THE CHEMISTRY OF MERCURY

ASPECTS OF INORGANIC CHEMISTRY
General Editor: C. A. McAuliffe

TRANSITION METAL COMPLEXES OF PHOSPHORUS, ARSENIC
AND ANTIMONY LIGANDS
Edited by C. A. McAuliffe

TECHNIQUES AND TOPICS IN
BIOINORGANIC CHEMISTRY
Edited by C. A. McAuliffe

The Chemistry of Mercury

Edited by

C. A. McAuliffe
University of Manchester Institute of Science and Technology

MACMILLAN OF CANADA / MACLEAN-HUNTER PRESS

First published 1977 by
The Macmillan Press Ltd
London and Basingstoke

First published in North America 1977 by
THE MACMILLAN COMPANY OF CANADA LIMITED
70 BOND STREET
TORONTO M5B 1X3

ISBN: 0-7705-1469-3

Printed in Great Britain

Contents

PART 3   THE ORGANIC CHEMISTRY OF MERCURY

A. J. Bloodworth

PART 4   THE BIOCHEMISTRY AND TOXICOLOGY OF
         MERCURY
         K. H. Falchuk, L. J. Goldwater and B. L. Vallee

Preface

The aim of this series is to provide reasonably comprehensive coverage of a topic in inorganic chemistry. This volume meets this requirement quite well and, as far as I am aware, there is no more comprehensive treatise on mercury than is presented here.

Mercury is an *ancient* element, and the chapter by Farrar and Williams is a very readable account of its history. Because mercury has been used so much by man I felt that no complete account of its chemistry could ignore its history.

Apart from the accepted use of mercury in organic syntheses, there is tremendous attention being devoted at present to the interaction of mercury compounds and organomercurials with biological systems. The importance of such interactions was brought to light by the 'Minamata disaster' of 1953, though the dangers of mercury had been known for some time. Thus, chapters are included in this book that detail the general coordination chemistry of mercury; the general chemistry of mercury has stimulated only spasmodic interest among inorganic chemists because of the lack of 'd–d' spectra, but this situation is rapidly being improved upon. Dr Bloodworth presents a detailed review of mercury–carbon bond formation and reactivity, and Professor Vallee and his coworkers provide an extremely timely account of biochemical and toxicological aspects of mercury and its compounds.

In order to avoid duplication, work presented in recent reviews by other authors is not duplicated here, but adequate cross-referencing is provided.

I am grateful to individual authors who have collaborated with me in producing this book; as Editor I accept responsibility for any production errors which appear. I wish to thank William Levason and Margaret McAuliffe for help at all stages of production and Penelope Barber for her considerable aid in editing.

Charles A. McAuliffe
Manchester, November 1976

PART 1

A History of Mercury

W. V. Farrar and A. R. Williams

University of Manchester Institute of Science and Technology

1 Mercury in the Ancient World (Europe and the Near East)

The commonest ore of mercury is the red sulphide, cinnabar, from which the free metal can be obtained by simple heating in air. Droplets of mercury are also sometimes found in the veins of cinnabar. As far as can be ascertained, a knowledge of cinnabar and mercury first appears in the Mediterranean world about the fifth century B.C. Claims for earlier dates have so far proved insubstantial, though there is nothing impossible about them, since—as we shall see—cinnabar ink was used in China from 1100 B.C.

There is no evidence that mercury was known in ancient Egypt, and it is not mentioned in the Ebers Medical Papyrus.[1] Specimens contained in small glazed earthenware bottles shaped like pineapples have been found in tombs of the eighteenth Dynasty (sixteenth century B.C.),[2] but were probably put there at a much later date, perhaps by the Arabs.[3] Cinnabar was used as a pigment in Ptolemaic Egypt, from the third to the first centuries B.C.

It has been suggested that mercury was known to the Hittites (sixteenth to thirteenth century B.C.), since the remains of an ancient cinnabar mine near Iconium (a Hittite site in Asia Minor) contained about fifty skeletons with stone hammers and lamps, entombed by the collapse of a gallery.[4] Without further evidence, however, this does not prove Hittite use of mercury, because the mines were still being worked in the Arab period, perhaps up to the tenth century A.D.[5] Similarly, attempts have been made to identify the drug 'sandu', referred to in Assyrian texts, with cinnabar;[6] but there is no real evidence that they were identical, and a red pigment excavated from Khorsabad (tenth to eighth century B.C.) proved on analysis to be iron oxide.[7]

The enterprising Phoenicians may well have been responsible for the first exportation of cinnabar from Spain.[8] Although there is no direct evidence of their involvement with mercury, they were active in exploiting the mineral resources of Spain from their trading centre at Tartessus from the eighth to the

third century B.C.; and the Almadén cinnabar mines were certainly worked well
before the arrival of the Romans at the latter date.[9] Cinnabar was used as a
pigment in Greece in the fifth century B.C.,[10] and the first literary mention
in Europe of the metal itself is a passing reference by Aristotle (384–322 B.C.)[11]

> The bodies which do not admit of solidification are those which contain no
> aqueous moisture . . . and those which do possess some water but have a pre-
> ponderance of air, like oil and quicksilver, and all viscous substances such as
> pitch and birdlime.

Theophrastus (?368–288 B.C.), a pupil of Aristotle, described both the purification
of cinnabar for use as the pigment vermilion, and a method of preparing metallic
mercury[12]

> Spanish cinnabar, which is extremely hard and strong, is natural, and so too is
> that of Colchis, which is said to be found on precipices and shot down by
> arrows. The manufactured variety comes from one place only, which is a little
> above Ephesus . . . Here a sand which glows like scarlet kermes-berry is collected
> and thoroughly pounded to a very fine powder in stone vessels. It is then washed
> in copper vessels, and the sediment is taken and pounded and washed again.
> There is a knack in doing this, for from an equal quantity of material some
> workers secure a great amount of cinnabar, and others little or none . . . The
> process is said to have been invented and introduced by Callias, an Athenian
> from the silver-mines, who collected and studied the sand, thinking that it
> contained gold owing to its glowing appearance. But when he found that it
> contained no gold, he still admired its fine colour and so came to discover the
> process, which is by no means an old one, but dates back only some ninety
> years before the archonship of Praxibalus at Athens. From these examples it is
> clear that Art imitates Nature, and yet produces its own peculiar substances,
> some for their utility, some merely for their appearance, like wall-paint, and
> some for both purposes, like quicksilver; for even this has its uses. It is made by
> pounding cinnabar with vinegar in a copper mortar with a copper pestle. And
> perhaps one could find several things of this kind.

Since Praxibalus was archon (a sort of chief magistrate) from 315–314 B.C.,
this dates the purification of crude cinnabar to about 400 B.C. Presumably the
Spanish cinnabar, which needed no such purification ('natural') was used from a
somewhat earlier date. The metal was evidently made by displacement from a
solution of its acetate, but it is surprising that a copper amalgam was not formed.
    The extraction of mercury by roasting the ore, and the property of amalgama-
tion, are first described by Roman writers; Vitruvius, writing near the end of the
first century B.C., is a little vague about the extraction process[13]

> When the ore has been collected in the workshop, because of the large amount
> of moisture, it is put in the furnace to dry. The vapour which is produced by
> the heat of the fire, when it condenses on the floor of the oven, is found to be
> quicksilver. When the ore is taken away, the drops which settle because of their
> minuteness cannot be gathered up, but are swept into a vessel of water; there
> they gather together and unite.

He goes on to say that four *sextarii* of mercury weigh 100 *librae*, which (taking the accepted values of these units) gives a specific gravity of about 14.5 instead of 13.6. He also notes that a stone, however heavy, will float on mercury, whereas even a small piece of gold will sink, and concludes 'that the gravity of bodies depends on their species and not on their volume.'

The extraction was carried out in Rome, the workshops being 'between the temples of Flora and Quirinus.' The ore was imported almost entirely from the Almadén (Sisapo) mines in Spain, which were worked by a private company under government inspection.[14] As in other mines in the Roman Empire, the miners would be slaves and criminals. There may also have been cinnabar mines at several places in present-day Yugoslavia, but the Roman dates proposed for these workings have been disputed, and medieval dates are equally likely.[15] Production from the cinnabar mines at Monte Amiata in Tuscany, which had been worked from pre-Roman times,[16] was discontinued in the first century B.C., probably as a consequence of the Servile Wars, for the Romans did not wish to have any colonies of industrial slaves within the borders of Italy.[17] As much as 2000 *librae* of mercury were extracted annually, and sold for gilding, for recovering gold from worn-out cloth-of-gold, or simply as a curiosity. The price was fixed by law at seventy sesterces per *libra*. The present-day unit of mercury production, the 'flask' of 76 lbs, is of Roman origin; 76 lb = 100 *librae* almost exactly.

Caius Plinus Secundus (Pliny the Elder), who died in the famous eruption of Vesuvius in A.D. 79, compiled an enormous Encyclopedia of Natural History dealing with the whole of knowledge then available, though often in an uncritical and half-understood fashion. His account of gilding is recognisable, but not easy to follow[18]

> The most convenient method for gilding copper would be to employ quicksilver, or, at all events, hydrargyros [a distinction between native mercury and that obtained from cinnabar] . . . To effect this mode of gilding, the copper is . . . heated by fire, in order to enable it, when thus prepared, with the aid of an amalgam of pumice, alum, and quicksilver, to receive the gold leaf when applied.

(The pumice and alum would clean the surface of the metal, so that the mercury could form a surface amalgam, on which the gold leaf could stick.)

Pliny's account of refining gold by amalgamation is similarly confused[19]

> There is a mineral also found [in Spain] which yields a humour that is always liquid, and is known as quicksilver. It acts as a poison upon everything, and pierces vessels even, making its way through them by the agency of its malignant properties. All substances float upon the surface of quicksilver, with the exception of gold, this being the only substance that it attracts to itself. Hence it is, that it is such an excellent refiner of gold; for, on being briskly shaken in an earthen vessel with gold, it rejects all the impurities that are mixed with it. When once it has thus expelled these superfluities, there is nothing to do but to separate it from the gold; to effect which, it is poured out upon skins that have been well tawed, and so, exuding through them like a sort of perspiration, it leaves the gold in a state of purity behind.

The confusion here is obviously between the cleaning of mercury by squeezing through a soft leather bag, and the recovery of gold from its amalgam, which must be done by heating to boil off the mercury.

Amalgams of copper and tin were probably known from an earlier date, under the name of 'asem'. There is a recipe for making 'asem' attributed to Phimenas the Saite of the late third century B.C.[20]

2 Mercury in Alchemy

'Alchemy' is the name given to attempts to make gold or silver from base metals. Its origins are obscure and undoubtedly complex, but it was the main activity of proto-chemists for at least a millenium and a quarter, A.D. 250–1500 in Europe and the Near East, and for even longer in China. Needham[21] has introduced the useful terms 'aurifiction' and 'argentifiction' to denote the making of a plausible imitation of gold or silver, whether or not for purposes of deception; and 'aurifaction' and 'argentifaction' to denote processes purporting to make real gold or silver from base metals. Both techniques existed side by side throughout the history of alchemy, and were sometimes confused, one suspects, even by the operators themselves.

The earliest known alchemical texts are the Leiden and Stockholm Papyri, probably written in Alexandria in the late third century. They include the following recipes[22]

> Take some mercury and some leaves of gold, and make up into the consistency of wax; taking the vessel of silver, clean it with alum, and taking a little of the waxy material, lay it on with the polisher and let the material fasten itself on. Do this five times. Hold the vessel with a genuine linen cloth in order not to soil it. Then taking some embers, prepare some ashes [and with them] smooth [it] with the polisher and use a gold vessel. It can be submitted to the test for regular gold . . .
>
> Manufacture of silver. Buy charcoal which the smiths use and soften it in vinegar one day. After that, take one ounce of copper, soak it thoroughly in alum, and melt it. After that, take eight ounces of mercury, but pour out the mercury thus measured into a secretion of poppy juice. Take also one ounce of silver. Put these materials together and melt; and when you have melted them, put the lumps so formed into a copper vessel with the urine of a pregnant animal and iron filing dust [for] three days.

The first recipe is a straightforward gilding by means of gold amalgam. The second, though confused, seems to be the preparation of a copper-silver amalgam which looks like silver. Both are clearly examples of aurifiction and argentifiction. Complicated recipes for aurifaction began to appear a little later, and in these techniques too, mercury often played a great part. But from the beginning there was also an 'esoteric' side to alchemy, in which the materials and processes described were symbolic rather than real, and the final aim was the spiritual ennoblement of the alchemist himself. It is often difficult to tell whether a real or symbolic meaning is intended for an alchemical text, especially since in either case the language is likely to be made intentionally obscure in order to confuse the uninitiated. Some kind of theoretical basis for the practical side of alchemy can be found in the 'Jabirian corpus', a collection of writings in Arabic put together in the ninth century under the name of a certain Jabir ibn Hayyan, apparently an historical personage, though a shadowy one. According to 'Jabir', all metals are made up of mercury, sulphur, and arsenic;[23] to change a base metal into gold it was necessary to alter the proportions of mercury, sulphur, and arsenic in the metal concerned. This was to be achieved by repeated distillations, to obtain the 'pure elements' or 'spirits', rather than the corrupt, earthy materials normally available.

If the authors of the Jabirian corpus did in fact manipulate (real) mercury and sulphur in the way that their theories would suggest, they must have discovered the synthesis of cinnabar by heating the two elements together. This has some importance as being the first deliberate synthesis of a naturally occurring compound. As far as we know, the Jabirian corpus does not actually describe this, but in the next century a fairly clear account was given by the Persian al-Razi (864–923), who became known in Europe as Rhazes. Part of his work, translated into Latin in the thirteenth century as *Razis de aluminibus et salibus*, contains the following passage[24]

> Another coagulation of mercury with the vapour of sulphur. There is another method, and it is the coagulation of [mercury] with the odour [vapour] of sulphur. That is, you put it in a piece of thick cloth, tie it, and suspend it in a vessel. Place sulphur in the bottom of the vessel and seal up the junction [between vessel and lid]. Kindle a fire under it for a whole day, and then take it off the fire so that it may cool. Then you will find a red salt *uzifur* [Arabic *zanjufur* = cinnabar]. You may use this in "The work of redness" [an aurifaction] and you will rejoice with it, because in it are many marvels which cannot be numbered.

The same text has a confused but recognisable account of the preparation of mercuric chloride.[25] This is perhaps the first synthesis of a definite chemical compound that does not occur in nature, and one which has, moreover, some remarkable properties ('corrosive sublimate').

> A sublimation of mercury. Likewise, there is a sublimation of it with salt and atrament [blue or green vitriol]. That is, you take the amount of [mercury] you wish, and pound it with its equal [weight] of atrament and its equal [weight] of salt. Let the salt be dissolved in a pot or dish, and then grind it with the other [solids] and afterwards dry [all] so that it is mingled with them. Then put this in an aluthel [that is, a tube] without moisture and keep it in a baker's oven overnight. Then, when you may take it out by hand, you pour from it the dissolved salt, return it to the oven, and pour from it several times until the

residue is darkened. You cause it to be put in the oven as many times [as it needs] until it is coloured red. Then place it in an aluthel and fasten a joined vessel from it [that is, a receiver]. Place a fire under it, and so you have put it upon the oven in the form of an athanor [that is, set up for a slow distillation]. Kindle a fire below it for three hours in the summer time, and until nightfall in winter time. Leave it until it becomes cold, open the aluthel, and you will now find in the upper part of it a clear white sublimate, with the help of God and the power of this device.

(It is, of course, likely that the sublimate also contained calomel, but—outside China and Japan—this does not seem to have been recognised as a distinct substance until much later.)

At about the same time as al-Razi, a compilation called *Mappae Clavicula*† appeared in Europe, containing such remnants of Greco–Roman chemical techniques as survived the fall of the Western Empire.[26] Of its 263 chemical recipes, about forty-five utilise mercury. The following may be taken as typical

To make very much gold. Take quicksilver 8 oz; sheets of gold 4 oz: sheets of good silver 5 oz: sheets of copper-bronze 5 oz: sheets of auricalcum [brass?] 5 oz: feathery alum and flowers of bronze which the Greeks call calcantum [blue or green vitriol?] 12 oz: orpiment with the colour of gold 6 oz: elidrium [?] 12 oz: and then you must mix all the sheets with quicksilver and make [them] like wax. Then put in the elidrium and orpiment together, afterwards casting in the flowers of bronze and alum. Put all [these] in a dish upon the fire and heat gently, sprinkling into [it] from [your] hand crocus [iron oxide] steeped in vinegar and a little nitre [natron, sodium carbonate]. In fact you sprinkle 4 oz of the crocus little by little until it is dissolved, and you let [it] stand so that it may absorb. Then, when it has become coagulated, lift it up, and, with heating, you will have gold.

This sounds like a corrupt and unworkable version of an original aurifiction recipe to make some sort of amalgam that looked like gold. Not all of the *Mappae Clavicula*, however, is alchemical. The last chemical recipe describes how excess mercury (amalgam) can supposedly be removed from gilded ware.

If quicksilver sticks to gold work, which cannot be placed in the fire, take the urine of a man, mix together atrament [blue or green vitriol] and salt, and make a thick paste of this. Put [some] of this upon the quicksilver, which is adhering to the gold work, and leave it on a good while. Then wipe [it] so that it disappears. After this, rub down with an iron [tool] as you know how.

Despite the *Mappae Clavicula*, alchemy was almost unknown in Europe until translations of Arabic writings into Latin began to appear in the twelfth century, mostly originating in Spain. Among the works translated were parts of the Jabirian corpus, and the name was latinised to Geber. Few of the numerous books attributed to 'Geber', however, have been shown to have Arabic originals, and it

† 'Little key to painting.'

is now believed that the five most important of them (at least) are Latin composi-
tions of the thirteenth and fourteenth centuries. Although they repeat much of
the substance of Arabic alchemy, both practical and esoteric, some of them
contain new observations of importance. The author of the thirteenth-century
Geber book *Summa Perfectionis* knew that prolonged heating of mercury in air
produced a red ash or calx (mercuric oxide); and the *De inventione veritatis* of
similar date has a much clearer and more plausible preparation of 'white sublimate'
(mercuric chloride) by heating together mercury, salt, alum, and the essential
oxidising agent, saltpetre.[27]

The German Dominican, Albertus Magnus (1193–1280), in his *De lapidibus*
written about 1260, knew enough about the Jabirian theory to expound a modified version of it in which mercury is the 'mother-principle' and sulphur the 'father-
principle', which together make up all metals. He was evidently familiar with the
mercury–sulphur reaction, and even implied that it was used on a commercial
scale to make vermilion pigment: 'manufacturers of minium make it by sub-
liming sulphur with mercury'.[28] ('Minium' in early writings can mean either
cinnabar or red lead, but the meaning here is clear from the context.)

Despite the saintly Albert, however, western alchemy soon gained a reputation
for dishonesty—though nobody was quite sure that what the alchemists were
claiming to do was impossible. One way of securing the confidence of the simple
was to take some mercury (actually a dilute silver amalgam) and manipulate it over
the fire with much ceremony, so that the mercury boiled away, leaving a small
button of silver, which any silversmith would pronounce genuine. By about 1390,
when Chaucer wrote *The Canon's Yeoman's Tale*, the alchemist in popular
imagination was already a fraud and a trickster.

> "Sir", he addressed the priest, "send out your man
> For quicksilver, as quickly as you can;
> Let him bring several ounces, two or three,
> And when he's back I promise you shall see
> A miracle you never saw before."
> "Certainly", said the priest, "he's at the door,
> It shall be done at once". The man was sent,
> Ready to do his bidding; off he went
> And, to tell truth, returned immediately
> Bringing three ounces of the mercury,
> And gave them to the canon for a start.[29]

And soon they were launched on the Great Work—which ends with the Canon
riding off into the sunset with his disgruntled Yeoman, leaving the priest a sadder
and poorer man. Eventually the only serious alchemists left were those of the
esoteric tradition, who did no experiments and so could not be easily exposed.

3   Mercury in China

It will not be possible to give a proper account of the history of mercury in China until Joseph Needham and his collaborators have completed the publication of their monumental *Science and Civilisation in China*. The following is a mere sketch, based on such material as is at present available.[30]

Among the earliest evidences of Chinese civilisation are the numerous 'oracle-bones', believed to date from before—though not very long before—1100 B.C. The ancient Chinese foretold the future from the pattern of cracks produced by applying a hot metal rod to an animal's shoulder-blade (scapulimancy); the forecast was then often written on the bone in a primitive form of the ideographic script. Some of the writing is in a red ink made from cinnabar,[31] which is not uncommon in China, especially in the province of Hunan. (China is still a major world producer of mercury.) Cinnabar ink has a long tradition of use in China, at times being the prerogative of the Emperor himself.

Apart from this long-standing use of cinnabar ink, a knowledge of mercury and its compounds seems to arise in China at about the same time as in the Mediterranean world, and apparently independently. Mercury itself is mentioned, together with cinnabar, in the *Classical Pharmacopoiea* (*Shen Nung Pên Tshao Ching*), the compilation of which began about 200 B.C., though it did not reach its final form until the second century A.D. (*The Classical Pharmacopoiea* itself is lost, but it has been reconstructed from copious quotations in later works.) A mercury still is known dating from the later Han dynasty (first or second century A.D.). The earliest mirrors also date from this dynasty; they consist of a polished layer of tin amalgam on a bronze plate.

Alchemy was practised in China from the Han dynasty onwards, though, as else-where, its beginnings are shrouded in legend. It was in full flower with Ko Hung (A.D. 261–323), a completely historical personage approximately contemporary with the Alexandrians who compiled the Leiden and Stockholm papyri. To use Needham's terminology, both aurifiction and aurifaction were attempted, though to a Chinese alchemist a still more important aim was the achievement of personal immortality, or at least longevity. This was to be attained by the drinking of an elixir, the preparation of which was not clearly described, but which probably contained compounds of arsenic and mercury. Several alchemists seem to have met their deaths in this way. It was expected that the bodies of these 'immortals' would remain uncorrupted, and in this connection a dramatic discovery was made in 1972. A princely tomb excavated near Changsha disclosed an air- and water-tight coffin containing the body of a woman who must have died about 186 B.C. This body was 'like that of a person who had died only a week or two before'; when the skin was pressed it returned to normal on release of the pressure. This preservation had been achieved neither by embalming, mummification, tanning, nor freezing, but apparently (few details are yet available) by immersion in an aqueous liquid containing mercury salts.

Some time later in the first millennium A.D., Chinese alchemists prepared both corrosive sublimate and calomel by heating various mixtures containing mercury, salt, and alum; unlike their Western contemporaries, they distinguished clearly between them. Calomel was used in Chinese medicine many centuries earlier than in the West, and a small factory in Japan was described in the 1890s as still making calomel by the traditional process.[32] Calomel is also one of the eighty pharmaceutical specimens preserved in a temple at Nara (Japan) since A.D. 756; the sample is 99.5 per cent pure. The preparation of both chlorides is clearly set out in the *Great Pharmacopoiea* (*Pên Tshao Kang Mu*) compiled in 1596.

From an unknown date in the first few centuries B.C., mercury was used in China (as in the Mediterranean world) to extract native gold and silver by amalgamation. A tin–silver amalgam was also used for dental purposes, the first reference being in A.D. 659.

As in many other subjects, knowledge of mercury and its compounds seems to have advanced at about the same speed and in the same directions in China and the West, with only minimal contact between them until the seventeenth century. Chinese proto-chemistry must, however, have influenced—and been influenced by— that of India, for there was much coming and going between the two countries. Unfortunately, it is as yet impossible to say much about early chemistry in India because of the lack of hard facts and firm dates. It is certain, however, that mercury was well known in India in the first millennium A.D., and it may well have been from the Indian tradition that the Arabs learned to use mercury in medicine.

4   The Mercury Industry and the Amalgamation Process

Ores of mercury, of which cinnabar (HgS) is the only important one, are widely distributed, though it is probable that all the richest sources were exhausted long ago, since the ore-bearing veins are usually on the surface or at no great depth. Spanish ores still contain up to (exceptionally) 7 per cent mercury, but in other parts of the world mercury contents of 0.3 per cent or even less have been profitably worked because of the simplicity of the extraction process. This is still basically the same as in ancient times; the cinnabar, concentrated if practicable by washing or flotation, is roasted in a current of air, with or without the addition of desulphurising substances such as lime or iron oxide. The condensate of crude mercury has then to be freed from soot, dust, and other impurities by a variety of *ad hoc* methods including distillation and filtration. From medieval times until quite recently, by far the largest tonnage usage of mercury has been in the extraction of gold and (especially) silver by amalgamation. Accounts of the mercury industry and the amalgamation process must therefore run together.

In the most primitive process known, cinnabar and brushwood were simply piled up in heaps and set on fire; when the ashes were raked over, pools of mercury could be collected. Sometimes green branches were put on the top of the heap to act as a crude condenser. The wastefulness and danger of this method needs no emphasis. By the late Middle Ages, however, this seems to have been replaced by the process known as *'per descensum'*; essentially that described by Agricola in 1556.[33] Two earthenware pots were adjusted over each other. The upper pot, filled with ore and closed at the top, was covered over with burning fuel. The mercury vapour passed down through small holes in the bottom of the pot to be condensed in the vessel below.

The first evidence of an amalgamation process being worked on a large scale comes from the twelfth-century Moslem geographer al-Idrissi, who reported that mercury was an important article of trade in Egypt at that time, some being re-exported for the extraction of gold in the countries in the south.[34] By contrast, the first great European compendium of chemical technology—Biringuccio's *De la pirotechnia*, published in 1540—treats amalgamation mainly as a method of recovering scraps of precious metals from the sweepings of the workshops of goldsmiths and silversmiths. Even for this technique (which, as we have seen, was known to Pliny) Biringuccio 'gave to the one who taught it to me a ring with a diamond worth 25 ducats, and I also pledged myself to give him the eighth part of whatever profit I should gain from this operation.'[35] His book has a woodcut of a simple mill for carrying out the amalgamation, but he mentions the possibility of working up slags and ores only in passing, almost hypothetically. He still believed, as did the ancient authors, that squeezing the amalgam through a leather 'purse' would leave the precious metal behind, though he wrote of distillation as an alternative. Agricola, writing a few years later, though he knew Biringuccio's book and copied much from it, seems to have felt the amalgamation process for silver not worth mentioning; though he described the corresponding process for gold in some detail, and made the same mistake about the leather bag.[36]

The amalgamation process was introduced in the silver mines of New Spain (Mexico) about 1554. Various names are cited in this connection; they were probably all partners in the same enterprise. One of the partners, Gaspar Lomann, was obviously a German, and it seems likely—though there is no direct evidence—that the process originated in the silver mines of the Tyrol, using mercury from Idria (Yugoslavia). If so, it had to be re-introduced into Central Europe in the eighteenth century, as 'the Mexican process'. In its earliest form, the 'patio' (courtyard) process, it was described by the Jesuit Juan de Acosta in 1590.[37] A mixture of ground and roasted silver ore with mercury and various additives (water, salt, vinegar, copper sulphate) was spread out on a paved area and mules, sometimes drawing rollers, were walked round and round on it. The amalgam was separated from the gangue by washing—there was much loss at this stage— and concentrated by squeezing through linen bags. Acosta knew, however, that the resulting near-solid product was not pure silver, but contained five parts of mercury to one of silver; and he described the distillation: 'the silver remaining without changing the forme, but in weight it is diminished five partes of that it was, and is spungious'. The 'loaves' of spongy silver were hammered into bars, and needed no further refining.

The mercury was theoretically recoverable, but inefficient working made a constant supply necessary; about one-and-a-half pounds of mercury were needed to produce a pound of silver. This at first caused difficulties, since there was little available locally (useful deposits of mercury were not discovered in Mexico until the eighteenth century). In 1559, aware that his treasury depended heavily on American silver, Philip II of Spain ordered that virtually the whole output of the Almadén mines should be sent to Mexico, and that the output should be increased. These mines were a royal monopoly, though mortgaged to the Fuggers, an Augsburg banking house who had paid the enormous bribes for the election of Charles V, Philip's father, as Holy Roman Emperor. For over a century, the 'Plate Fleet' set out from Spain every year laden with mercury, and returned with silver.

About 1545, what proved to be the world's richest silver mine was discovered at Potosí (now in Bolivia), high up in the Andes. Transport of mercury from Spain, though attempted, was almost impracticable; it involved two long and hazardous sea voyages, a land crossing of the Isthmus of Darien, and a final long haul from the Pacific coast over 13 000-foot passes to Potosí. The mines could therefore not be much worked until after 1563, when a fortunate discovery of mercury ore was made at Huancavelica, where the Indians were using the cinnabar as body-paint. This place, though also in the Andes and in the same Spanish viceroyalty of Peru, is about a thousand miles from Potosí, and the mercury had to be carried on pack-llamas along the network of roads originally made for the royal runners of the Incas. Potosí was then independent of European supplies. The silver mine was worked almost to exhaustion in two centuries by the forced labour of the native population. But conditions at Huancavelica were particularly bad, for the method used was to heat the rock-face by fire and crack it by splashing with cold water. The life expectancy of a miner at Huancavelica was about six months—an aspect of life there not commented on by Acosta. These two mines, of silver and mercury, worked with so much suffering, were the main pillars supporting the precarious Spanish economy during 'the golden century'; and the resulting flood of silver into Europe was one of the main causes of the disastrous inflation that followed.[38]

In the 1620s and 1630s the mines at Potosí came under the acute observation of Alvaro Alonso Barba (1569–?1640), who combined the offices of parish priest and superintendent of the mines. He is an interesting figure of whom little is known. His only book, *El arte de los metales*, was written in Potosí in 1637 and published in Madrid in 1640. (We have used a rather poor translation, *The art of mettals*, made by—or for—Lord Sandwich in 1670.) Although the book attempts to deal with all the metals then known, its longest and most important chapters, as might be expected, are those on silver and mercury. Barba's science was a little old-fashioned for his day; he mixed sound practical observations with medieval theory and quotations from the classics. Instead of the messy 'patio' process he recommended that the amalgamation be carried out in closed vessels with mechanical (water-powered) stirring; and instead of relying on traditional rule-of-thumb, he advocated laboratory tests of the silver ores to determine the best additives, and their optimum quantity, especially with regard to the danger of 'flouring' (formation of fine droplets that will not coalesce).

Despite his belief that mercury entered into the composition of all metals, Barba denounced as 'a great error' the widely held opinion 'that because for so many years the best Refiners in these Kingdomes have wasted at the least so much Quicksilver, as they have gotten Plate, therefore the Quicksilver is really and truly consumed in the operation'. And he roundly condemned the prodigality with which the whole operation was being conducted

the first and fundamental remedy whereof, is in my opinion, that the Mettals
be Refined by one that understands the Art, and is Authorized thereunto by
Publick License, after strict examination of his sufficiency, which is required
before the admission unto divers callings in the Commonwealth, without comparison of much less importance than this is. The Masters of Refining Works
have taken no Care at all in this matter, because how negligently soever they
Refine their own Oar, they lose nothing . . . It is a very great Trust that is put
into the Refiners, the whole Riches which this most prosperous Country
produceth, being put into their hands without account, or any obligation of
the quantity they are to return; their word and honesty only, without reply,
or appeal from their sentence is the only security of the truth, of what the Oar
hath yeilded; and it has need to be a strong security, when the violent incitation
of private interest is to deceive. He therefore that liveth continually amongst
these occasions, had need to be well furnished with the honor of a Christian,
lest having his fingers perpetually kneading in the paste, a good deal do not
stick unto them; there ought to be a great deal of circumspection in chusing
this Officer, for no mischief that hinders the Refining of the Oar, or extravagant
consumption, or loss of Quicksilver, can occasion so great prejudice as a Refiner
of a wicked conscience.

Barba was far ahead of his time in his insistence that a chemical process could
and should be costed throughout, not only for fuel, raw materials, and labour but
for less obvious things such as plant depreciation. It is not, of course, known to
what extent these recommendations were put into effect, but they are remarkable
as being made in that time and place. His improved mercury stills for Huancavelica
were certainly widely (though slowly) adopted, under the name of 'Bustamonte
furnaces'—Bustamonte was apparently a later superintendent of the Huancavelica
mine, who wrote a memorial to Philip IV of Spain on the advantages of Barba's
invention. At Almadén too, the old '*per descensum*' process was proving in-
adequate, and larger installations were put up in which the mercury vapour was
condensed in long brickwork tunnels or 'galleries'. Even so, a nineteenth-century
writer, who had seen most of the European mines, used such adjectives as 'rude'
and 'barbarous' in describing their manner of working.[39] But apart from some
well-conducted modern mines where the cinnabar is burned in a continuous
process, this still seems to be the state of the art over much of the world today.
The same writer mentioned a disastrous fire in the Idria workings in 1803, which
could only be extinguished by flooding. Nine hundred people suffered toxic
symptoms from the resulting fumes, and the mines were virtually unproductive
for many years.

The exhaustion of the seemingly inexhaustible 'silver mountain' at Potosí
by the late eighteenth century greatly reduced the importance of the Huancavelica
mine. During the wars of independence in the 1820s, output dropped almost to
nothing; after a slow recovery, it is again an important world source of mercury,
though producing less than half its maximum under the Spaniards. Also during
the eighteenth century, worthwhile cinnabar deposits were discovered in Mexico,
and in the (then) Spanish province of California, at a place hopefully called

New Almaden. This latter source was not seriously exploited until 1850, when California became one of the United States; it came just in time for the great 'gold rush' of that decade, but there also the output had declined steeply at the end of the century. By that time, however, the amalgamation process for silver was being phased out all over the world (though it is still used for gold in some places) and demand for mercury declined.

World production today is fairly steady at around a quarter of a million tons per annum. The main tonnage uses are in electrical equipment and in the Castner-Kellner alkali process. Although there is no immediate shortage, proven world resources are not enormous. Before many decades have passed, it will almost certainly be necessary to restrict mercury to uses where no substitute can be found, and to be scrupulous on recovering that which must be used—measures equally desirable for reasons both of economy and public health.

5 Mercury and the New Chemistry

Although mercury was a very real substance to the miner, the assayer, and the apothecary, the chemical theorists of the seventeenth century usually referred to it in pseudo-mystical language. According to the followers of Paracelsus, who were still numerous, all bodies were composed of the *tria prima*, salt, sulphur, and mercury—though these were not the substances ordinarily known by these names but symbols for qualities or principles. There was presumably some relationship between real mercury and Paracelsan 'mercury', but no one was willing to define it. Robert Boyle, in his *Sceptical Chymist* (1661) wished to sweep away all these confused ideas

> And certainly he that takes notice of the wonderful operations of quicksilver, whether it be common, or drawn from mineral bodies, can scarce be so inconsiderate as to think it of the very same nature with that immature and fugitive substance which in vegetables and animals chymists have been pleased to call their mercury.

Boyle, however, was not able to put anything definite in the place of these ideas, except the assertion that mercury was a 'mixt body', which nevertheless retained its identity through many chemical changes.

The nature of mercury was much clarified when J. A. Braun of St Petersburg froze it to a malleable solid (1759), using a mixture of ice and nitric acid to obtain the requisite low temperature. Mercury was therefore, as had long been suspected, a metal like any other except for its low melting point.

Later in the same century, mercury, and the red ash or 'calx' made by heating it for a long time in air (*mercurius calcinatus per se*) played a great part in the experiments of Priestley, Lavoisier, and others, which led to the establishment of modern chemistry. In 1774, Priestley was living at Calne in Wiltshire, as librarian to Lord Shelburne and tutor to his two sons. He spent his ample spare time in 'philosophical experiments' relating to electricity, or to the study of gases, which he called 'airs'. To use his own words

Having procured a lens of twelve inches diameter, and twenty inches focal distance, I proceeded with great alacrity to examine, by the help of it, what kind of air a great variety of substances, natural and factitious, would yield, putting them into [a vessel] which I filled with quicksilver, and kept inverted in a bason of the same. With this apparatus, after a variety of other experiments . . . on the 1st of August, 1774, I endeavoured to extract air from mercurius calcinatus per se; and I presently found that, by means of this lens, air was expelled from it very readily. Having got about three or four times as much as the bulk of my materials, I admitted water to it, and found that it was not imbibed by it. But what surprized me more than I can well express, was, that a candle burned in this air with a remarkably vigorous flame . . .[40]

Priestley had, of course, discovered oxygen, though he did not give it this name, and he did not at first understand the significance of his discovery. It was not until March 1775 that he found that a mouse would live in his new 'air' much longer than in an equal volume of ordinary air. Also, he had developed a test for the 'goodness' of air, by reaction with nitric oxide. Fresh air would react with about half its volume of this reagent, while 'foul air' (from respiration or combustion) reacted with much less or none. He was astonished to find that, on this test, his new 'air' was five times as 'good' as fresh air.

Meanwhile, in October 1774, Priestley had visited Paris, and in his imperfect French 'at the table of M. Lavoisier, when most of the philosophical people of the city were present' told the company of his discovery. For several years Lavoisier had been working on the problem of combustion; it would be wrong to suppose that Priestley's tale made the whole matter clear in his mind at once, but he evidently realised that an experiment of crucial significance had been performed. Without delay he set in train some laboratory work, of which the first, still confused, results were published as the famous 'Easter Memoir' of 1775. Two years later he described his classical experiment.[41] Mercury was heated just below its boiling point for twelve days in 50 in.[3] of air; red calx was formed, and the residual gas (42 in.[3]) would support neither life nor combustion. The red calx was collected, and on heating alone furnished mercury and 8 in.[3] of Priestley's remarkable new gas. (Actually, as Lavoisier admitted, the experiment as published was an 'ideal' one, constructed from the results of several real ones.) Mixing the two gases re-formed the original 50 in.[3] of a gas indistinguishable from common air. Combustion was therefore combination of the burning body with one constituent of the atmosphere, present in about 1/5 by volume; the remaining 4/5 was inert. In these memoirs the work of Priestley, though not ignored, was mentioned in a rather grudging and ambiguous fashion, which caused some offence.

This was the beginning, rather than the end, of Lavoisier's reconstruction of chemistry. The later stages are not intimately connected with mercury and will not be detailed here. The keystone of this new chemistry was Lavoisier's concept of *element* as the last point that chemical analysis is capable of reaching (*le dernier terme auquel parvient l'analyse*). Mercury figures in the list of elements that appears in his *Traité élémentaire de chimie* (1789).

The foundation of the new chemistry was complete when John Dalton discovered, in 1803, how to calculate the relative weights of the atoms of Lavoisier's elements from a knowledge of chemical analyses. In part 2 of the first volume of his *New System of Chemical Philosophy*, published in 1810, the rough analyses of mercuric oxide available to him led him to an atomic weight for mercury of 'about 167' (H = 1). A more accurate value of 202.5 was calculated by Berzelius in 1826 (modern value 200.7). Dalton was also probably the first to state clearly that mercury formed two series of salts, derived from two oxides (red and black), one containing exactly twice as much oxygen as the other per unit weight of mercury (1827).[42]

6 Mercury–Nitrogen Compounds

Certain compounds containing mercury–nitrogen bonds, though apparently simple in composition, have a long and confused history, which need only be summarised here. Treatment of a solution of mercury in nitric acid with 'sal ammoniac and salt of tartar' (that is, ammonium chloride and potassium carbonate) gave, under some conditions, a 'fusible white precipitate'. This preparation is said to be described in a *Testament of alchemy* bearing the name of Raymond Lully (Ramòn Lull, 1235–1315), a Catalan scholar and mystic, but actually written in the mid-fourteenth century by an anonymous author. The composition of 'fusible white precipitate' is $HgCl_2 . 2NH_3$.

By varying the conditions of preparation, an 'infusible white precipitate' can be obtained, first noted as being different from the other by Lemery in 1663.[43] (Often, of course, the precipitate formed must have been a mixture.) This substance still appears in modern pharmacopoeias as *mercurius praecipitatus albus.* Its composition was later found to be NH_2HgCl.

In 1846, a third important compound was discovered by Millon in the reaction of yellow mercuric oxide with aqueous ammonia.[44] It was basic ('Millon's base') and formed a series of salts; its composition can be written as $(HgOH)_2NH_2OH$.

These substances proved difficult to fit into any theory of valency, and they were the subject of desultory controversies into which it would be pointless to enter. Their structures were not settled until the advent of X-ray crystallography.[45] Only the 'fusible white precipitate' proved to be a simple monomeric compound $(H_3N–Hg–NH_3)^{2+} 2Cl^-$. The 'infusible' precipitate is a linear polymeric salt $(–NH_2–Hg–)^+_n Cl^-_n$; Millon's base is a three-dimensional network polymer.

It is perhaps appropriate at this point to mention the curious 'ammonium amalgam' discovered by T. J. Seebeck in 1808,[46] in the wake of Humphry Davy's isolation of sodium and potassium by electrolysis. Seebeck found that if a pool of mercury were placed on a block of ammonium carbonate, and an electric current passed with the mercury as cathode, the metal would froth up to a buttery mass, which slowly decomposed, giving ammonia and hydrogen. This discovery aroused tremendous interest and the new 'amalgam' was studied by many other chemists, including Berzelius and Davy himself. There was still much confusion about the nature of chemical elements and the status of ammonium salts. The existence of 'ammonium amalgam' enabled Ampère (1816) to propose a solution to this problem, with 'ammonium' (NH_4) as what became known later as a 'compound radical'

> This difficulty [the great similarity between ammonium salts and those of sodium and potassium] would disappear if we suppose that—like cyanogen, which although a compound body, has all the properties of the elements which combine with hydrogen to form acids—the compound of one volume of nitrogen with four volumes of hydrogen which combines with mercury in the amalgam discovered by M. Seebeck, and with chlorine in ammonium chloride, behaves in all the compounds which it forms like the metallic elements.[47]

The true nature of this substance seems still be to obscure, and it is seldom mentioned in recent chemical texts.

This is perhaps the most appropriate place to mention the toy known as 'Pharaoh's serpent' or more correctly 'Pharaoh's serpent eggs', made of mercury thiocyanate. It was introduced in Paris in 1865. 'An ingenious piece of parlour magic has lately been introduced by the conjurer Cleverman . . . something like a pastille is placed on a plate, a light is applied, and in a moment the pastille swells up and seems to uncurl itself, and something resembling a snake appears on the plate.'[48] They were all the rage for a time, but there were the inevitable accidents, both because the pastilles looked like sweets, and because the fumes evolved were highly toxic. Non-mercurial substitutes were devised, but probably none was as effective as the original.

7   Mercury in Industry

The employment of mercury in the extraction of gold and silver has already been mentioned. The trades of hat-making and gilding will be dealt with in chapter 10. Another industrial outlet stemmed from the discovery in 1800 by Edward Charles Howard (1774–1816) of mercury fulminate and its detonating properties.[49] Howard's method is essentially that used ever since—the action of ethanol on a solution of mercury in nitric acid. He carried out various tests of the explosive powers of his new compound (in which he was fortunate to escape serious injury) and recognised that it was an entirely different kind of explosive from gunpowder, and useless for firing bullets. In modern terms, it was 'brisant' rather than 'propellant'. E. Goode Wright of Hereford (1823) published a description of a detonating 'cap' for cartridges that contained mercuric fulminate: this was soon followed by regular manufacture by the chemist F. Joyce of Soho, London.[50]

A considerable increase in the scale of manufacture began in the 1860s, due partly to the introduction of the metal cartridge (with a larger detonating charge) at the time of the American Civil War and partly to Alfred Nobel's use of mercury fulminate as the detonator of choice for his new 'dynamite'. The making and handling of this substance on the large scale was, of course, hazardous in the extreme and claimed many victims. The well-known chemist Henry Hennell was blown to pieces at Apothecaries' Hall in London, while mixing six pounds (!) of fulminate to make a bomb for use in the first Afghan War in 1842.[51] At about the same time, the cartridge-maker William Eley was killed when his laboratory was destroyed in an explosion caused by fulminate. Nor was this the only hazard; workers in the detonator factories suffered severely from 'fulminate itch', which began as a troublesome irritation, but could eventually develop into deep incurable ulcers or 'powder holes' on the hands and fingers.[52] Several countries have made fulminate detonators illegal during the present century as less-dangerous substitutes have been introduced.

The largest tonnage use of mercury in industry is in the Castner–Kellner process for making chlorine and caustic soda by the electrolysis of brine, using a rocking mercury cathode. The idea was patented independently in 1892 by Hamilton Y. Castner, an American working in England, and by the Austrian Karl Kellner. An amicable agreement was arrived at, and the process was operated at Niagara Falls and at Runcorn, Cheshire, from 1897. Its success overturned the economics of the existing alkali industry, and speeded the decline of the Leblanc process, which was already living on its former by-products. It is still very widely used; the products inevitably contain traces of mercury, and, since they have so many outlets, have added to the mercury burden of the environment.

Mercury salts are also important as catalysts in several reactions of industrial importance. There is the famous case of the oxidation of naphthalene to phthalic anhydride catalysed by mercuric sulphate, said to have been discovered in the 1880s when a thermometer was accidentally broken in the reaction mixture. Of more importance is the hydration of acetylene to acetaldehyde catalysed by a solution of mercuric sulphate in 25 per cent sulphuric acid. This reaction (which proceeds via some ill-defined organomercury compounds) was first described by M. Kutscherow of St Petersburg in 1881.[53] The availability of cheap calcium carbide after 1900 made it of industrial interest, and it was worked by two German firms (*Consortium für Elektrochemische Industrie* and *Chemische Fabrik Griesheim Elektron*) from about 1912, and by the Canadian Electro-products Co. at Shawinigan a few years later. It is still one of the basic reactions of the petro-chemical industry, and played a part in the 'Minamata disaster' to which reference will be made later.

Since the introduction in the 1930s of mercury-vapour discharge tubes for street lighting—later extended to office and domestic lighting—there has been heavy usage of mercury by the electrical industry.

Here may also be mentioned the brief vogue for protecting wood against dry rot (the fungus *Merulius lachrymans*) by soaking it in a bath of aqueous mercuric chloride. Though not entirely new, this treatment was popularised by an impoverished Irish inventor, John Howard Kyan (1774–1850) who had studied the rotting of the timber supports in his father's copper mines in County Wicklow. The Admiralty tested pieces of treated wood in the 'fungus pit' at Woolwich from 1828 to 1831—an early example of biological testing—and the results were so satisfactory that Kyan took out patents (E.P. 6253 and 6309/1832), and in 1836 sold his rights for a considerable sum to the Anti-Dry Rot Co. 'Kyanising' became a fashionable word, and popular songs were even written about it. When the railway system began to expand, the kyanising of wooden sleepers seemed to offer a new and profitable market, but it was soon found that the mercury salt promoted the corrosion of the iron rails and bolts. On grounds of cost also it was inferior to 'creosote', available cheaply from gas-works, and in the 1850s the treatment fell out of favour. It is interesting as the first industrial use of mercury compounds as fungicides.

8 Mercury in Pharmacy

It is by no means certain that mercury or any of its compounds were used for medicinal purposes in the ancient world. The question hinges on the correct interpretation of various technical pharmaceutical terms. At any rate mercury and mercurial preparations were not an important part of *materia medica*, and Galen (A.D. 130–200) regarded them as too dangerous to be of any use. Since Galen came to be looked on as the prime authority on medical matters for over a thousand years, mercurials can have played no part in Western medicine before the end of the Middle Ages.

The Arabic medical tradition, on the other hand, though respectful towards Galen, was also open to other influences—in this case perhaps from India. As early as the ninth century, Arabic physicians were making ointments by pulverising mercury in a vehicle such as animal fat, which prevented the droplets from recombining. These ointments were prescribed for skin and eye infections. They became known in Europe, partly through the Crusaders who brought back a knowledge of *unguentum saracenum* and partly through the writings of ibn Sina or Avicenna (980–1037), some of which were available in Latin translation in the thirteenth century.

However, the great vogue of mercury in medicine began with the appearance in Europe, just before 1500, of an apparently new and usually fatal disease, spread by sexual contact. It was called by many names, both vulgar and learned, until 'syphilis' (first used by Fracastoro in 1530)[54] became generally accepted. It was widely believed—though it cannot be proved, and the question has been inconclusively debated ever since—that the sailors of Columbus brought back the disease from the Americas. Certainly it was spread throughout Europe by the French army and its camp-followers after the campaign against the Spaniards in Italy in 1495; and it had the typical characteristics of a new disease ravaging a susceptible population, being more quickly fatal, and with more dramatic symptoms, than are usual nowadays.

No doubt every conceivable remedy was tried by the first sufferers, and the belief soon gained ground that *unguentum saracenum* was effective. There are many conflicting claims for the first use of mercury in the therapy of syphilis, but it became well established in the early years of the sixteenth century; countless medical men, reputable and disreputable, made fortunes from this treatment— it was jested that at last a way had been found to turn mercury into gold. The patient was smeared from head to foot with this unpleasant grease ('inunction') for a period of weeks or months; but this was the least unpleasant part of the 'cure', for the usual distressing symptoms of mercury intoxicaton—salivation, loose teeth, swollen gums, bladder irritation, psychological disturbance—had to be endured as well as the original disease. A little later in the same century, inunction was supplemented or sometimes replaced by exposure of the whole body to mercury vapour. This was the 'tub', often referred to by Elizabethan and Restoration playwrights. It was typically a large wooden box, originally of the kind used for salting down meat for the winter, in which the patient sat or stood naked, with only his head projecting through a hole in the top; cinnabar would be roasted in a separate retort, and the vapours admitted to the box by a tube. This treatment, for several hours every day, was continued until the patient could stand no more, or could no longer pay.

It must not be supposed that these heroic measures against syphilis were adopted without opposition. Throughout the sixteenth century and later, a continual and often bitter controversy raged between 'mercurialists' and 'anti-mercurialists', not only within the medical profession, but among the interested public. For reasons that are not clear, the anti-mercurialists frequently had the support of the Church, which favoured the rival remedy, guaiacum, sometimes called 'holy wood' (though many prominent churchmen submitted to the mercury treatment). Guaiacum is a resin collected from the Caribbean trees *Guajacum officinale* and *G. sanctum*; it contains phenols such as guaiacol, which are no doubt bactericidal, but for the treatment of syphilis are wholly ineffective. But at least it could not be said of guaiacum, as it was of mercury, that the cure was worse than the disease; and the anti-mercurialists numbered many distinguished men, such as the royal physician Jean Fernel, who refused to treat François I with mercury[55] (the king died, supposedly of syphilis, in 1547).

It was appreciated that absorption of mercury through the skin, whether by inunction or the 'tub', was a fairly slow process, and might not be effective in time to save the patient if the disease had reached an advanced stage. Oral administration of mercury and its compounds was therefore attempted, and the practice had the vociferous support of Paracelsus (1493–1541), a well-known exponent of mineral, as against herbal, medicines. A short, sharp 'cure' of, for example, soldiers on campaign was also very desirable. Mercury compounds, however, tend either to be spectacularly poisonous (as corrosive sublimate, $HgCl_2$), or inactive because of almost complete insolubility (as cinnabar, and to some extent, mercury itself). The discovery of calomel (Hg_2Cl_2), which is sparingly soluble, *relatively* non-toxic, and a powerful purgative, was regarded as a great step forward. Calomel, as we have seen, was known to Chinese pharmacy in much earlier times, but it was probably discovered in Europe about 1600, and was at first a secret remedy. It is said[56] to be referred to obscurely by Oswald Croll, a German physician, in his *Basilica Chymica* (1609); and it is openly treated in the *London Pharmacopoeia* of 1618. The two methods of preparation given there would be those used today—addition of common salt to a solution of mercury in nitric acid, and sublimation of a mixture of mercury and corrosive sublimate. The name of 'calomel' was given to it by Turquet de Mayerne, physician to James I of England another royal sufferer from syphilis, according to report. Other relatively safe mercurials, less popular than calomel, were 'Turpeth mineral' (a basic mercury sulphate) and 'infusible white precipitate' (chapter 6).

During the eighteenth and nineteenth centuries the mercurial treatment of syphilis had no serious rival. Physicians lost faith in guaiacum; another American drug, sarsaparilla (from *Smilax* spp.) had a brief vogue. A fourth method of administration was introduced, in the form of intramuscular injections of 'grey oil', a suspension of mercury droplets in a liquid medium. Many complex organic mercurials were made in the hope of reducing toxicity, and usually rejected after a short trial. The first really hopeful alternative to mercury came with the introduction of Salvarsan by Ehrlich in 1911. Even then mercury continued in diminishing use for another twenty years, and in some medically backward places may still be used. During this period of more than four centuries, the majority of the medical profession would have agreed with Nicholas Lemery[43]

> Hitherto there is no Remedy found out to be so sovereign for the Cure of Venereal Maladies, as Mercury; wherefore its greatest Enemies have been forced to fly to it, after they had tried a long Time, to no Purpose, to drive out the Poison by other Remedies. And in Truth if we knew any milder ones that were able to terminate the Accidents of the Pox as well as this does, 'twould argue much Rashness to make use of *Mercury*, because it is not always conducted according to our Desires, and sometimes very scurvy Consequences do happen upon it; but we know no other that can be esteemed to approach it in Vertue for all Venereal Diseases, and especially the Universal Pox.

The question of whether mercury *really* cured syphilis is still unsettled and likely to remain so. Experimental programmes that could be devised to settle the matter would be both unethical and impracticable. The problem is complicated by the following factors.

(i) Until the introduction of the Wassermann reaction (1906) it was impossible to be certain of a diagnosis of syphilis, a disease whose symptoms can resemble those of many others. In particular, until the work of John Hunter, towards the end of the eighteenth century, it was often confused with the other common venereal disease, gonorrhea. This means that many of those who were treated with mercury and survived, apparently cured, can never have had syphilis at all.

(ii) The 'typical' case of syphilis (there are many atypical ones) progresses in three phases, with remissions between the phases during which the symptoms may be minor or absent. Many supposed cures must therefore have been merely remissions. The interval between the second and third phases can last several years, in which time the patient may well die from some entirely different cause. Hunter

himself had syphilis (from accidental self-inoculation), took the mercury treatment
for three years, and regarded himself as cured; then died of a heart attack during
a violent dispute with his colleagues.

Therefore, many (and possibly all) of those who went through the horrors
of the mercury treatment must have suffered in vain—and the 'cure' itself must
have caused many deaths from poisoning. If so, it must rank as one of the worst
examples of 'iatrogenic illness' in the history of medicine. Yet its apparent success
prompted the extension of mercury therapy to a host of other complaints during
the seventeenth and eighteenth centuries. This was the age of the 'blue pill' (a pill
of mercury and chalk), the standby alike of the fashionable physician of London
or Paris, the army surgeon, and the country apothecary.

This vogue for the widespread prescribing of mercury received its literary
impetus from Augustin Belloste, physician to the ducal household of Savoy. As a
sequel to his widely read treatise on surgery, *Le chirurgien d'hôpital*, he published
a collection of tracts,[57] one of which referred to mercury as a therapeutic agent
in terms of extravagant praise. This was taken up in England, notably by Thomas
Dover (1660–1742), 'the quicksilver doctor' who graduated in medicine as a young
man, spent most of his life at sea as a privateer (or, less politely, a pirate) and
settled down to medical practice in London at the age of sixty-one. Mercury was
to him an almost universal medicine, though 'Dover's powder', for which he is
remembered, is non-mercurial (essentially a mixture of opium and ipecacuanha).
His book, *The ancient physician's legacy to his country*, published in 1733, was
a best-seller, and ran finally into eight editions. Dover claimed that the book was
'what he has collected himself in Forty-nine Years Practice; or, an account of the
several Diseases incident to Mankind, described in so plain a Manner, that any
Person may know the Nature of his own Disease. Together with the several
Remedies for each Distemper, faithfully set down.'

The remedy was usually mercury, drunk neat, or taken in a variety of other
ways

> What an admirable Medicine then is Mercury! . . . the Pox, the Piles, scorbutic
> and scrophulous Ulcers: Inflammations and Fluxions of the Eyes; the Itch, the
> Leprosy, and all cutaneous Foulnesses; internal Ulcers, White Swellings, Tumours,
> Sharp Humours in the Stomach and Guts; Stone, Gout, and Gravel; What an
> Army of the most terrible Foes! against the major Part of which this Friend to
> Nature is a Specific, and the best Remedy against the rest that is yet known.[58]

His defences against the charge that mercury was poisonous ranged from sweet
reason

> Much may be said to shew the Impossibility of Quicksilver doing any Damage
> to the Patient: what gives offence to Nature is, what we term Spiculae, Points,
> or Edges. Now Quicksilver always retaining a globular Figure, together with the
> Softness of the Body, no harm can happen from the Use of it . . .

through rhetoric

> It is a difficult Matter to remove vulgar Errors; they are as strongly rooted as
> the most inveterate Disease, and Reason and Physic are frequently baffled by
> both. How high did the Cry run formerly against the Use of the Bark, one of the
> best Medicines in the *Materia Medica*? What Fears? What Apprehensions of its
> evil Consequences? What strange Misconstructions of its Effects, and downright
> absolute Perversion of its Properties? Nothing can be safe from such Miscon-
> structions; neither Merit in Man, nor Excellence in Medicine . . .

to simple abuse

> PROPHYLACTICUM: OR, a Preservative against the Miserable Consequences
> of the Venomous Bite of a MAD CREATURE. Being a Calm Reply to an Out-
> rageous Libel, intitled Remarks on the Review of the Quicksilver Controversy
> [the title of one of Dover's pamphlets].

Clearly Dover had his opponents; but, fantastic though his career had been, he
differed from most of his contemporaries only in the strength of his language and
the absoluteness of his opinions. So widespread was the use of mercury and its
compounds that by the middle of the nineteenth century anatomists and physio-
logists seriously debated whether mercury might be a normal constituent of the
human body; nearly all cadavers contained quantities detectable by their (not
very sensitive) tests. The effects on the physical and mental health of populations
can only be guessed at. (Animals, incidentally, were treated in the same way; a
Lincolnshire druggist is said to have sold twenty-five *tons* of mercurial ointment
to local farmers in a single year.[59]) Perhaps the most irresponsible prescribing was
of draughts of mercury to relieve constipation—a minor complaint for which many
more effective and safer remedies were known. Even a level-headed physician like
Lemery thought it better to take a lot than a little[43]

> Mercury is given in the Disease called *Miserere*, unto two or three Pounds, and
> is voided again by Siege to the same Weight; it is better to take a great Deal of
> it than a little, because a small Quantity might be apt to stop in the Circum-
> volutions of the Guts, and if some acid Humours should happen to join with it,
> a *Sublimate Corrosive* would be there made; but when a large Quantity of it is
> taken, there's no need of fearing this Accident, because it passes quickly through
> by its own Weight.

Fortunately, by the beginning of the nineteenth century, saner counsels were
being heard. A popular encyclopedia, for instance, sternly reproved the practice[60]

> Whether [mercury] may have been supposed to open the bowels by some self-
> moving power, or by its mechanical weight, the theory is equally absurd; and
> what is worse, the practice is not only useless, but dangerous. There are cases
> on record, in which the quicksilver, thus administered, had accumulated in the
> intestines, formed itself a sac by its weight, and at length produced death, by
> passing into the cavity of the abdomen, in consequence of the rupture of this
> sac. It is obvious, indeed, that the medicine could not ever force a passage by

its weight, along the course of the bowels, since it must ascend occasionally, if the body is erect; and the whole passage must be nearly horizontal, if the patient be in a recumbent position.

But although this barbarous form of medication was eventually discontinued, it was probably the Victorian period that witnessed the most widespread use ever of mercurials in medicine—especially in the form of calomel. This was given freely to infants, for example, as an ingredient in teething powders and aperients; 'pink disease' (acrodynia) was recognised as a clinical entity, and ascribed to a variety of causes, such as malnutrition, long before it was found (about 1950) to be due to calomel poisoning. Donovan's Solution (a mixture of the iodides of mercury and arsenic) was also freely prescribed, to be taken by mouth for skin eruptions, not necessarily of venereal origin. Thousands of quack remedies were on sale, especially in the United States, containing mercurials as the active constituents; no doubt their only virtue was that, after taking them, the patient felt that *something* was happening to him!

Edward Frankland (1825–1899), of whom more is said later, has given us a behind-the-counter glimpse of mercury in pharmacy from his apprenticeship to a Lancaster apothecary in the early 1840s[61]

> In a room on the first floor, there was a very large marble or serpentine mortar, about two feet internal diameter. The pestle was about nine inches in diameter and one foot long, with a wooden shaft about six feet long securely fixed into it, its other end working loosely in an iron ring fixed to a beam in the ceiling. Thus the pestle could be worked round and round and backwards and forwards in the mortar. For the preparation of mercurial ointment, about fourteen pounds of hogs' lard and five or six pounds of quicksilver were placed in the mortar, and had to be triturated until a magnifying glass failed to show any globules of mercury. This blending of mercury with lard is an exceedingly tedious operation; working, in the aggregate, two full days a week, it required about three months to complete it. Moreover, the resistance to the motion of the pestle in the lard is very great, making the labour very hard and the arms ache. Of course this operation is usually performed by mechanical power . . . but in order to save a few pence per pound, Mr X imposed this hard labour and risk of salivation on his junior apprentice. The home-made product also was by no means equal in quality to the machine-made article. Notwithstanding all these three months labour, I could never succeed in entirely "killing" the quicksilver; a lens always showed the ointment to be full of minute globules.

A more rational approach to the use of mercurials in medicine began with Robert Koch's claim (1881)[62] that mercuric chloride was a more effective bactericide than the 'carbolic acid' (phenol) popularised by Lister. This judgment was based on *in vitro* tests—of which Koch was a pioneer—and attempts to extend it to hospital practice proved disappointing owing to the unavoidable necrosis associated with 'corrosive sublimate'. Koch's high reputation, however, prompted the German chemical industry to investigate the bactericidal effects of more complex mercury compounds, such as the mercury salts of organic acids. These

began to come on the market in 1914, under such names as Afridol, Mercuro-
chrome, Merthiolate. These had a considerable vogue during the First World War
and afterwards as ingredients of 'antiseptic soaps' and 'germicidal ointments'. These
are still made and used, though there is some evidence that they are only marginally
more effective (if at all) than plain soap or Vaseline. A soap of this kind, containing
mercuric iodide, sells very well in Africa—not for its antiseptic properties but because
it has a bleaching action on black skin. Fildes (1920) showed that the pharmacological
activity of mercurials was due to the great affinity of mercuric ions for thiol (SH)
groups—a discovery unwittingly made in 1834 by the Danish chemist Zeise,[63] who
first prepared aliphatic thiols and named them 'merc-aptans' (*corpus mercurium
captans*; readily combining with mercury).

The main present-day use of mercurials in pharmacy is as diuretics. This property of 'blue pill' and of calomel, the two most popular mercurial medicines, was
well known to the great medical teacher Boerhaave of Leyden (1668–1738) and
perhaps earlier. It seems to have been thought of no interest until 1920, when
Vogl in Vienna noticed the diuretic effect of mercury salicylate prescribed for a
syphilitic patient, and found that a certain proprietary mercurial called Novasurol
had an even stronger effect. This led to many new mercurial formulations, and—
although there are now other and safer diuretics—the mercury compounds still
retain their place in modern Pharmacopoeias. The trend in chemotherapy is, however, strongly against heavy-metal compounds, and it may be doubted whether
mercury will have any function in pharmacy at all in thirty years' time.

9 Organomercury Compounds

Organometallic chemistry had its effective beginning when Edward Frankland, working in Bunsen's laboratory in Marburg, discovered that metallic zinc would react with methyl iodide (1849). On his appointment, two years later, to the chair of chemistry at the newly founded Owens College in Manchester, he used his opportunity to extend this reaction to other metals.[64] Mercury, he found, would react with methyl iodide when exposed to sunlight, and he prepared crystalline methylmercuric iodide (CH_3HgI) and some of its derivatives. He noted their nauseous taste, but was totally unaware of their extraordinary toxicity—though he survived this, as well as much other exposure to mercurials, to die at a normal age. It was in the same paper (1852) that Frankland put forward his idea of 'saturation capacity', which later grew into the concept of valency. Shortly afterwards, Zinin in Russia found that the more reactive allyl iodide readily formed C_3H_5HgI even without the assistance of sunlight.

The mercury dialkyls were accidentally discovered by George Bowdler Buckton (1818–1905) in 1858, while he was working in the Royal College of Chemistry in London. He was trying to make methylmercuric cyanide by double decomposition, but the reaction took an unexpected course

$$2CH_3HgI + 2KCN \rightarrow Hg(CH_3)_2 + 2KI + (CN)_2 + Hg$$

and he isolated mercury dimethyl as a heavy volatile liquid.[65] He too escaped unscathed, though when the investigation was resumed a few years later, there was a disaster that has passed into the folklore of chemistry. When Frankland left Manchester in 1857, he went to St Bartholomew's Hospital in London ('Bart's'), where he again took up the study of organomercurials. With his assistant B. F. Duppa, he discovered a new and more convenient synthesis of mercury dialkyls, in the reaction of sodium amalgam with an alkyl iodide in ethyl acetate—the latter acting not only as a solvent, but being for some reason essential for reaction to take place.[66] Frankland left Bart's in 1863, and the research was continued by his successor William Odling, together with a number of assistants, including Buckton and a young German, Dr Carl Ulrich. The unfortunate Ulrich, when 'preparing a large quantity of the mercuric methide in the midst of January 1865 . . . met with an accident, breaking one of the tubes which contained the preparation. According to his own statement, he inhaled a great portion, not having taken the necessary precautions'. The following day 'his countenance had attained a dull, anxious, and confused expression' and he was advised to seek medical attention. On 3 February he was admitted to Bart's in a very weak condition; on the 9th, he became noisy and had to be put under mechanical restraint. The next day, his breath and body began to smell offensively, and he was comatose, except that from time to time he raised himself and uttered incoherent howls. He died on the 14th.[67]

A young technician, identified only as 'T.C.' from the same laboratory was admitted to Bart's on 28 March. His symptoms, at first milder than Ulrich's, developed inexorably, and by the summer he was completely demented, with no control over his bodily functions. He continued in this pitiful state for many months, finally dying on 7 April 1866. A third assistant was also taken ill, but soon ceased to be mentioned in the reports, and may have recovered.

As a result of these tragic events, a publicity-seeking chemist of no great ability named T. L. Phipson waged a virulent press campaign against Frankland, although Odling admitted that the responsibility was his. Accusations of criminal carelessness and counter-accusations of distortion of facts were freely exchanged;[68] the episode may even have done a little good in improving the appallingly low standards of laboratory hygiene. Amidst all the fuss, Buckton (who had first prepared 'mercuric methide') wisely retired from chemistry, married Odling's daughter, and spent the rest of his long life as a country gentleman. The most inexcusable part of the whole affair is that Frankland contributed a detailed article on organomercurials to an important chemical handbook, *Watts' Dictionary of Chemistry* (1882), in which no hint is given of any possible hazard in working with these substances; indeed the *taste* of mercury dimethyl is described as 'faint but mawkish'!

It is understandable that relatively little further work was done on organo-mercury compounds for about thirty years after the events at Bart's; though the inoffensive mercury diaryls and arylmercury halides were prepared from aryl bromides and sodium amalgam.[69] Interest revived just before the end of the century, when the important 'mercuration' reaction was discovered by Otto Dimroth. Mercuric acetate and benzene were found to react to give phenyl mercuric acetate, and the reaction was even more facile with suitably substituted benzenes, such as phenols, phenol ethers, and derivatives of aniline.[70] Together with an equally general but somewhat less useful reaction discovered by Peters a few years later[71]

$$ArSO_2H + HgCl_2 \rightarrow ArHgCl + SO_2 + HCl$$

it made possible the preparation and study of a host of relatively non-toxic organomercurials. The first compound containing mercury as a member of a heterocyclic ring was described by von Braun in 1913, though the structure he put forward is incorrect.[72] It is also appropriate to mention at this point the 'mercarbides', first properly described in 1898,[73] though known long before. These are non-crystalline, heavily mercurated substances made, for example, by refluxing ethanol with a suspension of mercuric oxide in aqueous caustic alkali. The mercarbides are now believed to be mixtures of polymers of relatively low molecular weight. At about the same time (1900) Hofmann and Sand began the study of the complex reactions between mercury salts and olefins. For many years there was uncertainty as to whether the products of these reactions were best regarded as coordination complexes (like the olefin–platinum-metal compounds) or as true organomercury compounds containing a 'normal' C—Hg bond. The question has been decided in favour of the latter.[74]

Compounds of the heavy metals were, in 1900, the most effective fungicides known, and naturally it was not long before these now readily available organomercurials were being tested for this purpose. For one major application—the mildew-proofing of textiles—they have never proved suitable; but as agricultural seed-dressings they were very successful, and have remained so to this day. Seed-borne diseases caused by microscopic fungi (smut, bunt, and various kinds of rust) have always been responsible for damage, sometimes amounting to havoc, in the world's cereal crops. The first tests date from about 1912, when the German Agricultural Service reported favourably on a substance described as 'chlorphenolquecksilber' supplied by the firm of Bayer. In 1915, Bayer put this preparation on the market under the name of 'Uspulun' (from *Ustilago* and *Puccinia*, two genera of pathogenic fungi), and competitive products of similar composition appeared from other firms after the First World War. They all had the serious disadvantage that the seed had to be steeped in an aqueous solution of the fungicide, and then dried. These costly operations were eliminated by Bayer's 'Tillantin' (from *Tilletia caries*, the bunt fungus). This was a nitrophenol mercury compound, sold as a fine powder, which simply needed to be mixed with the grain in rotating drums (1924).

Phenyl mercuric acetate, by far the most important of this class of fungicide, was marketed by Bayer in 1929 under the name 'Ceresan'. Under various brand names and formulations, its use is now world-wide; its effectiveness is well proven enough to outweight the many hazards associated with it, both in production and use (it is highly dermatitic to some people). Toxic reactions among workmen treating the grain, or farmers sowing it, are now rare; but there have been several disasters caused by accident or ignorance, where treated grain has been used for food instead of planting. Even colouring the grain has not always proved sufficient warning. But mercurial fungicides have one great advantage, in that resistant strains of the target organisms never seem to have arisen yet.

From about 1950 onwards, phenyl mercuric acetate and related compounds have been used in enormous quantities to control 'slime', (a viscous microbial growth) in paper mills. The effluent from these mills, discharged into rivers, lakes, or enclosed seas such as the Baltic, gave rise to concentrations of mercury in mud and water that caused justifiable concern. No definite ill-effects were ever traced to this practice, but non-mercurial slimicides are now widely used. They tend to be less effective, but they are less objectionable from an environmental point of view.

A short account of the outbreak of organomercury poisoning at Minamata, Japan, in the 1950s, is reserved for the final chapter.

10　Mercury as Poison

Mercuric chloride, 'the corrosive sublimate', probably first made by Arabic alchemists in the tenth century, became notorious in medieval and Renaissance Europe as a violent poison and was no doubt responsible for many sudden and mysterious deaths. The most famous case in England was that of Sir Thomas Overbury, imprisoned in the Tower after a Court intrigue, and finally done to death by a bribed apothecary in 1613. Matters were much the same in China, where the Court physicians found a test for mercury at an early date, by pressing pieces of gold foil into the viscera, and looking for evidence of amalgamation. Half a grain was a fatal dose, and though antidotes such as milk and butter were hopefully prescribed, death from renal failure was almost inevitable. Lemery, in his *Cours de chimie*,[43] had an ingenious explanation for this dramatic toxicity and its corrosive action on tissue in terms of a crude atomic theory

> The *Corrosion* of *Sublimate* does proceed from the edged Acids which fix in the body of *Mercury*, and it may be said with great Probability, that this Metal always retaining a round Figure (let it be divided never so subtilly) does rarifie by the Heat of Fire into an abundance of little Balls, which the acid Spirits do fix into on all Sides, and so interlace themselves in it, that they hinder its rising higher, and do together make one Body, that is called *Sublimate*. But when this *Sublimate* is applied to Flesh, the Heat and Moisture of it do set in Motion the *Mercurial* Parts, and the Motion of the little Balls being once raised, they rowl about with great Fury, and tear the Flesh with the Edges they contain, which are like so many little Knives cutting wherever they touch; from whence it comes to pass, that if the *Sublimate* should be taken inwardly, it kills in a very little Time . . .

Mercuric chloride, or mercury itself, are the most plausible candidates for the 'leperous distilment' that Claudius poured into the ear of his sleeping brother, King Hamlet. Apart from corrosive sublimate, however, most writers, though they knew that mercury and its compounds were poisonous, dealt with the matter in an astonishingly casual way. The notorious ill-health of those who worked in mercury mines and roasted the ore was ample proof of the dangers of mercury vapour; but Agricola, an acute observer and a humane man, discussing the distillation of crude mercury, merely remarked

The pots, lest they become defective, are moulded from the best potters' clay, for if there are defects the quicksilver flies out in the fumes. If the fumes give out a very sweet odour it indicates the quicksilver is being lost, and since this loosens the teeth, the smelters and others standing by, warned of the evil, turn their backs to the wind, which drives the fumes in the opposite direction; for this reason, the building should be open around the front and the sides, and exposed to the wind.[75]

Lemery, of course, was not at a loss for an explanation of the effects of mercury vapour

Those who draw it out of Mines, or work much with it, do often fall into the Palsie, by Reason of Sulphurs that continually steam from it; for these Sulphurs consisting of gross Parts, do enter through the Pores of the Body, and fixing themselves rather in the Nerves, by Reason of their Coldness, than in the other Vessels, do stop up the Passage of the Spirits, and hinder their Course.[43]

Father Acosta knew of a wonderful antidote[76]

. . . for that the fume of Mercurie is mortall . . . the workmen preserved themselves from this venome, by swallowing a double duckat of gold roled up; the which being in the stomacke, drawes unto it all the quicke-silver that enters in fume by the eares, eyes, nostrilles, and mouth, and by this meanes freed themselves from the danger of quicke-silver, which the gold gathered in the stomacke, and after cast out by the excrements; a thing truly worthy of admiration.

There was thus a certain casualness about the known toxic properties of mercury metal. Part of this indifference can be ascribed to the widespread use of mercury in the treatment of syphilis, to which reference has already been made; the inevitable symptoms of mercury intoxication were confused with the symptoms of the disease. The heavy and distressing salivation, for instance, was implicitly taken by some practitioners as confirmation of a correct diagnosis of syphilis, and by others as a sign that the treatment was 'working'. Sometimes, however, the effects were too disastrous to be ignored, as in the case of *H.M.S. Triumph*.[77] In 1810, this ship salvaged 130 tons of mercury from the wreck of a Spanish vessel off Cadiz. With the usual lack of concern, the containers were stored in the breadroom; and being wet with sea water, soon rotted, and the mercury quickly contaminated the whole ship. Almost all the officers and men suffered severely from 'ptyalism' (salivation), most became unfit for service, and there were some deaths. The ill-named *Triumph* was brought back to England, cleaned and ventilated at great expense, but this did little to help matters, and the ship became virtually unserviceable.

Apart from such random disasters, the main classes at risk from mercury were the medical profession, workers in mercury mines, mirror-makers, and scientific workers—to which were later added hatters. The hazards were first clearly documented by Bernardino Ramazzini (1633–1714) in his book *De morbis artificum* (1700, 1713), which is regarded as the foundation of industrial medicine (though to a modern reader it seems disappointingly uncritical).[78] The use of a tin amalgam to form a reflecting surface was known to the Chinese in the early centuries of our era, though the glass mirror with a reflecting back made of tin or silver amalgam does not appear until the European middle ages.[†] Ramazzini reported serious symptoms of poisoning among the craftsmen who worked with the amalgam, though nothing much seems to have been done about it as long as the process was used—it was gradually made obsolete by the 'silver mirror' discovered by Liebig in 1835 (a solution of a silver salt in a glass vessel, reduced by an aldehyde, deposits silver on the walls as a brilliant reflecting layer).

It is quite difficult to ascertain the true history of the use of mercury compounds in the hat-making trade. Hats are made of felt, a non-woven textile of animal hairs oriented at random, which interlock because of scales or roughnesses on their surfaces. Wool is scaly, and felts naturally; but other hairs, including the abundant rabbit and the once fashionable beaver, have to be artificially roughened with an acid liquor (usually dilute nitric acid) before they will felt satisfactorily. The addition of mercury salts to this liquor was found to give a much better felt, and this modified process became known as 'carrotting' because the fibres were changed in colour to a dull orange. The danger arises at a later stage in making hats, when the felt is dried and a dust strongly charged with mercury pervades the workshop, which was usually ill-ventilated. The workpeople then suffer from salivation, erethism, and 'hatter's shakes', a tremor whose consequences can be seen most clearly in the subject's handwriting. It has been both asserted and denied that the expression 'mad as a hatter' (which can be traced back to the early nineteenth century) derives from this industrial hazard. What is at issue is *when* this mercurial treatment was introduced. There are no contemporary references before the middle of the nineteenth century (for example reference 39), but some writers believe it to have been in use in the eighteenth or even the seventeenth century. There was certainly a secret process (actually called 'secretage') ascribed to Huguenot hatters who left France after the revocation of the Edict of Nantes in 1685, but it seems impossible now to determine whether or not this involved mercury. Perhaps the process was discovered and re-discovered several times. The use of mercury in felting continued well into the present century, but is now forbidden in most countries; there are acceptable substitutes.[52]

Laboratory workers of all kinds have, like the medical profession, always been at particular risk from mercury vapour. Apart altogether from the compounding and administering of mercurial medicines, the number and variety of pieces of scientific and medical equipment employing the unique properties of mercury is very large, and continually increasing. The following is an admittedly incomplete list; mercury barometer (Torricelli, 1644)[‡]; mercury-in-glass thermometer (Fahrenheit, 1714); mercury trough for collection of water-soluble gases (Priestley,

[†] The first account of making this kind of mirror is in Giambattista della Porta's *Natural Magick* (1558), a strange medley of technological recipes and conjuring tricks.

[‡] This was not the first scientific instrument to use mercury. During the Middle Ages, 'mercury clocks' were made both in Europe and China, on the principle of the well-known water-clock or clepsydra. It is not likely that they were very successful, since the flow-properties of mercury are much altered by traces of dirt.

1772); the eudiometer (late eighteenth century); the sphygmometer (Poiseuille, 1828) and numerous other applications of manometry; the mercury-vapour pump (Sprengel, 1865); and the McLeod gauge (1874). There is also the extensive use of mercury in electrical experiments from 1800 onwards. It was used, for instance, by Humphry Davy in his isolation of the alkali metals by electrolysis (1807), and continually by such electricians as Faraday, Wheatstone, Joseph Henry, and Ohm. The use of mercury as an electrical contact is particularly dangerous in the long term, since minute sparks can cause local boiling, and thus higher concentrations of vapour. Even in non-electrical work, however, globules of mercury carelessly spilt, or escaping from broken apparatus, can lodge unobserved in sinks, under benches, or between floorboards. Faraday himself showed that gold leaf, suspended over mercury in an enclosed space, would rapidly amalgamate even at room temperature, owing to the appreciable vapour pressure of the metal. (The effect can be shown quite spectacularly by suitable illumination of a beaker of mercury with ultra-violet light.)

A paradoxical situation thus arose, in that laboratory workers were aware, on an intellectual level, of the dangers of mercury vapour, but did not apply this knowledge to themselves. The hazards were first forcibly pointed out as late as 1926, in an understandably emotional article by Alfred Stock (1876–1946).[79]

> My decision to publish a frank account of my personal sufferings [he began] . . . is prompted by a keen desire to warn, in the strongest terms, all who have to deal with metallic mercury of the dangers they face, and to spare them the wretched experiences which have cast a shadow over a great part of my life.

Stock, as a physical chemist, was continually exposed to mercury vapour, especially from vacuum pumps, for a period of twenty-five years. While still a young man he suffered from irritation of the mucous membrane of the nose, for which he had to undergo much painful medical treatment, including surgery, without relief. He also had violent headaches, tremor of the hands, and a socially troublesome inflammation of the bladder, as well as loss of memory and 'slow and pedantic' mental processes. Even when assistants and long-staying visitors to the laboratory were affected in the same way, the cause was not suspected. 'I felt stupid all the time I was in Germany', said one of them. Eventually, in 1924, a piece of research had to be completed in a hurry because a member of staff was leaving; cleanliness was abandoned and mercury droplets were everywhere. On this occasion an assistant developed the symptoms of gross mercury poisoning, which was correctly diagnosed by his brother, a physician. When the place was properly cleaned up, Stock's own symptoms disappeared, and his health apparently remained good for the rest of his life.

The ravages caused among scientists and laboratory technicians (as well as physicians and apothecaries) by mercury vapour can never be estimated. Mild cases would be indistinguishable from minor indisposition and psychological trouble; one is led to speculate about the continual ill-health of certain eighteenth- and nineteenth-century scientists and medical men, and their unnecessary rancour in debate. Stock himself pointed to Pascal and to Faraday as possible sufferers, though more recent evaluation of their known symptoms has led to a 'not proven' verdict, since in both cases their medical history was quite complex. A plausible case has been made out that the unexpected death of Charles II was due to mercury poisoning—he spent much time towards the end of his life amusing himself with amateurish 'alchemical' experiments.[80] It may be significant that Newton (on the evidence of his unpublished notebooks) was boiling several pounds of mercury just before his period of temporary insanity in 1692-3.

Stock also wrote some rather alarmist papers implicating the use of amalgams in dental fillings as a cause of widespread sub-acute mercury poisoning.[81] Silver amalgam had been introduced for this purpose by a Parisian dentist called Taveau in 1826, though there were technical objections to its use that were only slowly overcome; from 1843 to 1850, for example, the American Association of Dental Surgeons required members to sign a pledge not to use 'any amalgam whatever' (copper amalgam is the one usually used nowadays). The matter is still not entirely cleared up, but it is now believed that the absorption of mercury from dental fillings is too small to represent a health hazard.

The problem of dental fillings is of course only one aspect of the problem of mercury in the environment. There can be no doubt that the background concentration of mercury has increased enormously during the historical period, especially during the last few centuries. Since the beginning of mercury mining, millions of tons of the element have been taken from beneath the earth where it was 'locked up' (usually as cinnabar) and liberated finally after many adventures into the air or the sea. Patients liberally medicated with mercury have been buried or cremated; Spanish galleons laden with mercury have sunk on their way to the silver mines of the Indies; the mercury used in the amalgamation process is now in the soil, the sea, or the air; every shot fired from a fulminate-detonated cartridge adds a little more mercury to the atmosphere; the Castner–Kellner process ensures that traces of mercury find their way into almost every product made by the chemical industry. There is no known terrestrial process that will lock up much of this mercury again in an inert form; and there are biological processes, both known and suspected, that will sequester environmental mercury to a higher (though still small) concentration. As with other heavy metals such as lead and cadmium, it is very difficult to discover whether or not these minute but ubiquitous concentrations of mercury have any effect on health. (Unlike lead, mercury is fairly rapidly eliminated by the body, and is not a cumulative poison; but, unlike arsenic, there is no evidence of acquired tolerance to continued small doses.)

These remarks refer only to *inorganic* mercury. Recent events have shown, however, that organomercury compounds in the environment constitute a much more serious threat. This is the aftermath of the 'Minamata disaster', a distressing and often fatal illness, which attacked a poor fishing community on the island of Kyushu, Japan.[82] The first cases were noted in 1953, and three years later the 'epidemic' had assumed such grave proportions as to demand a full investigation. Possible causes, such as malnutrition or infectious disease, were eliminated; but it was eventually found that fish and shellfish from local waters, the staple diet of most of the population, contained abnormally high amounts of mercury. When fishing was banned, there was economic and social distress, but no further cases of illness. Those already affected, however, did not regain their health, as would be expected from ordinary mercury poisoning. The mercury clearly came from a nearby chemical works, where (among other processes) acetaldehyde was made from acetylene using a mercuric salt as catalyst; the plant was rather inefficient and about a kilogram of mercury was lost for every ton of aldehyde produced. It went into the shallow waters of what was almost an inland sea. The firm believed that the mercury so lost was inorganic, but the symptoms of 'Minamata disease' were not those of inorganic mercury intoxication, nor were the amounts of mercury ingested by the victims particularly high. There was, for example, no salivation, but the effects, especially on the brain, were much more like those observed in the Bart's tragedy of the 1860s, and in the few later accidents with alkylmercurials.

Closer examination of the factory effluent showed that it did indeed contain small quantities of the methylmercury ion CH_3Hg^+ and that shellfish in the bay sequestered this as CH_3HgSCH_3. This confirmed the diagnosis of organomercury poisoning, but the matter did not end there. Independent work in Sweden[83] showed that bacterial action in bottom sediments or in rotting fish could convert inorganic mercury into both the methylmercury ion (soluble in water) and dimethylmercury (volatile). The anaerobic mud of the shallow seas off Minamata could perform this methylation, so that even if the effluent had contained only inorganic mercury, the disaster could still have happened. The implications of this deeply disturbing discovery are now being studied, and the health authorities of many countries are watching the situation closely.

It is perhaps fortunate that the inevitable depletion of the world's mercury resources will eventually make the metal so expensive that it will no longer be possible to cast it needlessly into the environment beyond hope of recovery.

Bibliography

The only recent study of the history of mercury is Leonard J. Goldwater's, *Mercury: a history of quicksilver* (Baltimore, 1972), which came into our hands while this chapter was being written. Far from being a recital of the same facts in different words, it forms a useful complement to our work. Goldwater is a medical man, and the book is frankly weak on chemical matters, as well as confusing in its arrangement; but it is recommended to readers who require more information on the medical use of mercury, on mercury in the environment, and on the economics of mercury mining.

Special topics are dealt with in the following books:

A. P. Whitaker, *The Huancavelica Mercury Mine* (Cambridge, Massachusetts, 1941).

L. A. G. Strong, *Doctor Quicksilver* (London, 1955)—a biography of Thomas Dover in which the small amount of available material is interestingly extended by the novelist's art.

W. Eugene Smith and Aileen M. Smith, *Minamata* (New York, 1975)—was published too recently to be used in the present study.

This applies also to an important article by M. Teich, *Ann. Sci.*, **32** (1975) p. 305, dealing with the amalgamation process in Central Europe in the eighteenth century, but with interesting references to Spanish–American practices.

References

1. W. Wreszinski, *Der Papyrus Ebers* (Leipzig, 1913).
2. E. O. von Lippmann, *Entstehung und Ausbreitung der Alchemie* (Berlin, 1900) p. 600.
3. J. R. Partington, *Origins and development of applied chemistry* (London, 1935), 84.
4. W. Gowland, *Archaeologia*, 69 (1920) p. 157.
5. Ref. 3, p. 381.
6. R. C. Thompson, *Chemistry of the ancient Assyrians* (London, 1925) p. 59.
7. Ref. 3, p. 292.
8. A. Weiskopf, *Z. angew. Chem.*, 14 (1910) p. 429.
9. Pauly-Wissowa, *Real-Encyclopädie*, vol. 9, 'Hydrargyrum' p. 55.
10. E. R. Caley, *J. chem. Educ.*, 23 (1946) p. 314.
11. Aristotle, *Meteorologica*, Book IV, 385b.
12. Theophrastus, *De lapidibus*, ed. and trans. D. E. Eichholz (Oxford, 1965) p. 81.
13. Vitruvius, *De architectura*, Book VII.
14. O. Davies, *Roman mines in Europe* (Oxford, 1933) p. 138.
15. Ref. 14, p. 139; R. J. Forbes, *Studies in ancient technology*, Vol. 8, p. 173; M. M. Vassits, *Rev. archéologique*, 43 (1954) p. 60.
16. T. Haupt, *Berg- und Hüttenmännische Z.*, 1888, p. 41, says that mercury has been found in an Etruscan tomb (?fourth century B.C.)
17. Ref. 14, p. 67.
18. Pliny, *Natural History*, Book XXXIII, chap. 20.
19. Ref. 18, chap. 32.
20. M. P. E. Berthelot, *Introduction à la chimie des anciens et du moyen âge* (Paris, 1889) p. 23.
21. J. Needham, *Science and civilisation in China*, vol. 5, part 2 (Cambridge, 1974).
22. E. R. Caley, *J. chem. Educ.*, 3 (1926) p. 1149; 4 (1927) p. 979.
23. M. P. E. Berthelot, *Les origines de l'alchimie* (Paris, 1885), p. 207.
24. R. Steel, *Isis*, 12 (1930) p. 10 (recipe 23).
25. Ref. 24, recipe 20.
26. T. Phillips, *Mappae Clavicula; a treatise on the preparation of pigments during the Middle Ages* (1846) p. 183.
27. J. W. Mellor, *A comprehensive treatise on inorganic and theoretical chemistry*, vol. 4 (London, 1922–37) p. 796.
28. *Book of minerals* (trans. D. Wyckoff) (Oxford, 1967) p. 212.

29. G. Chaucer, *The Canterbury Tales* (modernised N. Coghill) (Harmondsworth, 1951).
30. Li Ch'iao Ping, *The chemical arts of old China* (Easton, Penn., 1948); N. Sivin, *Chinese alchemy; preliminary studies* (Cambridge, Mass., 1968); J. Needham and Wang Ling, *Clerks and craftsmen in China and the West* (Cambridge, 1970); and ref. 21.
31. R. S. Britton, *Harvard J. Asiatic studies*, 2 (1937) p. 1.
32. E. Divers, *J. Soc. chem. Ind.*, 13 1894 p. 108.
33. H. C. Hoover and L. H. Hoover, *Georgius Agricola: De re metallica* 2nd edn. (New York, 1950) pp. 426–32.
34. Quoted in ref. 3 p. 193.
35. C. S. Smith and M. T. Gnudi, *The Pirotechnia of Vannoccio Biringuccio* (New York, 1959) p. 384.
36. Ref. 33, pp. 295–9.
37. J. de Acosta, *Natural and moral history of the Indies* (reprint of English trans., 1604) (London, 1880), pp. 211–21.
38. E. J. Hamilton, *American treasure and the price revolution in Spain* (Cambridge, Mass., 1934).
39. A. Ure, 'Mercury' in *Dictionary of Arts, Manufactures, and Mines* (London, 1853).
40. J. Priestley, *Experiments and observations on different kinds of air*, vol. 2 (1775) p. 33.
41. A. L. Lavoisier, *Traité élémentaire de chimie* (Paris, 1789) pp. 19–20.
42. J. Dalton, *A new system of chemical philosophy*, vol. 2 (Manchester, 1827) pp. 19–20.
43. N. Lemery, *Cours de chimie* (Paris, 1663). We have used an English translation published in 1720.
44. A-N-E. Millon, *Ann. chim.*, 18 (1846) p. 397.
45. A. F. Wells, *Structural inorganic chemistry* 3rd edn. (Oxford, 1962).
46. T. J. Seebeck, *Ann. chim.*, 66 (1808) p. 191.
47. A. M. Ampère, *Ann. chim.*, 2 (1816) p. 16.
48. *Chem. News*, 12 (1865) pp. 158, 170, 307
49. E. C. Howard, *Phil. Trans. Roy. Soc.*, 90 (1800) p. 204.
50. *The rise and progress of the British explosives industry* (London, 1909).
51. W. T. Brande, *A manual of chemistry* 6th edn. (London, 1848), vol. 1 p. 1003.
52. D. Hunter, *Diseases of occupations* (London, 1969).
53. M. Kutscherow, *Ber. dtsch. chem. Ges.*, 14 (1881) p. 1540.
54. G. Fracastoro, *Syphilis, or the French Disease* (English trans., London, 1935).
55. C. Sherrington, *The endeavour of Jean Fernel* (Cambridge, 1946).
56. J. R. Partington, *A history of chemistry* (London, 1961), vol. 2, p. 168.
57. A. Belloste, *Suite du Chirurgien d'hôpital, contenant differens traitez, du mercure; des maladies des yeux, et de la peste* (Paris, 1725).
58. 'A Gentleman of Trinity College, Cambridge' [T. Dover], in *Encomium argenti vivi* (London, 1733).
59. A. W. Blyth and M. W. Blyth, *Poisons: their effects and detection* 5th edn. (London, 1920) p. 683.
60. 'Constipation' in *Rees' Cyclopaedia* (1819).
61. M. N. W. and S. J. C. (eds.), *Sketches from the life of Sir Edward Frankland* (London, 1902) p. 28.
62. R. Koch, *Mittheilungen aus dem Kaiserlichen Gesundheitsamt* (1881) p. 234.
63. C. W. Zeise, *Liebig's Ann.*, 9 (1834) p. 1.

64. E. Frankland, *Phil. Trans. R. Soc.*, 142 (1852) p. 417.
65. G. B. Buckton, *Phil. Trans. R. Soc.*, 148 (1858) p. 163.
66. E. Frankland and B. F. Duppa, *J. Chem. Soc.*, (1863) p. 415.
67. *Chem. News*, 12 (1865) p. 213; 13 (1866) p. 59.
68. *Chem. News*, 13 (1866) pp. 7, 23, 35, 47, 59, 84.
69. E. Dreher and R. Otto, *Liebig's Ann.*, 154 (1870) p. 93.
70. O. Dimroth, *Ber. dtsch. chem. Ges.*, 31 (1898) p. 2154.
71. W. Peters, *Ber. dtsch. chem. Ges.*, 38 (1905) p. 2567.
72. J. v. Braun, *Ber. dtsch. chem. Ges.*, 46 (1913) p. 1792; 47 (1914) p. 490;
    S. Hilpert and G. Grüttner, *ibid.*, p. 186.
73. K. A. Hofmann, *Ber. dtsch. chem. Ges.*, 31 (1898) p. 1904.
74. J. Chatt, *Chem. Rev.*, 48 (1951) p. 7.
75. Ref. 33, p. 428.
76. Ref. 37, p. 212.
77. W. Barnett, *Phil. Trans. R. Soc.*, 113 (1823) p. 402.
78. B. Ramazzini, *Diseases of workers* (English trans., New York, 1964).
79. A. Stock, *Z. angew. Chem.*, 39 (1926) p. 461.
80. M. L. Wohlbarscht and D. S. Sax, *Notes Rec. Roy. Soc.*, 16 (1961) p. 154.
81. A. Stock, *Z. angew. Chem.*, 39 (1926) p. 984; 41 (1928) p. 663.
82. R. Hartung and B. D. Dinman (eds.), *Environmental mercury contamination* (Ann Arbor, 1972); A. Tucker, *The toxic metals* (London, 1972).
83. S. Jensen and A. Jernelöv, *Nature*, 223 (1969) p. 753.

PART 2

The Coordination Chemistry of Mercury

W. Levason† and C. A. McAuliffe

Department of Chemistry, University of Manchester Institute of Science and Technology

† Present address: Department of Chemistry, The University, Southampton
SO9 5NH.

11 Introduction

The coordination chemistry of mercury has received rather less attention than that of the later transition elements, due in part to the fact that as a d^{10} metal ion Hg^{2+} exhibits neither paramagnetism nor 'd–d' spectra, the study of which have provided much of the impetus in other areas of coordination chemistry. However, the current interest in the biological properties of mercury clearly points to the need for a more complete understanding of the ability of mercury to bind various donor atoms and of the resulting stereochemistry.

This part aims to review comprehensively the compounds of mercury with the exception of organomercurials. Thus, all complexes with a direct Hg–C bond are excluded. The comprehensive review of mercury chemistry by Gmelin[1] covers the literature up to 1960 (and up to 1965 in later parts), and emphasis in the present work is thus placed on the more recent developments. Other relevant review articles on particular aspects are noted in the appropriate sections. The literature has been covered up to May 1975.

12 General

Mercury has the electronic configuration $[Xe]4f^{14}5d^{10}6s^2$, and its first three ionisation potentials are 10.43, 18.65, and 34.4 eV, respectively, with the result that no more than two electrons are removed under chemically significant conditions.

The general aspects of mercury chemistry have been described by Roberts[2] and will not be repeated in detail here. The chemistry of mercury in relation to that of other B metals has also been discussed.[3] The thermodynamic properties, chemical equilibria, and standard potentials of mercury in solution have recently been reviewed,[4] and so the solution chemistry is generally not discussed here.

Mercury differs from other metals in readily forming a polycation, the mercurous ion, Hg_2^{2+}. Consideration of the standard potentials[5] yields

$$Hg_2^{2+} = Hg_{(\ell)} + Hg^{2+} \qquad E^0 = -0.115 \text{ V}$$

and

$$K = [Hg^{2+}]/[Hg_2^{2+}] = 1.15 \times 10^{-2}$$

which show that the mercurous ion is stable to disproportionation, but only just so, and any ligand that appreciably reduces the activity of the Hg^{2+} ion either by complexation or precipitation will bring about disproportionation. Since many ligands bond Hg^{2+} quite strongly the number of stable mercurous complexes is limited.

13 Mercurous Mercury Hg_2^{2+}

Although the Hg_2^{2+} ion is no longer the only polycation[6] known, it is by far the most common and readily obtained. The evidence for the formulation Hg_2^{2+} as opposed to Hg^+ has been discussed by numerous authors (references 1, 2, 7 and references therein), and will not be recounted here. The first report[8] of $\nu(Hg-Hg)$ in aqueous solutions of mercury(I) nitrate was published as long ago as 1934, and more recently Raman spectral data have been reported for a series of mercury(I) compounds (table 13.1). A considerable quantity of X-ray structural data on mercury(I) compounds has become available recently, and is tabulated in table 13.2.

Mercury(I) fluoride is usually prepared from the carbonate and 40 per cent aqueous hydrofluoric acid. It is decomposed by water, and on heating disproportionates into mercury and HgF_2. The other mercury(I) halides are insoluble in water and are usually obtained by precipitation of aqueous mercury(I) nitrate solution with an alkali halide.[33] All four halides are light sensitive. The vapour density of mercury(I) chloride corresponds to 'HgCl', which has been shown to be due to disproportionation into mercury and $HgCl_2$. Halogens (X_2') convert the Hg_2X_2 compounds into mercury(II) but there has been much controversy over the nature of the products when $X' \neq X$ (q.v.).

The original X-ray data on the mercury(I) halides, Hg_2X_2, was of low accuracy, but based on the available values it was proposed that the $d(Hg-Hg)$ was related to the electronegativity of the X group.[34] Dorm[14,35] has re-investigated the structures and shown that there is no obvious trend in the Hg-Hg distances (table 13.2) and that the Hg-Hg distance is not significantly different in Hg_2F_2 or Hg_2Br_2.

Mercury(I) cyanide does not exist; addition of cyanide ions to a mercury(I) solution produces $Hg(CN)_2$ and mercury, as would be expected in view of the strong $Hg^{II}-CN$ interaction. However, mercury(I) thiocyanate,[36,37] azide,[38] and cyanate[39] are known, while the case of selenocyanate[37] does not seem to have been examined recently. There is no evidence that mercury(I) oxide, sulphide, hydroxide, or peroxide can be isolated, despite claims in the older literature.[1]

Table 13.1. Mercury–Mercury Stretching Frequencies, $\nu(\text{Hg}-\text{Hg})$ cm^{-1} (from references 9–13)

Hg_2F_2	solid	185.9
Hg_2Cl_2	solid	166.5
Hg_2Br_2	solid	132.2
Hg_2I_2	solid	112.5
$Hg_2(NO_3)_2 \cdot 2H_2O$	solid	179.8
$Hg_2(NO_3)_2$	aqueous solution polarised	171.7
$Hg_2(ClO_4)_2 \cdot 4H_2O$		181.6
$Hg_2(ClO_4)_2$	aqueous solution polarised	173.3
Hg_2SO_4	solid	172.2†
Hg_2CO_3	solid	174.5
$Hg_2(phen)_2(NO_3)_2$		158
$[Hg_2(phen)(NO_3)_2]_n$		172
$Hg_2(phen)(NO_3)_2$		182
$[Hg_2(pyrazine)(NO_3)_2]_n$		207
$Hg_2(pyrazine)_2(NO_3)_2$		162
$Hg_2(2\text{-pyrazincarboxamide})_2(NO_3)_2$		165
$Hg_2(4\text{-methylquinoline})_2(NO_3)_2$		162, 124
$Hg_2(4\text{-aminobenzonitrile})_2(NO_3)_2$		179
$Hg_2(3\text{-amino-4-methoxybenzenesulphonyl-}$ $\text{fluoride})_2(NO_3)_2$		164, 140
$Hg_2(3\text{-fluoroaniline})_2(NO_3)_2$		160
$Hg_2(2,5\text{-difluoroaniline})_2(NO_3)_2$		177, 154
$Hg_2(\text{aniline})_6(NO_3)_2$		162
$Hg_2(p\text{-phenoxyaniline})_2(NO_3)_2$		187
$Hg_2(4\text{-aminobiphenyl})_2(NO_3)_2$		169, 132
$Hg_2(4\text{-}(CF_3)\text{-aniline})_2(NO_3)_2$		174
$Hg_2(p\text{-aminobenzoic acid methyl}$ $\text{ester})_2(NO_3)_2$		185, 163
$Hg_2(p\text{-aminobenzoic acid ethyl}$ $\text{ester})_2(NO_3)_2$		182, 132
$Hg_2(2\text{-naphthylamine})_2(NO_3)_2$		182
$Hg_2(\text{quinoline})_2(NO_3)_2$		151
$Hg_2(\text{acridine})_2(NO_3)_2$		212, 192
$Hg_2(o\text{-phenylenediamine})_2(NO_3)_2$		171
$Hg_2(p\text{-phenylenediamine})_2(NO_3)_2$		180
$\{Hg_2[N_2(COMe)_2]\}_n$		181
$Hg_2[NRSO_2F]_2$	$R = SO_2F$	192.5
	$R = CO_2Me$	183.5
$Hg_2(BrO_3)_2$		128, 183
Hg_2SeO_4		120, 181
$Hg_2(IO_3)_2$		120, 181
$Hg_2(CH_3CO_2)_2$		134, 168
$Hg_2(CCl_3CO_2)_2$		113

† Also quoted as 142, 193.

Table 13.2. Structural Data on Hg(I) Compounds

Compound	d(Hg−Hg) nm	Remarks (nm)	Reference
Hg$_2$F$_2$	0.2507(1)	2F at 0.214, 4F at 0.2175	14
Hg$_2$Cl$_2$	0.2526(1)	2Cl at 0.243, 4Cl at 0.321	14
Hg$_2$Br$_2$	0.249(1)	2Br at 0.271, 4Br at 0.332	14
Hg$_2$I$_2$	0.269	Hg−I = 0.268 (low accuracy)	15
Hg$_2$(NO$_3$)$_2$. 2H$_2$O	0.254(1)	Hg−O = 0.215; H$_2$O−Hg−Hg−OH$_2$ unit	16
Hg$_2$(BrO$_3$)$_2$	0.2507(6)	Hg−O = 0.216	17
Hg$_2$SO$_4$	0.250	Hg−Hg−O = 164.9°, Hg−O = 0.224	18
Hg$_2$SeO$_4$	0.251	Hg−Hg−O = 160°, Hg−O = 0.221	18
Hg$_2$(ClO$_4$)$_2$. 4H$_2$O	0.250	Hg−O ≈ 0.21	19
Hg$_2$[C$_6$H$_4$(COO)$_2$]$_2$	0.2519(4)	Hg$_2$$^{2+}$ linearly bonded to O, O of diff. phthalate groups	20
Hg$_2$(CH$_3$COO)$_2$		monoclinic structure	21
Hg$_2$(phen)(NO$_3$)$_2$	0.2516	phen coordinated N,N to *one* Hg	22
Hg$_2$(4-CN-pyridine)$_2$(ClO$_4$)$_2$	0.2498	Hg−Hg−N = 176°; ligand bonded via py N	24
Hg$_2$(OPPh$_3$)$_6$(ClO$_4$)$_2$	0.2522(2)	approx. td coordn. about Hg	25
Hg$_2$(py-*N*-O)$_4$(ClO$_4$)$_2$	0.2523(2)		26
Hg$_2$(3-Cl-py)$_2$(ClO$_4$)$_2$	0.2487	Hg−N = 0.221; Hg−Hg−N = 167.4°	27
Hg$_2$(1,8-naphthyridine)$_2$-(ClO$_4$)$_2$	0.2511(1)	Hg−N = 0.203(2); Hg−N^{II}= 0.278(3)	28
Hg$_2$[N$_2$(COMe)$_2$]$_2$	0.29(1)		29
Hg$_2$(1,4-diazine)(NO$_3$)$_2$	0.2499(1)	chain structure	30
Hg$_2$SiF$_6$. 2H$_2$O	0.2495(3)	Hg−O = 0.220; Hg−Hg−O = 170.9°	31
(Hg$_2$)$_3$(AsO$_4$)$_2$	0.2535(4)	polymeric	32

A range of stable oxysalts of mercury(I), viz: nitrate,[16] chlorate,[17] bromate,[17] iodate,[40] sulphate,[18] selenate,[18] perchlorate,[19,41] carbonate[42] are readily prepared, and there are a number of others, reported many years ago,[1] which almost certainly exist, but have not been studied more recently, for example periodate, chlorite, tellurate. Mercury(I) fluoroborate[43] and fluorosilicate[31] and several carboxylates are well characterised—acetate,[21] oxalate,[44] phthalate,[20] and mono-, di-, and tri-chloroacetate.[36] A recent attempt to repeat the preparation of mercury(I) nitrite was unsuccessful.[45]

The structures of these mercury(I) salts that have been determined (table 13.2) generally show the presence of a O−Hg−Hg−O grouping, essentially linear in structure, the terminal oxygens being provided either by water molecules, as in Hg$_2$(NO$_3$)$_2$. 2H$_2$O, which is better formulated as [H$_2$O−Hg−Hg−OH$_2$](NO$_3$)$_2$,[16] or by the oxyanion as in Hg$_2$SO$_4$ or Hg$_2$SeO$_4$.[18] Dehydration of Hg$_2$(NO$_3$)$_2$. 2H$_2$O yields a yellow solid of composition Hg$_2$(NO$_3$)$_2$, which decomposes on heating.[45] The infrared spectrum[45] indicates coordinated nitrate groups, but the structure does not appear to have been further examined. Interestingly, while both the fluoroborate and fluorosilicate (containing [H$_2$O−Hg−Hg−OH$_2$]$^{2+}$)

can be crystallised; the tetraphenylborate ion brings about disproportionation due to the insolubility of mercury(II) tetraphenylborate.[45]

Until relatively recently it was supposed that Hg_2^{2+} would not form any complexes, due to the tendency to disproportionate in the presence of ligands that coordinate strongly to Hg^{2+}. In 1959 Anderegg[46] reported evidence for the complexation of mercury(I) by phenanthroline in aqueous nitric acid solution and the isolation of $Hg_2(phen)_2(NO_3)_2$. Potentiometric studies established the formation of mercury(I) complexes of polyphosphate and dicarboxylate ions in solution,[47] and subsequently the same method was used to show that aniline forms a complex with mercury(I) in solution, which although less stable than the corresponding mercury(II) compound, is much more stable than those formed by Ag(I) or Ni(II).[48]

Potts and Allred[45] prepared a series of complexes with oxygen donor ligands— Ph_3PO, Ph_3AsO, pyridine-N-oxide, dimethylsulphoxide—but found that triphenylphosphine, triphenylstibine, triphenylphosphite, triphenylphosphine sulphide, sodium dimethyldithiocarbamate, diethyldithiophosphate, urea or thiourea cause disproportionation. These authors point out that no compound of mercury(I) appears to be stable in the presence of ligands higher than water in the spectrochemical series, and there are some anomalies such as urea, which is lower than water (as an oxygen donor) in the spectrochemical series, that decompose Hg_2^{2+}, but this may be due to N-bonding by the urea.[45]

The structure[22] of $Hg_2(phen)_2(NO_3)_2$ shows the phenanthroline to be chelated to one mercury atom, the other nearest neighbours being oxygen atoms of the nitrate groups. Kepert and Taylor[23] prepared mercury(I) compounds of stoichiometry $HgL_2(ClO_4)_2$ for L = 2-chloropyridine, 5-nitroquinoline, quinoline, 2-picoline, acridine, 2-methylquinoline, 2,6-lutidine, and 2,4,6-trimethylpyridine. The ligands 2-ethylpyridine and 4-benzylpyridine afforded $Hg_2L_4(ClO_4)_2$.

Kepert *et al.*[24] prepared $Hg_2(4\text{-CN-pyridine})_2(ClO_4)_2$ as white crystals by reaction of mercury(I) perchlorate tetrahydrate and the ligand in ethanol. The structure consists of the expected linear ($Hg\text{--}Hg\text{--}N = 176°$) N--Hg--Hg--N unit, with the cyanopyridine ligand bonded via the ring nitrogen rather than the cyano group. The corresponding complex of 3-chloropyridine has an essentially similar structure[27] although there is a greater distortion from linearity attributable to Cl . . . Cl repulsion between neighbouring molecules. In view of the differing modes of coordination produced by phenanthroline and the substituted pyridines, Dewan *et al.*[28] examined the complex with 1,8-naphthyridine, since although the latter is potentially a bidentate ligand, it has an unusually small 'bite'. The structure of the complex (figure 13.1) shows the naphthyridine to be coordinated in an asymmetric manner and predominantly through one nitrogen atom (Hg--N = 0.203(3) nm) with a much weaker interaction with the second nitrogen (Hg--N = 0.278(1) nm).

Brodersen *et al.*[10,11] extended the known Hg(I)--nitrogen compounds by preparing $Hg_2L_2(NO_3)_2$, where L = p-phenoxyaniline, 4-aminobiphenyl, 4-trifluoromethylaniline, the methyl and ethyl esters of p-aminobenzoic acid,

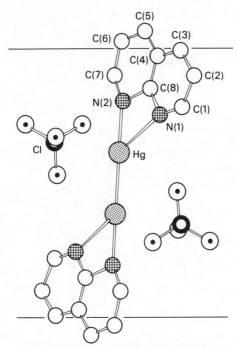

Figure 13.1. Hg$_2$(1,8-naphthyridine)$_2$(ClO$_4$)$_2$

2-naphthylamine, quinoline, benzoquinoline, acridine, 4-phenylpyridine, pyridazine, 2-pyrazinecarboxamide, 4-methylquinoline, 4-aminobenzonitrile, 3-amino-
4-methoxybenzenesulphonylfluoride, 4-fluoroaniline, 3-fluoroaniline and 2,5-
difluoroaniline; Hg$_2$L$_4$(NO$_3$)$_2$ for L = 4-fluoroaniline, and Hg$_2$(aniline)$_6$(NO$_3$)$_2$;
all of which on the basis of infared and Raman spectroscopy were assigned discrete
molecular structures [L$_n$–Hg–Hg–L$_n$]$^{2+}$. The ligands *o*-phenanthroline, pyrazine,
o-phenylenediamine, and *p*-phenylenediamine form [Hg$_2$L$_2$(NO$_3$)$_2$]$_n$ with
polymeric structures. This was subsequently confirmed in the case of the pyrazine
complex by an X-ray structure determination.[30] These workers suggest that in
order to stabilise the mercury(I) complex the electron density on the nitrogen
must be within certain (as yet undefined) limits; too low a density and substitution
in the [H$_2$O–Hg–Hg–OH$_2$]$^{2+}$ group does not occur, too high a density and
disproportionation to mercury(II) and mercury(0) is promoted.

In contrast to the nitrogen donors, pyridine-N-oxide does not coordinate
axially[26] to Hg$_2$$^{2+}$ in Hg$_2$(pyNO)$_4$(ClO$_4$)$_2$. Three of the four pyNO ligands bridge
adjacent dimers (Hg–O = 0.219–0.277 nm) producing 4- or 5-coordinate mercury
(figure 13.2).

Kepert *et al.*[25] obtained a hexakis(triphenylphosphine oxide)dimercury(I)-
bisperchlorate from Hg$_2$(ClO$_4$)$_2$ and Ph$_3$PO, which compares with Hg$_2$(Ph$_3$PO)$_4$-
(ClO$_4$)$_2$ and Hg$_2$(Ph$_3$PO)$_5$(ClO$_4$)$_2$ obtained previously.[45] The structure of

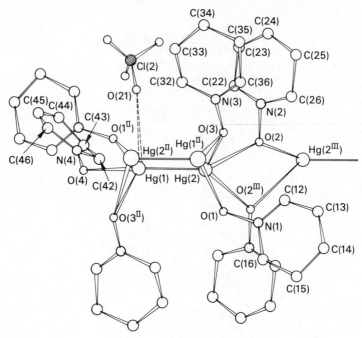

Figure 13.2. $Hg(pyNO)_4(ClO_4)_2$

$Hg_2(OPPh_3)_6(ClO_4)_2$ consists of essentially tetrahedrally coordinated mercury atoms to a second mercury and three oxygen atoms of $OPPh_3$ groups.

It is apparent from table 13.2 that the Hg—Hg bond length is somewhat sensitive to other ligands present, but does not vary in any simple manner as a function of ligand. (It must also be stressed that several of the determinations are of low accuracy.) While the electronegativity correlation[34] has been shown to be incorrect, it would appear that further studies are necessary before all the data can be rationalised.

It has been suggested[49] that trithiodiacetylacetone (L) forms a mercury(I) complex $Hg_2Cl_2L_2$, which would appear to be the only example with a sulphur donor ligand, although the data reported do not rule out the possibility that the complex contains mercury(II). Mercury(I) is also present in the mineral eglestonite (Hg_6Cl_4O) and in $2HgO \cdot Hg_2Cl_2$,[50] while $Hg^I_2Hg_2^{II}S_2(ClO_4)_2$ has been reported.[51]

The reported formation of mercury(I) nitrogen compounds on reaction of mercury(I) salts with ammonia or alkylamines[1] has been shown to be incorrect (references 52, 53 and references therein) the products being mercury(0) and mercury(II). However, 'N–H' acids[52] form stable mercury(I) compounds, the first example being the diacetylhydrazine compound prepared in 1958.[29] The orange $Hg_2[N_2(COCH_3)_2]$ was originally obtained from $N_2H_2(COCH_3)_2$ and $Hg_2(NO_3)_2$ in weakly acidic solution and on the basis of a calculated radial distribution Hg–Hg was estimated as 0.29 nm,[29] but this was subsequently corrected to 0.246 nm.[12] More recently the compound has been prepared from mercury(I) carbonate in benzene solution, along with the trifluoro- and trichloroacetyl analogues,[12] but attempts to prepare the bis(ethoxycarbonyl) or bis(*p*-toluenesulphonyl)hydrazine analogues were unsuccessful.

The $RNH(SO_2F)$ ($R = SO_2F$, CO_2Et, CO_2Me, $CONEt_2$) ligands react with mercury(I) carbonate to form $Hg_2[RN(SO_2F)]_2$, which are believed to be discrete complexes (figure 13.3) in contrast to the hydrazine derivatives, which are polymeric.[12] On the basis of the $\nu(Hg-Hg)$ frequencies observed in the Raman spectra of these complexes and by application of the Badjer formula, Hg–Hg was estimated[12] as 0.244–0.247 nm.

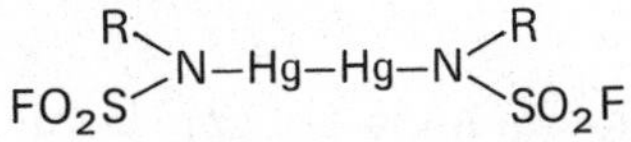

$$FO_2S \diagdown \overset{R}{\underset{}{\diagup}} N-Hg-Hg-N \overset{R}{\underset{SO_2F}{\diagdown}}$$

Figure 13.3

The $[Hg(NS)]_x$ and $Hg_2(S_7N)_2$ obtained from $S_4(NH)_4$ and S_7NH, respectively, on treatment with $Hg_2(NO_3)_2$ in N,N-dimethylformamide are almost certainly of similar constitution.[54]

14 Other Polycations

The reduction of Hg^{2+} in molten $AlCl_3$–NaCl produced evidence for a Hg_3^{2+} ion, as well as for Hg_2^{2+},[55] and $Hg_3(AlCl_4)_2$ is obtained from a $1:2:2$ molar ratio of $HgCl_2 : Hg : AlCl_3$ at 240 °C after six days.[56,57] The crystal structure of the latter revealed an almost (174.4°) linear $[Hg–Hg–Hg]^{2+}$, with Hg–Hg = 0.2551(1) nm and 0.2562(1) nm, rather longer than in mercurous compounds.[58] The Hg . . . Cl distance is only 0.254 nm, indicative of considerable interaction with the anion. The yellow hexafluoroarsenate(V), $Hg_3[AsF_6]_2$, has been formed from mercury and arsenic pentafluoride or mercury(I) hexafluoroarsenate.[58,59] The Raman spectrum in liquid sulphur dioxide exhibits ν(Hg–Hg) at 118 cm^{-1} (strong polarised), indicating a linear structure, and this was subsequently confirmed by an X-ray, which showed that the Hg_3^{2+} ion was linear and symmetrical with Hg–Hg = 0.2552(4) nm.[60] Dissolving mercury in SbF_5/liq.SO_2 forms $Hg_3(Sb_2F_{11})_2$.[59] The reaction of mercury with AsF_5 in liquid sulphur dioxide also yields dark red $Hg_4(AsF_6)_2$, which contains the centrosymmetric Hg_4^{2+} ion[61] (figure 14.1). An excess of mercury reacts with AsF_5 in liquid sulphur dioxide at room temperature to yield an insoluble golden solid, originally[62]

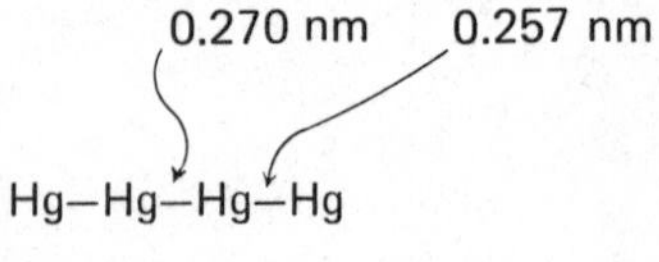

0.270 nm   0.257 nm

Hg–Hg–Hg–Hg

Figure 14.1

formulated 'Hg_3AsF_6', but shown by X-ray crystallography to be the remarkable $Hg_{2.85}AsF_6$, which contains infinite chains of mercury atoms running through the cubic close-packed fluoride lattice in mutually perpendicular directions (figure 14.2). The Hg–Hg$_{(av)}$ bond is 0.264(1) nm and the formal charge is +0.35. The compound has a conductivity approaching that of a metal, and each mercury chain is essentially a 'one-dimensional metal'.[63]

Booth et al.[64] have obtained e.p.r. evidence for the formation of Hg_2^{3+} or Hg_2^+ in frozen solutions of mercury salts exposed to ^{60}Co γ-rays at 77 K.

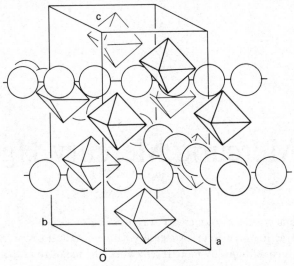

Figure 14.2. An isometric view of $Hg_{2.86}AsF_6$ showing the infinite chains of mercury atoms (circles) running through the lattice of AsF_6^- ions (octahedra)

15 Mercuric Mercury Hg^{2+}

The structural chemistry of mercury was reviewed by Grdenic[50] in 1965. Although the section on mercurous mercury has been shown to contain a number of erroneous conclusions by more recent work, the majority of the review devoted to mercury(II) is an invaluable introduction to the stereochemistry of this metal. In the ten years since the appearance of this work, many more X-ray structure determinations have been reported, and these are described in the appropriate sections of the present work. However, the classification of Grdenic in terms of 'characteristic' and 'effective' coordination is still valid, and thus reference 50 is recommended as an introduction to the stereochemistry of mercury(II).

15.1 Halogen Compounds

Mercury(II) fluoride, HgF_2, is a white crystalline solid, which darkens in colour on exposure to air, due to hydrolysis. It is obtained by heating Hg_2F_2, by chlorination of Hg_2F_2 when the $HgCl_2$ formed simultaneously can be sublimed away on heating, or by fluorination of mercury or $HgCl_2$.[65,66] It has the cubic CaF_2 structure, Hg . . . F = 0.240 nm, and is one of the few mercury(II) compounds in which the bonding is believed to be predominantly ionic.[67] A dihydrate $HgF_2 . 2H_2O$ is obtained by dissolving yellow HgO in 40 per cent aqueous HF;[68] on heating it decomposes into Hg(OH)F, which contains mercury in an irregular octahedral environment (O_3F_3), and which may be better described as a polymeric mercurionium fluoride, $[Hg(OH)]_n^{n+} . nF^-$.[69] There is also a report of a second hydroxyfluoride, $Hg_3F_4(OH)_2 . 3H_2O$, formed by hydrolysis of mercury(II) fluorosulphonate.[70]

The other three mercury(II) halides, HgX_2 (X = Cl, Br, I), are readily obtained from reactions of the elements. The structure of $HgCl_2$ consists of mercury linearly coordinated to two chlorines (0.225 nm) with four other chlorines (2 x 0.334, 2 x 0.363 nm) completing a distorted octahedral arrangement.[71,72]

Mercury(II) bromide has a deformed brucite layer structure,[73] once again with two near (0.248 nm) and four more distant bromine atoms (0.323 nm). The structural chemistry of mercury(II) iodide is more complex. The red form, which is the stable form at room temperature, consists of corner-linked HgI_4 tetrahedra ($Hg–I = 0.2783$ nm),[74] while the yellow form, produced by precipitation in aqueous solution or by heating the red form above 127 °C,[75] contains distorted octahedrally coordinated mercury(II).[73] There is also an unstable orange form, which contains Hg_4I_{10} units formed from four corner-linked HgI_4 tetrahedra connected into HgI_2 layers.[76]

The behaviour of the mercury(II) halides in solution in various solvents, and the unusual properties of fused mercury(II) bromide have been reviewed in detail elsewhere.[65,77]

The nature of the mixed halides $HgXY$ ($X \neq Y$), formed on treatment of Hg_2X_2 with Y_2 has been a cause of controversy for some time; considerable doubt remains as to their chemical individuality. Rastogi *et al.*[78-80] prepared HgClI, HgBrI, HgFBr, HgFI, HgClBr, and also $Hg(CH_3COO)X$ ($X = Br, I$) by reaction of the heavier halogen with the appropriate Hg_2X_2. X-ray powder patterns appear to rule out the possibility that these products are mixtures of the corresponding symmetrical mercury(II) halides. Evidence for the formation of mixed halides in solution (references 80–83 and references therein) and in vapours and melts[84,85] containing HgX_2 and HgY_2 has also been reported.

Numerous basic halides (oxyhalides) have been described,[64] and the structures of several of them have recently been determined. The known oxychlorides are $Hg_3O_2Cl_2$, Hg_3OCl_4, $Hg_4O_3Cl_2$, and $Hg_5O_4Cl_2$.[86] The $Hg_3O_2Cl_2$ compound was originally formulated $Hg(OHgCl)_2$[87], but more recent work[88] has described the structure as containing two-coordinate ($Hg–O = 0.207$ nm), and three-coordinate ($Hg–O = 0.217, 0.232, 0.233$ nm) mercury. Trigonal $O(HgCl)_3^+$ cations and Cl^- anions ($Hg–O = 0.204$ nm, $Hg–Cl = 0.231$ nm) exist in Hg_3OCl_4.[89,90] It has been suggested that $Hg_5O_4Cl_2$ should be formulated $[Hg_4Cl_4]^{4+}[Hg_6O_8]^{4-}$.[91]

The structure of only one oxybromide, $Hg_5O_4Br_2$, has been determined and consists of infinite $–Hg–O–Hg–O–$ chains ($Hg–O = 0.205$ nm) with other mercury atoms, four coordinate, (O_4) lying between the chains. The bromines are essentially ionic.[92] This raises doubt about the formulation given above for $Hg_5O_4Cl_2$, since the bromine analogue was previously formulated in a similar way. Little seems to be known about the occurrence and constitution of mercury(II) oxyiodides.

Mercury(II) oxide also forms compounds with other halides, for example, $(HgO)_2NaI$.[93]

Fluoromercurates(II) have been little studied. The colourless, extremely moisture-sensitive $MHgF_3$ ($M = K, Rb, Cs$) are formed by fluorination of $HgCl_2 . 2NH_3 + MCl$.[94] The potassium salt is orthorhombic ($a = 0.620$, $b = 0.628, c = 0.881$ nm); the other two are cubic [$a = 0.457$ nm (Cs), 0.447 nm (Rb)]. The thallium analogue is also cubic ($a = 0.4475$ nm).[95] The reaction of

$HgF_2 . 2H_2O$ with pyridine and HF gives $[pyH]_2HgF_4 . 2H_2O$, which becomes anhydrous when recrystallised from methanol.[68] Similar complexes are known with protonated α-picoline or 2,6-lutidine as cations.[68]

The halomercurates(II) containing Cl, Br, or I are well known and have been reviewed by Deacon.[96] The stability of the halomercurate(II) ions in aqueous solution is $(F \ll)Cl < Br < I$,[97,98] and it was suggested that the major species present in solutions of $HgX_2 + X^-$ are HgX^+, HgX_2, HgX_3^-, and HgX_4^{2-}. More recently studies have been extended to a range of non-aqueous solvents, for example MeCN, DMF, ethyl acetate, methyl methacrylate, and ethanol.[99-105] Much of the earlier work reported little more than the stoichiometry of the various products obtained from solutions of HgX_2 and metal halides. The reported compounds are listed in Gmelin[106] (alkali and alkaline earth cations), and by Gmelin[106] and Deacon[96] (sulphonium, phosphonium, arsonium, etc., cations). Phase diagrams for the alkali halide/mercury(II) halide systems have been constructed.[107,108] Although complexes of various $MX : HgX_2$ ratios are known it is apparent from a study of the sources listed above that the commonest stoichiometries are M^IHgX_3 and $M^{II}HgX_4$, with $M^IHg_2X_5$ and $M^IHg_3X_8$ much less common. Many of the reported complexes are formulated as hydrates.[66,106] Mixed halomercurates(II), for example $[HgCl_2I_2]^{2-}$, $[HgBrI_2]^-$, and $[HgBr_2I]^-$, and $[HgX_2X'X'']^{2-}$ $(X \neq X' \neq X'')$ have been reported,[80,82,109-112] and their individuality established by the presence of lines in their Raman spectra not present in the spectra of the simple halomercurates(II).

The spectroscopic properties of halomercurates(II) have been examined in detail recently with a view to establishing the nature of the anions present.[113-117] The spectra of HgX_3^- and HgX_4^{2-} ions have been discussed in terms of the structural units present, and it is proposed that $Hg_3X_8^{2-}$ ions are best formulated $(HgX_3^-)_2 . HgX_2$, while $Hg_2Cl_5^-$ may be $HgCl_3^- . HgCl_2$.[115] Raman spectral data for acetone solutions of HgX_3^-, HgX_4^{2-}, HgX_5^-, and $Hg_3X_8^{2-}$ ions suggests that in all cases the dominant species in solution is HgX_3^-, although evidence for $Hg_2I_5^-$ was found in the solution spectra of $Me_4NHg_2I_5$.[114] Electronic spectra of a number of halomercurates(II) have been recorded.[100,101,118,119] The X-ray structural data available on halomercurate(II) complexes shows that stoichiometry is a poor guide to structure (table 15.1).

The only discrete $HgCl_4^{2-}$ ion appears to be that found in the complex with the alkaloid perloline.[125] The other chloro complexes generally contain the mercury in a distorted $HgCl_6$ environment,[50] which may be linked into chains (figure 15.1) as in $K_2HgCl_4 . H_2O$[120,121] or into layers as in α-NH_4HgCl_3[123] (figure 15.2) or to form two-fold ribbons of $(Hg_2Cl_6)_n^{2-}$ as in $NaHgCl_3 . H_2O$[127] (figure 15.3).

The 'compound' $NaHgCl_3$ does not exist; dehydration of the hydrate yields a mixture of NaCl and $HgCl_2$,[114] and the reported structure[137] appears to be that of the hydrate. A second form of NH_4HgCl_3 has been identified[124] with the same structure as NH_4CdCl_3.

The structures of bromo- and iodomercurates(II) are less well known. In

Table 15.1. Halomercurate(II) Complexes

Compound	Structure (nm)	Reference
$K_2HgCl_4 . H_2O$	$HgCl_6$ octahedra in chains; Hg–Cl = 0.2383(2), 0.2979(2), 0.3249(2)	120, 121
$CsHgCl_3$	Cubic distorted $HgCl_6$ octahedra in chains; Hg–Cl = 0.229(2 Cl), 0.297(4 Cl)	122
NH_4HgCl_3 α	Tetragonal $HgCl_6$; Hg–Cl = 0.234(2 Cl), 0.296(4 Cl)	123
β	Isomorphous with NH_4CdCl_3	124
$[Perloline]_2[HgCl_4] . H_2O$	Tetrahedral; Hg–Cl ≈ 0.250	125
$[Me_4N] HgCl_3$	as $[Me_4N]HgBr_3$	126
$NaHgCl_3 . 2H_2O$	Orthorhombic $HgCl_6$; Hg–Cl = 0.235, 0.240, 0.281(2 Cl), 0.327(2 Cl); $[HgCl_6]_n$ chains	127
$[NH_4]_2HgCl_4 . H_2O$	Orthorhombic, similar to K analogue	128
$KHgBr_3 . H_2O$	Irregular tetrahedra of $HgBr_4$, sharing corners	129
$[Me_4N]HgBr_3$	Monoclinic $HgBr_4$; Hg–Br = 0.249, 0.252, 0.255, 0.292	130
Tl_4HgBr_6	Isolated $HgBr_6$; Hg–Br = 0.254(12)(2 Br), 0.3109(11)(4 Br)	131
$CsHgBr_3$	Cubic $CaTiO_3$ structure; $HgBr_6$ environment	128
$KHgI_3 . H_2O$	HgI_4 tetrahedra sharing corners; Hg–I = 0.220(5), 0.273(1), 0.283(4), 0.290(4)	132, 133
Cs_2HgI_4	Distorted HgI_4 tetrahedra; Hg–I = 0.291, 0.278, 0.271, 0.286	134
$[Me_3S]HgI_3$	3-coord. Hg; trigonal planar anion; Hg–I = 0.270	135
$[Me_4N]HgI_3$	as $[Me_4N]HgBr_3$; Hg–I = 0.272 (av.)	126, 132
Ag_2HgI_4	Two forms cubic zinc blende; high-temp. form Hg–I = 0.274; tetragonal low-temp. form	128
Cu_2HgI_4	Two forms similar to Ag analogue; tetragonal Hg–I = 0.264	128
$[A] Hg_2I_6$	Discrete Hg_2I_6 units Hg–I = 0.270, 0.304 (A is the 3,5-bis(N,N-diethylimonium)-1,2,4-trithiolane cation)	159

$CsHgBr_3$ the mercury is six-coordinate,[128] but the $KHgBr_3$. H_2O and $[Me_4N]HgBr_3$ species contain the metal in a very distorted tetrahedral environment.[129,130] In the latter one bromine is considerably farther from the mercury than are the other three, and the structure seems best described as intermediate between discrete $HgBr_3^-$ anions and a polymeric chain structure. Isolated $HgBr_6^{4-}$ octahedra exist in Tl_4HgBr_6,[131] and $[NH_4]_4HgBr_6$, Tl_4HgI_6, and $Tl_4HgCl_2Br_4$ are isomorphous. The Tl_4HgCl_6 compound could not be obtained, attempted preparations producing tetragonal $Tl_{10}Hg_3Cl_{16}$. The TlX/HgX_2 systems have been examined in some detail.[131,138]

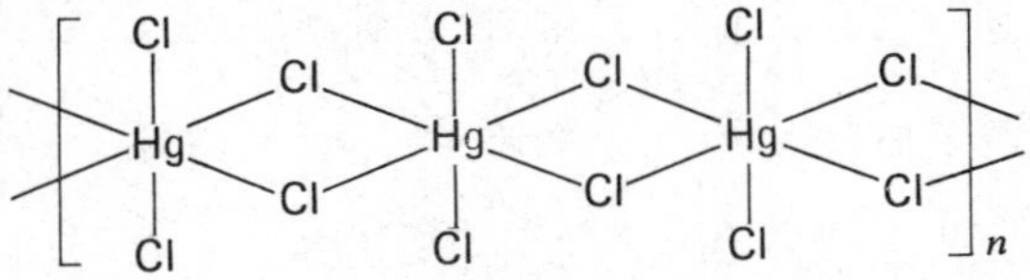

Figure 15.1

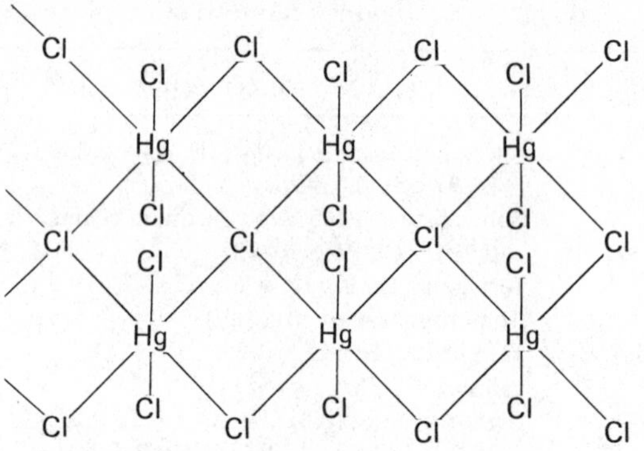

Figure 15.2

Discrete trigonal planar HgI_3^- ions are contained in $[Me_3S]HgI_3$,[135] but in $KHgI_3 . H_2O$, Cs_2HgI_4, and $[Me_4N]HgI_3$ distorted HgI_4^{2-} species are present.[126,132–134] The 'complex' $K_2HgI_4 . 3H_2O$ has been shown to be a mixed halide—$KHgI_3 . KI . 3H_2O$—by an X-ray study.[132] The Ag_2HgI_4 and Cu_2HgI_4 compounds display unusual structures: a high-temperature form and a low-temperature tetragonal structure.[128,139] Discrete $Hg_2I_6^{2-}$ ions are present in the compound with 3,5-bis(N,N-diethylimonium)-1,2,4-trithiolane cations.[159]

The reactions of HgI_2 with CdI_2, PbI_2, or BiI_3 in DMF or DMSO yield compounds $CdHg_2I_6 . 6DMF$, $CdHg_2I_6 . 8DMSO$, $PbHg_2I_6 . 8DMSO$, and $BiHg_3I_8 . 8DMF$; the structures of these species are still undetermined.[140]

Unusual $[HgI]^+$ moieties appear to be present in $[HgI]_2MF_6$ (M = Ti, Sn), produced from the reaction of HgO, K_2MF_6, and iodide ions in 40 per cent HF.[141] An X-ray study has revealed the presence of infinite $_\infty^1[Hg_{2/2}I]^+$ chains linked by MF_6 octahedra.[136]

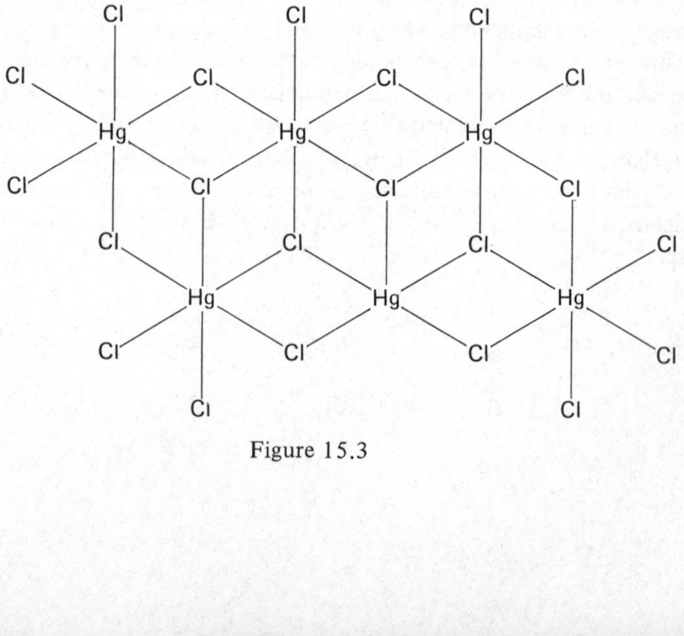

Figure 15.3

15.2 Mercury(II) Pseudohalides

Mercury(II) cyanide is produced by reaction of mercury(II) salts with most metal cyanides, but it is best prepared from aqueous hydrocyanic acid and HgO.[142] Mercury(I) salts disproportionate when treated with alkali metal cyanides yielding $Hg(CN)_2$. In aqueous solution mercury(II) cyanide is little-ionised (dissociation constant to $Hg^{2+} + 2CN^- \approx 10^{-35}$) and thus exhibits few reactions of Hg^{2+} or CN^-, although HgS can be precipitated by addition of H_2S. The structure consists of almost linear $Hg(CN)_2$ molecules, Hg–C = 0.2015(3), 0.2019(3) nm, interaction occurring between neighbouring molecules, Hg . . . N = 0.2742(3) nm, two nitrogen atoms completing a distorted tetrahedron about the mercury.[143,144] In solution Raman spectroscopy suggests a linear $Hg(CN)_2$ structure.[145] On heating, $Hg(CN)_2$ decomposes into mercury, cyanogen, and a brown polymer, 'paracyanogen', $(CN)_x$.[146]

In solutions of $Hg^{2+} + CN^-$ spectroscopic, potentiometric and polarographic evidence shows that $Hg(CN)_2$, $Hg(CN)_3^-$, and $Hg(CN)_4^{2-}$ all exist,[147-149] and solid complex cyanides $M^I_2Hg(CN)_4$ are known for M = Na, K, Rb, Cs, Tl.[142,148,150,151] The $Hg(CN)_4^{2-}$ is tetrahedral,[152] although accurate structural data appear to be lacking. Solid cyanomercurates(II) of different stoichiometry have been claimed, principally $M^IHg(CN)_3$,[142] although $KHg(CN)_3$[153] has been shown to be a mixture of $Hg(CN)_2$ and $K_2Hg(CN)_4$.[148] X-ray crystallographic results show that $CsHg(CN)_3$ contains corner-linked $Hg(CN)_3^-$ units with a distorted tetrahedral coordination about the mercury.[154]

A reappraisal of the published data for the existence of mononuclear complexes $Hg(CN)_x^{(x-2)-}$ has shown that complexes with $x > 4$ do not exist.[155] Numerous complexes of type $M^IHg(CN)_2X$ (X = Cl, Br, I) can be obtained from $Hg(CN)_2$ and M^IX. The structure of $KHg(CN)_2I$ consists of $Hg(CN)_2$ molecules with each mercury weakly bonded to four iodines (0.338 nm).[156] In aqueous solution infrared spectroscopy suggests that $Hg(CN)_2$ is not present, a ν(CN) absorption at lower frequency [than that in $Hg(CN)_2$] may be ascribed to $[Hg(CN)_2I]^-$.[148] The structure of $Zn(NO_3)_2 . 2Hg(CN)_2 . 7H_2O$, obtained by evaporation of an aqueous solution of the constituents, shows it to be $\{Zn(H_2O)_4-[Hg(CN)_2]_2\}(NO_3)_2 . 3H_2O$, in which the zinc is six-coordinate (N_2O_4) with one nitrogen arom of each $Hg(CN)_2$ molecule functioning as a donor.[157] Numerous other 'double salts' of $Hg(CN)_2$ are known, and further X-ray studies may reveal further examples of this type of bonding. Mercuric oxide is soluble in aqueous $Hg(CN)_2$ solution to form $[Hg(CN)]_2O$, which has a molecular oxo-bridged structure in the solid state, but exists as Hg(CN)(OH) in solution.[158]

Mercury(II) cyanate is formed from AgNCO and mercuric chloride in methanol.[19] The potassium salt of the tetracyanatomercurate(II) is formulated as O–bonded on the basis of infrared measurements,[160] but ^{14}N n.m.r. measurements on the tetraalkylammonium analogues suggest N-bonding.[161] The isomeric fulminate ligand (CNO) affords the well-known mercury(II) fulminate, $Hg(CNO)_2$, usually obtained by the strange reaction of mercury, nitric acid, and ethanol.[162] The product is violently explosive and is widely used as a detonator. In view of its technical importance, and the extensive studies of its physical properties,[1,164] it is surprising that its structure has not yet been determined.

Mercury(II) fulminate reacts with alkali-metal fulminate solutions to form the highly explosive $M^I_2[Hg(CNO)_4]$ (M = K, Rb, Cs),[165,166] but with larger cations, for example Ph_4As^+, more stable complexes result.[165] Mercury fulminate reacts with PPh_3 or PCy_3 to form the non-explosive $Hg(CNO)_2(PR_3)_2$.[165] There are also some mixed complexes, $[Hg(CNO)_2(N_3)]^-$, $[Hg(CNO)_2X]^-$ (X = Cl, Br), and $[Hg(CNO)_2(CN)_2]^{2-}$.[165,167,168]

Mercury(II) thiocyanate is usually obtained by metathesis of KNCS with mercury(II) nitrate. The structure consists of *trans* octahedral HgS_2N_4 coordination with two short Hg–SCN bonds (0.2318(6) nm) and four longer Hg–NCS bonds (0.281 nm), all the thiocyanate anions being bridging.[169,170] The compound burns in air to yield a voluminous spongy ash of unknown composition ('Pharaoh's serpents').

In solutions containing excess NCS^- ions, complex thiocyanatomercurates(II), $Hg(SCN)_3^-$ and $Hg(SCN)_4^{2-}$, are formed.[171,172] The structures of several alkali metal, NH_4^+, and PPh_4^+ salts of both anions have been determined.[173-176] In the tetrathiocyanatomercurates(II), tetrahedral $Hg(SCN)_4^{2-}$ ions are present, Hg–S = 0.249(2), 0.255–0.257(2) nm (Ph_4P^+ salt), while the $[PPh_4][Hg(SCN)_3]$ complex contains infinite chains of $Hg(SCN)_3^-$ with one –SCN– bridging to produce a flattened tetrahedron about the mercury atom, Hg–S = 0.246 (2 × S) and 0.259 (bridge), Hg–N = 0.240 nm (bridge).[175] The rubidium and caesium trithiocyanatomercurates(II) contain trigonal planar $Hg(SCN)_3^-$ groups,[154] and since the K^+, NH_4^+, and Rb^+ salts are isostructural,[173] the earlier reported structure[178] appears to be in error.

The $Hg(SCN)_4^{2-}$ moiety functions as a ligand towards a number of metal ions, for example $[Co(NCS)_4Hg]$,[178] which contains CoN_4 moieties linked by thiocyanate bridges to HgS_4 tetrahedra; and $Co(NCS)_6Hg_2 \cdot C_6H_6$,[179] which contains tetrahedral HgS_4 groups. The $Co(NCS)_4Hg$ complex readily adds two further ligands at the cobalt, which thus achieves six-coordination, for example THF, dioxan, pyridine.[180] The $Co(THF)_2(NCS)_4Hg$ complex undergoes substitution readily with a variety of P- or N donors to form complexes containing ionic structures with octahedral cobalt(II) and tetrahedral $Hg(SCN)_4^{2-}$ ions, except for the PPh_3 complex, which is believed to be $(PPh_3)_2Co(NCS)_2Hg(SCN)_2$, both metals being in a pseudotetrahedral environment.

Similar complexes $ML_2(NCS)_4Hg$ containing tetrahedral HgS_4 and octahedral MN_4L_2 are known for M = Mn^{2+}, Fe^{2+}, Co^{2+}, Ni^{2+}, Zn^{2+}, Cu^{2+}, Cd^{2+}; L = THF or pyridine.[181] There are a number of other compounds of the type $M(NCS)_4Hg$ where M = Mn^{2+}, Fe^{2+}, Ni^{2+}, Cu^{2+}, Zn^{2+}, Cd^{2+},[182-185] that generally contain bridging thiocyanate groups producing a four-coordinate HgS_4 and a four- or six-coordinate M^{2+} ion.

Numerous complexes of mercury thiocyanate with O, N, P, etc. donors are known (q.v.). It would appear that all contain Hg–SCN or bridging Hg–SCN–

groups, the preference for thiocyanato over isothiocyanato coordination being expected in view of the soft or class-B character of Hg(II).

Mixed halide–thiocyanate HgX(SCH) (X = Cl, Br, I) are formed from equimolar amounts of the constituents; they contain octahedral mercury, achieved via bridging X and SCN groups (X_3SN_2 donor sets).[186]

Mercury(II) selenocyanate, $Hg(SeCN)_2$, and the $Hg(SeCN)_4{}^{2-}$ ion are known.[27,188] The stability constant of $Hg(SeCN)_4{}^{2-}$ is greater than that of $Hg(SCN)_4{}^{2-}$.[189,190] There are selenocyanate bridges present in $Hg(SeCN)_2$.[188] Though they have been less studied than the sulphur analogues, selenocyanate-bridged complexes $MHg(SeCN)_4$ (M = Cu, Pb,[188] Zn, Co, Cd[191–193]) have been prepared. Although the nature of the bridge is not always clear, it is reasonable to assume that mercury is predominantly Se-bonded and the M^{2+} metal is N-bonded. In $ZnHg(SeCN)_4$ it appears that a pseudotetrahedral environment exists about both zinc and mercury.[194] The compound Hg(SeCN)Cl and some $Hg(SeCN)_xI_y{}^{(x+y-2)-}$ ions have been reported.[195]

Mercury(II) azide is formed from HgO and aqueous HN_3 or by metathesis in aqueous solution between NaN_3 and $HgCl_2$. It exists in two modifications, the very explosive α- and the more stable β-form.[196,197] The infrared spectrum[198] suggests the presence of $Hg(N_3)_2$ units, and this has been confirmed by an X-ray study,[199] which showed an essentially linear N–Hg–N grouping, the arrangement in the crystal being such that each mercury is coordinated to five nitrogen atoms of neighbouring molecules to produce a distorted capped trigonal prismatic environment. The triazidomercurate(II) ion is also known in $[Ph_4P][Hg(N_3)_3]$.[167]

15.3 Oxygen Donors

15.3.1 Oxides

Mercuric oxide is formed by heating mercury in oxygen or air at approximately 300 °C, by heating mercury(II) nitrate, or by precipitation of aqueous mercury(II) salts with alkali hydroxide. The dry methods produce the red form, the wet method yields yellow HgO, the two forms differing in particle size; the yellow is more finely divided and chemically more active.[200–202] The rhombic structure consists of planar O–Hg–O–Hg zig-zig chains, Hg–O = 0.203 nm,[203] with four adjacent oxygen atoms in neighbouring chains at 0.282 nm. There is also a rhombohedral form,[204,205] produced by slow precipitation from dilute solutions of K_2HgI_4 and KOH, which differs in that the O–Hg–O–Hg chains are spiral, not planar, and the nearest neighbours are at 0.279(90) and 0.290(20) nm. Mercuric oxide decomposes on heating and is converted ultimately to HgX_2 by halogens.

A mercury(II) peroxide, HgO_2, is claimed to be formed from $HgCl_2$, KOH and H_2O_2 in alcohol, or from HgO and H_2O_2 in aqueous solution at 0 °C. Vannerberg reports two forms, α-HgO_2, probably rhombohedral, and β-HgO_2, which is orthorhombic, with the mercury and peroxide groups forming infinite zig-zag chains, Hg–O = 0.206(2 x O) and 0.266–0.268(4 x O), O–O = 0.32–0.35(4 x O–O) and 0.15 nm (1 x O–O).[206,207]

There is no evidence that a solid hydroxide exists; reaction of alkali metal hydroxide with mercury(II) salts precipitate HgO. Mercuric oxide is, however, slightly soluble in concentrated alkali although no complex hydroxo species seem to have been isolated.

Oxomercurates(II) of type $M_2Hg_2O_3$ (M = Na, K) were mentioned in 1960,[208] but nothing seems to be known about them. The M_2HgO_2 (M = Li, Na, K, Rb, Cs) are formed by heating HgO with MO_x in oxygen.[209,210] They form colourless tetragonal crystals, which are exceedingly moisture sensitive. The sodium compound contains linear $[HgO_2]$ units, Hg–O = 0.197 nm, each sodium being in a square pyramidal NaO_5 arrangement.[210]

15.3.2 Oxy salts

Mercury(II) oxide dissolves in all except weak acids to form mercury(II) salts, which are usually obtained as hydrates on evaporation. Since HgO is a weak base, excess acid is usually necessary to prevent hydrolysis and the formation of basic salts (q.v.).

There is no simple mercury(II) carbonate, the metathesis of mercury(II) salts with HCO_3^- or CO_3^{2-} produces either HgO or ill-defined basic carbonates. Anhydrous mercuric acetate crystallises on cooling from a hot solution of HgO in 50 per cent acetic acid.[211] It is soluble in water but the solutions are essentially non-electrolytes, and widely used in organic chemistry to mercurate aromatic moieties, and as an oxidising agent for secondary hydroxyl groups. Other mercuric carboxylates can be prepared, for example the oxalate, HgC_2O_4[12] and tartrate, $HgC_4H_4O_6$.[213] The synthesis and properties of mercury(II) trifluoro-acetate, and its use in the mercuration of aromatic compounds has been discussed in a recent review.[163]

Mercury(II) nitrite is said to form yellow crystals and to have a nitro, $O_2N–Hg–NO_2$, rather than a $Hg(ONO)_2$ structure,[214] but this probably needs re-examining. However, the long-known potassium mercury nitrite, variously formulated $2KNO_2 . Hg(NO_2)_2$ or $3KNO_2 . Hg(NO_2)_2$, obtained from potassium nitrite and mercuric nitrate has been shown by an X-ray study to be $K_3[Hg(NO_2)_4]$-NO_3.[215] The $Hg(NO_2)_4^{2-}$ ion contains eight-coordinate mercury (HgO_8) with a very distorted square antiprism arrangement of the oxygen atoms about the metal. The Hg–O distances lie in the range 0.234–0.258 nm.[215] The complex has also been examined by neutron diffraction.[216] The formation of $K_2Hg(NO_2)_4$ by treatment of mercury(II) chloride with silver nitrite in water and subsequently adding excess KNO_2 has been established, and the infrared spectrum suggests that the nitrite ligands are chelating.[217] A $Rb_2Hg[Hg(NO_2)_6]$ complex has been reported.[218]

Mercury(II) nitrate is normally obtained by dissolution of mercury in excess nitric acid, and on evaporation $Hg(NO_3)_2 . (H_2O)_x$ (x = 8, 2, 1) may be obtained. The anhydrous salt may be prepared by dissolving the dihydrate in molten mercuric bromide,[219] or by reaction of HgO with N_2O_4 to produce $Hg(NO_3)_2 . N_2O_4$ initially, and this readily loses N_2O_4 to yield the anhydrous compound.[220]

Double or complex nitrates are rare. Bullock and Tuck[221] isolated $[Me_4N]_2$-

$Hg(NO_3)_4$ from a solution of the constituents in ethanol, but alkali-metal analogues could not be obtained. Several mercury(II) phosphates are known: $Hg_3(PO_4)_2$,[222,223] $HgHPO_4$,[222] $Hg_2P_2O_7$[224] and some polyphosphates.[225] Mercury(II) arsenates have been reported,[226] but are generally ill defined.

Mercury(II) sulphite does not seem to have been isolated, and the double salts with alkali metals contain S-bonded sulphite (q.v.). Mercury(II) sulphate is obtained by dissolving HgO or mercury metal in concentrated sulphuric acid. The structure consists of mercury atoms in a distorted tetrahedral arrangement, Hg–O = 0.214(2 x O), 0.208, 0.228 nm.[227–229] The monohydrate, which crystallises from solutions of the anhydrous compound in moderately dilute acid has a structure based on $HgO_5(OH_2)$ octahedra with Hg–O = 0.217, 0.250(2 x O), 0.251(2 x O), 0.214 nm (H_2O), arranged in chains with sulphate groups linking the chains together.[230,231]

The very slightly soluble mercury(II) selenite, $HgSeO_3$, precipitates from mercury(II) nitrate solutions on addition of H_2SeO_3 or Na_2SeO_3.[232] It does not seem to have been established whether coordination is via the selenium or oxygen. The selenate, $HgSeO_4$, is isomorphous with $HgSO_4$,[228] and is prepared similarly. Several mercury(II) tellurates have been prepared,[232–234] viz. $HgTeO_3$, Hg_3TeO_6, $Hg_2H_2TeO_6$, and the structure of Hg_3TeO_6, which contains octahedral TeO_6 and tetrahedral HgO_4 moieties with each oxygen bonded to one tellurium and three mercury atoms, Hg–O = 0.233 nm, Te–O = 0.198 nm.

Mercury(II) salts of halogen oxyacids are well established—$Hg(ClO_3)_2 \cdot 2H_2O$, $Hg(BrO_3)_2 \cdot 2H_2O$, $Hg(ClO_4)_2 \cdot 6H_2O$, $Hg(IO_3)_2$, obtained by dissolving the oxide in the appropriate acid or, in the case of the iodate, by precipitation from aqueous solution. The perchlorate is the most important and crystallises as the hexahydrate from aqueous solution;[235] other hydrates have also been reported with 3 and 4 water molecules.[236]

Numerous basic salts are known, many of which were reported over fifty years ago and have not been examined by modern spectroscopic techniques. The more recent studies have revealed a number of interesting structures and only these structures will be discussed here.

Since mercury(II) oxide is only weakly basic most mercury salts hydrolyse to basic salts in aqueous solution unless acidified. Cooney and Hall obtained[237] Raman-spectroscopic evidence for the presence of polynuclear hydroxo-bridged species in aqueous solutions of $Hg(NO_3)_2$. X-ray scattering measurements[238] on acified mercury(II) perchlorate solutions indicate the presence of $Hg(H_2O)_6{}^{2+}$, Hg–O $\approx$ 0.24 nm. Hydrolysis produces polynuclear species that are thought to be of types $Hg_2OH(H_2O)_2{}^{3+}$, $Hg_3O(H_2O)_3{}^{4+}$ or $Hg_4O(OH)(H_2O)_3{}^{5+}$.[238]

Mercury(II) hydroxynitrate, $HgOH(NO_3)$, contains infinite chains of Hg–OH–Hg (figure 15.4), Hg–O = 0.2079 nm, 0.2093(5) nm, with the $NO_3{}^-$ ions between the layers.[239,240]

Mercury hydroxychlorate[241] and hydroxybromate[242] contain similar infinite $[Hg(OH)]_n{}^{n+}$ chains, with the mercury being further coordinated to six oxygens from $XO_3{}^-$ groups. The Raman spectra of $[Hg(OH)]_n{}^{n+}$ has been recorded.[243]

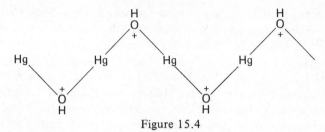

Figure 15.4

A different structure is present in basic mercuric sulphate (turnpeth mineral), $HgSO_4 . 2HgO$, which contains (figure 15.5) $(Hg_3O_2)_n{}^{2n+}$ ions, Hg–O = 0.203 nm,[244,245] with sulphate ions between the layers; $HgSeO_4 . 2HgO$ is similar.[244]

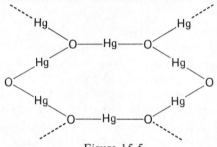

Figure 15.5

The $2HgSO_4 . HgO . 2H_2O$ of Bonifacic[245] has been re-formulated $Hg(OH)_2 . 2HgSO_4 . H_2O$ and contains $O_3SO–Hg(OH)Hg(OH)HgOSO_3$, the first established example of finite zig-zag chains, Hg–O = 0.210, 0.206 nm (in chain), $Hg–OSO_3$ = 0.214 nm; further more distant coordination of the sulphate oxygens to the mercury is also present.[246] Johansson[238,247] has examined the basic mercury per-chlorates $Hg(ClO_4)_2 . xHgO . yH_2O (3 > x > 1.5)$ obtained by dissolving HgO in perchloric acid. These were formulated $Hg_5O_2(OH)_2(H_2O)_x$, $Hg_7O_4(OH)_2-(ClO_4)_4$, and $Hg_2O(OH)(ClO_4)$, which contain the infinite Hg–O polymers shown in figure 15.6.

The basic mercury halides have already been discussed on p. 61.

15.4 Oxygen Donor Ligands

Pyridine-N-oxide forms 1 : 1, 2 : 1, and 6 : 1 complexes with mercury(II), the type of complex obtained depending on the anion and the preparative conditions. The 6 : 1 complexes $[Hg(pyNO)_6]Y_2$ (Y = ClO_4, BF_4, PF_6, SbF_6)[248,249] contain octahedral HgO_6 coordination, and this stoichiometry is achieved with weakly or non-coordinating anions. The structure of $[Hg(pyNO)_6](ClO_4)_2$ contains a regular octahedral cation with Hg–O = 0.235 nm, which is considerably longer than Hg–O in HgO,[250] and this compound is the first example of regular octa-hedral coordination for mercury(II). A similar structure has been assigned to $[HgL_3](ClO_4)_2$ (L = 2,2'-bipyridine-N,N'-dioxide).[251]

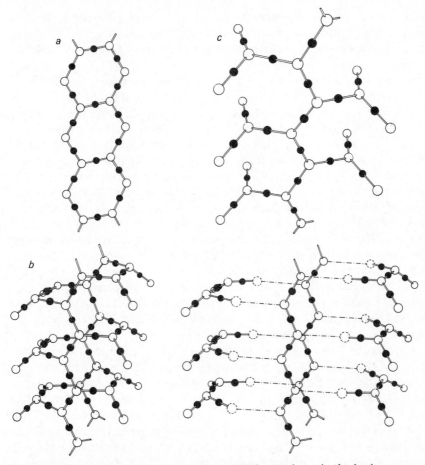

Figure 15.6. Part of the infinite one-dimensional complexes in the basic mercury(II) perchlorates. The smaller dark circles are mercury, the open circles, oxygen. (a) The triclinic $Hg_5O_2(OH)_2(ClO_4)_4(H_2O)_x$. (b) The orthorhombic $Hg_7O_4(OH)_2(ClO_4)_4$. To the right are drawn some characteristic structure elements of the complex. (c) The monoclinic $Hg_2O(OH)ClO_4$

The 1 : 1 complexes $[Hg(pyNO)X_2]_2$ (X = Cl, Br) are dimeric in solution,[248,249,252–256] and studies of the infrared and Raman spectra in solid state have been interpreted as consistent with halogen-bridged species, and not PyNO-bridged dimers as originally suggested.[249] An X-ray study of $Hg(pyNO)Cl_2$ has revealed the presence of bent $HgCl_2$ molecules, Hg—Cl = 0.2316, 0.2339 nm, held together by mutual interaction and by pyNO molecules, Hg—Cl = 0.3185, 0.3318 nm, Hg—O = 0.259, 0.260 nm.[257] Thus the mercury atom is in the characteristic six-coordinate environment, with two short Hg—Cl, two Hg—O, and two longer Hg—Cl bonds.[257] This contrasts with the essentially trigonal bipyramidal five-coordinate cadmium analogue.[257] Similar $[HgLX_2]_n$ complexes have been isolated with 2-, 3-, and 4-picoline-N-oxide, 4-cyanopyridine-N-oxide, lutidine-N-oxide, 2-ethylpyridine-N-oxide, and 4-methoxypyridine-N-oxide.[252,256] Several mercury(II) iodide analogues have been reported, but they appear to partially decompose in air.[254] The $[Hg(pyNO)_2Y_2]$ (Y = $CF_3CO_2^-$, $CCl_3CO_2^-$, NO_3) complexes may be hexa-coordinate and contain bidentate anionic ligands.[248,249] The $HgL(SCN)_2$ (L = pyNO, 2-, 3-, or 4-picNO, lutNO) and $HgL'(CN)_2$ (L' = picNO, lutNO) species are probably pseudohalide bridged,[258] while $Hg(4\text{-}CNpyNO)_2$-$(SCN)_2$ and $HgL''_2(CN)_2$ (L'' = pyNO, 4-picNO) may be tetrahedral monomers.[249,258] Mercuric cyanide does not complex with 3-picNO or 4-CNpyNO.[258] Bidentate chelation is observed for 2,2'-bipyridine-N,N'-dioxide in the pseudo-tetrahedral $HgLX_2$ (X = Cl, Br, SCN) species and in $Hg_2L(CN)_2$, which is polymeric with cyanide bridges.[259] Pappas *et al.*[260] have reported DTA studies on a range of mercury chloride adducts of pyridine-N-oxides, and Brill and Hugas[261] have obtained NQR data on mercury halide complexes of pyNO and 4-picNO. Quinoline-N-oxide forms[262] a 1:1 complex with $HgCl_2$, which contains mercury coordinated to two chlorines at 0.2299, 0.2304 nm, two oxygens at 0.256, 0.261 nm, and two further chlorines at 0.312, 0.335 nm. The 1:1 complex with 3,5-dibromopyridine-N-oxide, however, contains penta-coordinate, essentially square-pyramidal mercury, consisting of chains of $HgCl_2$ (0.231, 0.229 nm) with Br_2pyNO links between them (0.251 nm) and more distant coordination to further chlorines of neighbouring molecules.[263]

Carlin *et al.*[248] could not prepare $[HgL_6](ClO_4)_2$ with L = Me_3NO or Ph_3PO. Triphenylphosphine oxide forms colourless crystals of the 2:1 adducts with mercury halides in ethanol.[264,265] Other phosphine oxide complexes reported are $[(p\text{-}MeC_6H_4)_3PO]_2Hg_2X_4$ (X = Cl, Br, I),[266] $Hg(OPMe_3)Cl_2$, $Hg_5(OPMe_3)_2$-Cl_{10},[258,267] and $Hg(Ph_2vinylPO)Cl_2$.[268] A number of other complexes have been reported in the older literature but have not been re-examined. The structures of these complexes do not seem to have been definitely established. Triphenylarsine oxide forms 1:1 and 2:1 adducts with mercuric chloride.[269] The former is oxygen bridged,[262,270] while the $Hg(OAsPh_3)_2Cl_2$ complex has a distorted tetrahedral structure with Hg–Cl = 0.233 nm, Hg–O = 0.235 nm.[270,271] The far infrared spectrum of $[Hg(OAsPh_3)Cl_2]_2$ has beem measured.[271] Both $OPPh_3$ and $OAsPh_3$ form 4:1 complexes with mercuric perchlorate.[45]

Octamethylpyrophosphoramide, $(Me_2N)_2P(O)OP(O)(NMe_2)_2$,[272] and nonomethylimidodiphosphoramide, $(Me_2N)_2P(O)N(Me)P(O)(NMe_2)_2$,[273] form $[HgL_3](ClO_4)_2$ complexes, which probably involve HgO_6^{2+} coordination. Dimethylsulphoxide (DMSO) forms $[Hg(DMSO)_6](ClO_4)_2$ with mercuric perchlorate, and thioxan oxide (figure 15.7) and tetrahydrothiophene oxide form similar complexes.[248] Other DMSO complexes of stoichiometry DMSO:$HgCl_2$ 1:1, 1:2, and 2:3 are formed by direct reaction or by hydrogen peroxide oxidation of the corresponding dimethylsulphide complex.[275,276] Diethylsulphoxide gives similar results,[276] while R_2SO (R = Pr^n, Bu^n, Ph, Bu^i) form 1:1 $HgCl_2$ complexes and $R_2'SO$ (R' = Ph, Pr^i) form 1:2 species.[276,277] These complexes are all non-electrolytes and molecular-weight measurements indicate substantial dissociation in solution.[276]

$$O=S \overset{\displaystyle (CH_2)_2}{\underset{\displaystyle (CH_2)_2}{\diagdown \!\!\! \diagup}} O$$

Figure 15.7

Infrared and Raman spectra were interpreted as consistent with oxygen coordination of the DMSO, and possible structures have been discussed.[278,279] The X-ray structural determination[280] of $Ph_2SO \cdot HgCl_2$ confirms the $Hg-OSPh_2$ coordina-

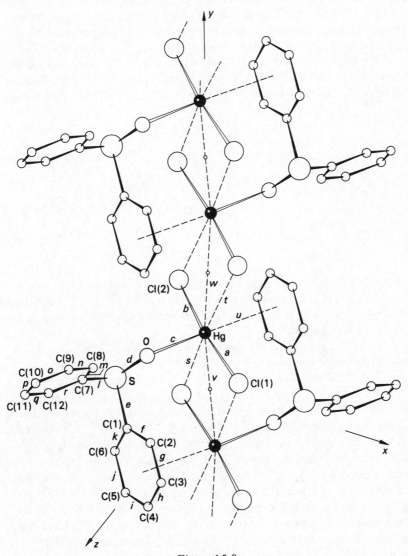

Figure 15.8

tion (figure 15.8) in which the mercury is coordinated to two chlorines (0.2291, 0.2289 nm) and two more distant chlorines (0.3230, 0.3284 nm) of neighbouring molecules, and the oxygen is at 0.258 nm with one phenyl group occupying the sixth coordination position (0.351 nm from the mercury). The structure of $HgCl_2(DMSO)_{2/3}$ (the 2:3 adduct) consists of $HgCl_2$ molecules and $[(DMSO)$. $HgCl_2]_2$ groups.[281] The $HgCl_2$ molecules have Hg–Cl = 0.2306 nm with long contacts to four other chlorines (0.3004–0.3081 nm), while in the dimeric units each mercury is coordinated to two oxygens (0.252, 0.256 nm), two chlorines (0.2309, 0.2320 nm), and two contacts to neighbouring molecules (Hg . . . Cl = 0.3302, 0.3372 nm), the dimeric units being held together by oxygens bridges. Mercury(II) thiocyanate and cyanide form 1:2 HgX_2:DMSO adducts.[278,282,283] *Cis*-4-*p*-chlorophenylthian-1-oxide forms a $HgCl_2$.2L complex in which the ligand is coordinated via the oxygen [Hg–O = 0.248, 0.297 nm (bridge), 0.270 nm (terminal), Hg–Cl = 0.2284, 0.2295 nm], (figure 15.9), the sixth coordination position on the mercury is occupied by the chlorine atom of the *p*-chlorophenyl group.[284]

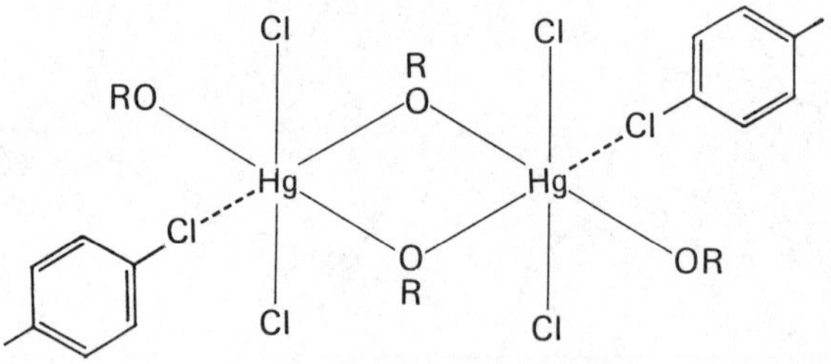

Figure 15.9

Much less is known about selenoxide ligands. Bis(4-ROphenyl)selenoxides (R = Me, Et) form 1:1 complexes with mercuric halides.[285] The 1:1 complexes of Ph_2SeO[286] and Me_2SeO[287] with $HgCl_2$ are coordinated via the oxygen, on the basis of their i.r. spectra.

Methanol and ethanol readily dissolve mercuric halides[288] and a number of adducts have been isolated. The 2:1 methanol/$HgCl_2$ complex has been X-rayed,[289,290] the structure consisting of mercury coordinated to two chlorines at 0.231 nm, to two oxygens (MeOH) at 0.282 nm, and to two further chlorines (at 0.307 nm) from neighbouring molecules completing the very distorted octahedron. Mercuric cyanide forms a 1:1 adduct with methanol.[291,292] Alkoxides have been little studied, but $Hg(OEt)_2$ is formed from NaOEt and HgX_2.[293] Mercury bis(trifluoromethyl)nitroxide, $Hg[ON(CF_3)_2]_2$, can be prepared from mercury and $ON(CF_3)_2$.[294,295] Similar $Hg(OSiR_3)_2$ (R = Me, Ph), are formed by treatment of mercury(II) bromide with the sodium silanolate.[296,297] $Hg(OSiMe_3)_2$ very easily decomposes into HgO and $Me_3SiOSiMe_3$.

Mercury(II) halides also complex with ethers. The structure of $HgBr_2$.THF consists of $HgBr_2$ units (Hg–Br = 0.2475 nm), oxygen-coordinated THF (Hg–O = 0.267 nm), and a very distorted octahedron being completed by contacts to three neighbouring bromines.[298,299] Mercuric cyanide forms a THF complex of stoichiometry $5Hg(CN)_2 \cdot 4THF$, the X-ray crystal structure of which reveals three types of mercury atoms. All have distorted octahedral coordination; one consists of $Hg(CN)_2$ units with four further $Hg-N$ contacts to neighbouring molecules; the second consists of $Hg(CN)_2$ with four THF ligands; and the third is $Hg(CN)_2$ with $2O(THF)$ and $2N(-NCHg)$ contacts.[298,300]

The mercury halide/1,4-dioxan systems have been thoroughly investigated. The known complexes are: $HgCl_2$ (dioxan), $HgX_2(dioxan)_n$ (X = Br, I, CN; n = 1, 2) and $Hg(SCN)_2(dioxan)$.[301-305] The structure of $HgBr_2(dioxan)_2$[306,307] consists of distorted octahedral $HgBr_2O_4$ (Hg–Br = 0.243 nm, Hg–O = 0.283 nm) units. A 1:1 adduct is formed between 1,3-dioxan and $HgCl_2$.[302] Preliminary structural data have also been reported for $HgCl_2 \cdot$ dioxan.[308]

Numerous other oxygen donor ligands complex with mercury(II), although in many cases the reported studies are pre-1950 and hence little evidence for the actual donor site is available. Among the ligands studied more recently and for which evidence (usually from i.r. spectral data) is available that indicates Hg–O coordination are: cyclohexane-1,4-dione,[309,310] γ-butyrolactam,[311] phenols,[312] tertiary amides,[313] uracil,[314] 1-methyl-2,3, or 4-pyridone,[315] and cumarin(5,6-benzopyrone).[316,317] X-ray crystallographic studies have been reported on the mercuric halide adducts of cyclohexane-1,4-dione,[310] azoxyanisole,[262] and uracil.[318]

The mercuric complexes of β-diketones have been examined by a number of workers.[319-321] In contrast to the complexes of these ligands with many other metals in which the β-diketone functions as a chelating O,O donor, with mercury(II) it appears that the coordination is via the central carbon atom[321] (figure 15.10).

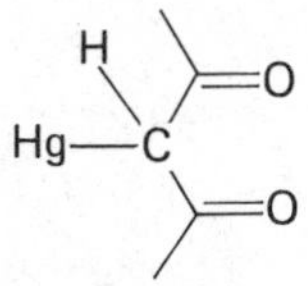

Figure 15.10

15.5 Sulphur Donor Ligands

Mercury(II) sulphide exists in two forms, a red hexagonal α-HgS (cinnabar) obtained from the elements or by passing H_2S into mercuric acetate in hot glacial acetic acid containing ammonium thiocyanate, and a black β-HgS (metacinnabar) precipitated from aqueous acidified mercuric chloride by H_2S.[322,323]

α-HgS is isostructural with hexagonal HgO and contains spiral chains, $(S-Hg-S-Hg)_n$, Hg–S = 0.236, 0.310, 0.330 nm, and is the stable form at ambient temperature.[324] The black β-HgS has a zinc-blende structure, Hg–S = 0.253 nm and is stable above 400 °C, although it exists indefinitely at room temperature despite its metastable condition.[325] Mercury sulphide is exceedingly insoluble in water, and unreactive chemically, being attacked only by concentrated HBr, HI or aqua regia. Oxygen converts it on heating to a mixture of $HgSO_4$, SO_2, Hg_2SO_4 and mercury. Mercuric sulphide dissolves in alkali-metal sulphide solutions, presumably to give complex anions,[326] and several complex sulphides, usually obtained by heating the components together, are known, for example M_2HgS_2 (M = Na, K), although no recent studies appear to have been reported.

Numerous mercury(II) complexes of thioether (R_2S) ligands are known. Many of those containing unusual thioether ligands were prepared for ligand-characterisation purposes, and little other than analytical and melting point data are reported—references to these are not included in this article. Thioether complexes are known with mercury(II) halides in ratios 1:1, 2:1, and 2:3, and are readily obtained by direct combination in aqueous alcoholic solutions.[267,327-337] Thus, complexes have been prepared with Me_2S,[327-329] Et_2S,[327,328,331,332] Pr_2S,[323-329,331,332] $Bu^n{}_2S$,[327,328,331] $Bu^i{}_2S$,[327,328,331] $Bu^s{}_2S$,[327,331] $Bu^t{}_2S$,[338] (isoamyl)$_2S$,[327,328] (n-amyl)$_2S$,[327,332] MeEtS,[327,333] $MePr^nS$,[327,329] $MeBu^n$-S,[327,329] $MeBu^tS$,[327] and EtPrS.[327,334] The structures of these compounds have been variously interpreted as containing sulphonium ions $[R_2SHgCl]Cl$[335] or as adducts $R_2S \cdot nHgX_2$. Branden[335] proposed the sulphonium-salt structure for $Et_2S \cdot (HgCl_2)_2$, that is, $[Et_2SHgCl]Cl \cdot HgCl_2$, which contains Hg–S = 0.241 nm, Hg–Cl = 0.271 nm. Biscarini *et al.*[327,338] investigated a wide range of thioether complexes by i.r. and electronic spectroscopy, and molecular weight and conductivity measurements, and concluded that they were better formulated as addition rather than substitution (that is, sulphonium) products. Thus, in the solid state the 1:2 ($R_2S \cdot 2HgX_2$) adducts consist of HgX_2 and $Et_2S \cdot HgCl_2$ molecules held together by weak chlorine bridges (figure 15.11). The 1:1

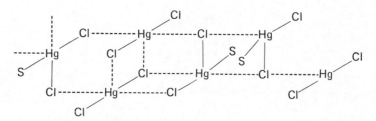

Figure 15.11

complexes are probably analogous to the $HgCl_2 \cdot L$[334] (L = tetrahydrothiophen) (figure 15.12). The 2:3 complexes are probably of intermediate structure since their i.r. spectra contain absorptions characteristic of the mercury environments in both the 1:1 and 1:2 complexes.[327] In solution[338] these complexes are dissociated to considerable extents, as evidenced by their molecular weights; since the solutions are non-electrolytes it is apparent that the dissociation is of the form

$$RR'S \cdot nHgCl_2 \rightleftharpoons RR'S + nHgCl_2$$

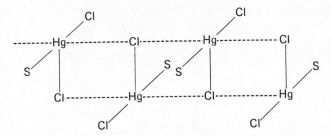

Figure 15.12

Hydrogen peroxide converts the thioether adducts to sulphoxide complexes, although some sulphones are also produced.[339] The thermal decomposition of $Bu^t_2S . HgCl_2$ produces $Hg_3S_2Cl_2$, $HgBu^tSCl$, Bu^tCl, HCl, and $Me_2C=CH_2$.[336]

Cyclic sulphides (thioethers) also complex readily with mercury(II) halides— $(CH_2)_3S$,[340] $(CH_2)_4S$,[332,335,340] $(CH_2)_5S$,[341] and the structure of the tetra-hydrothiophen, $(CH_2)_4S$, compound has been detemined.[335] 1,2-bis(methylthio)-ethane, $MeSCH_2CH_2SMe$, and its ethyl analogue form 1:1 adducts with mercury halides on mixing warm concentrated aqueous solutions of the halides and the dithioethers. The spirocyclic tetrathioether $C(CH_2SBu^t)_4$ forms Hg_2X_4L (X = Cl, Br, I) (figure 15.13), but the phenyl analogue did not complex with mercury(II).[346]

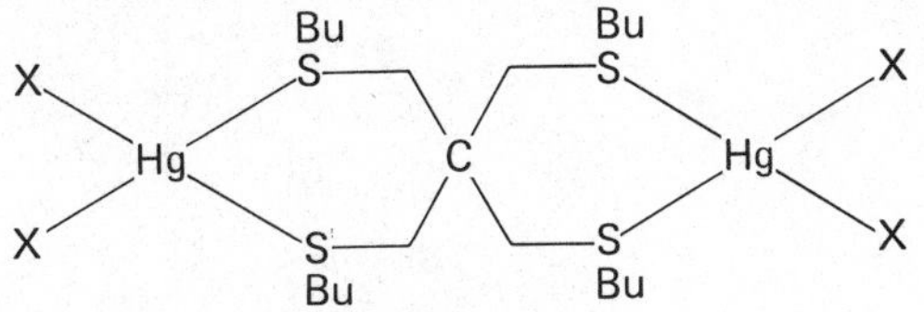

Figure 15.13

The dithioether-diolefins (figure 15.14) function only as disulphur donors towards mercuric halides, while but-3-enylbutylsulphide and *n*-butylpent-4-enylsulphide gave only uncharacterised oils.[347]

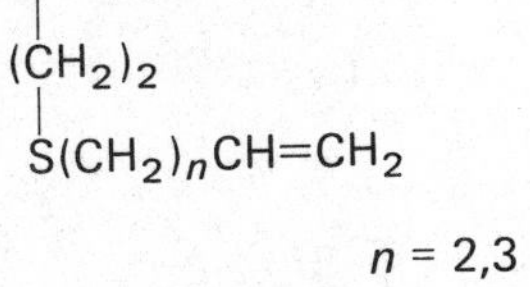

$$S(CH_2)_nCH=CH_2$$
$$(CH_2)_2$$
$$S(CH_2)_nCH=CH_2$$

n = 2,3

Figure 15.14

1,4-thioxan forms $HgCl_2 . 2L$,[348] the structure of which[349] shows the mercury to be in a distorted tetrahedral (S_2Cl_2) environment with the thioxan in a chair conformation, Hg–S = 0.257 nm, Hg–Cl = 0.248 nm. 1,4-

dithian forms $HgX_2 . L (X = Cl, Br, I, CN)$ and $Hg(ClO_4)_2 . 2L,$[350-352] which probably contains tetrahedral HgX_2S_2 moieties with dithian bridges. 1,3,5-trithian forms $HgX_2 . L (X = Cl, Br, I)$ complexes;[353-356] the chloro complex contains tetrahedral mercury (figure 15.15), $Hg–Cl = 0.244$ nm, $Hg–S = 0.261$ nm.

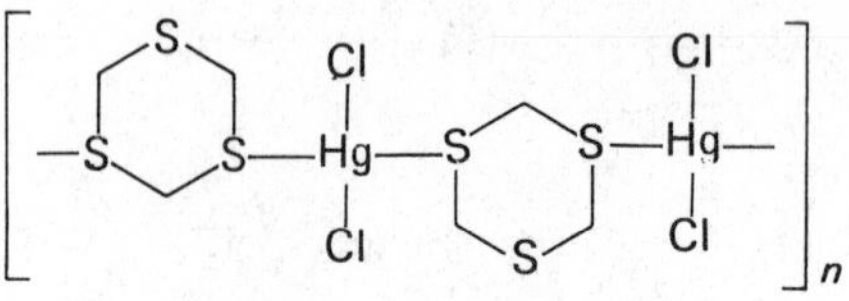

Figure 15.15

Phenoxanthin forms a 1 : 1 adduct with mercuric chloride, which has the structure[357] illustrated in figure 15.16; $Hg–Cl = 0.233, 0.308$ nm, $Hg–S = 0.312$ nm. The cyclic 1,6-dithiocyclodeca-*cis*-3-*cis*-8-diene forms a 1 : 2 complex with $HgCl_2$, which contains one approximately linear $HgCl_2$ unit ($Hg–Cl = 0.230$ nm, with further chlorines at 0.302, 0.293, 0.321, and 0.444 nm) and one

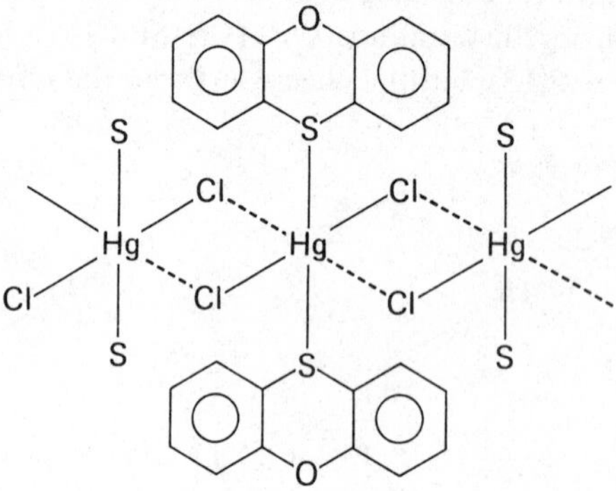

Figure 15.16

deformed $HgCl_2S_2$ environment ($Hg–S = 0.253$ nm, $Hg–Cl = 0.251$ nm)[358,359] (figure 15.17).

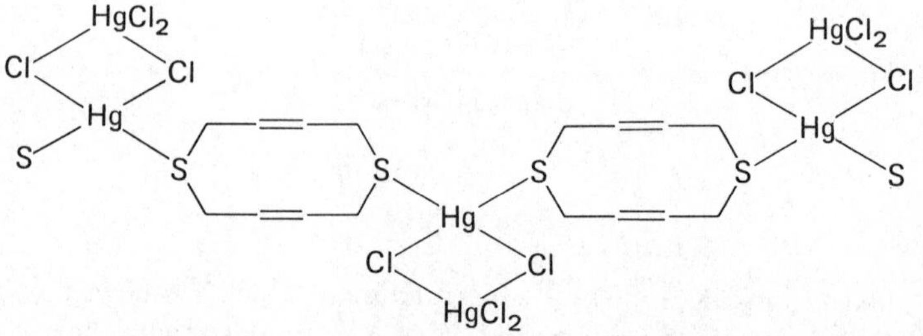

Figure 15.17

Mercury(II) mercaptides $Hg(SR)_2$ are formed from HgO and RSH. Infrared spectra suggest that the complexes with R = Me, Et, Pr^n, Pr^i, Bu^i, are linear S—Hg—S systems, but that the Bu^t analogue is polymeric.[187]

In $Hg(SMe)_2$ the mercury atom has two short bonds to sulphur (0.236 nm) and is coordinated to three other more distant sulphurs (0.326 nm), producing irregular tetragonal pyramidal coordination.[360] The ethyl analogue, $Hg(SEt)_2$, has a discrete molecular structure,[361,362] while the $Hg(SBu^t)_2$ is polymeric, each mercury being essentially tetrahedrally coordinated (Hg—S = 0.259, 0.266 nm).[361] The $Hg(SCF_3)_2$, obtained from HgF_2 and CS_2 in an autoclave[363,364] or photochemically from mercury and F_3CSSCF_3,[365,366] forms F_3CSHgX (X = Cl, Br, I) on reaction with HgX_2 and $M[Hg(SCF_3)_2X]$ with alkali metal halides (MX).

Reid *et al.*[367] found that tri-substituted thioureas react with $(PhC{\equiv}C)_2 Hg$ to

form $(RR'NS)_2 Hg$ (R = R' = Ph; R = Ph, R' = N⟮‾‾O⟯, —N⟮∶⟯ , —N⟮⬡⟯

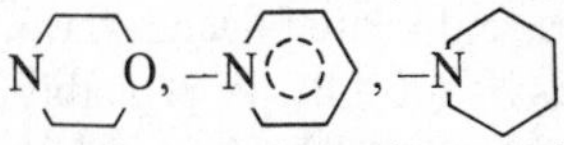

NMePh), which decomposed on heating to HgS. Other mercaptide complexes have been prepared from mercuric acetate and RSH, for example with R = Me_2CHCH_2-, $Me(CH_2)_8-$, Ph, while m-$C_6H_4(SH)_2$ behaves as a dinegative bridging ligand.[337] 1,3-dimercaptopropane and dihydrolipoic acid (figure 15.18) form 1:1 adducts with $HgCl_2$; lipoic acid (figure 15.19) itself forms a 1:1 adduct, but this is different to the complex with (figure 15.18), suggesting that the S—S bond remains intact.[368] It appears that RSHgX complexes can be prepared, but the structures are unclear.[369,370]

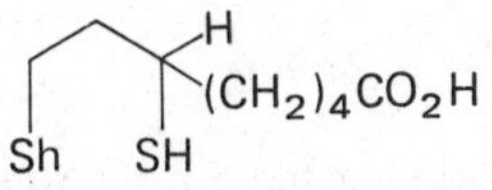

Figure 15.18 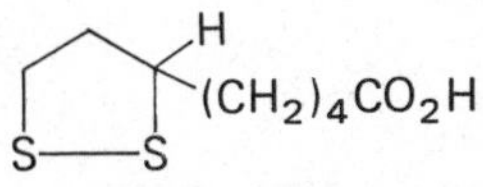   Figure 15.19

Bromine and iodine convert mercury(II) dialkyldithiocarbamates, $Hg(R_2dtc)_2$, into complexes analysing as '$HgX_2(R_2dtc)_2$' (R = Me, Et, Bu), which actually contain tetraalkylthiouram disulphide ligands.[371] These complexes can be made directly from the latter and HgX_2.[371] n.m.r. spectral data suggest that the complexes are tetrahedral with S_2X_2 donor sets,[372,373] and this was subsequently confirmed by an X-ray crystallographic analysis[374] of the tetramethylthiouram-disulphide—mercuric-iodide complex (figure 15.20); Hg—I = 0.2654, 0.2661 nm, Hg—S = 0.2651, 0.2882 nm. Tetramethylthiouram monosulphide, $Me_2NC(S)$-$C(S)NMe_2$ forms a similar complex.[372]

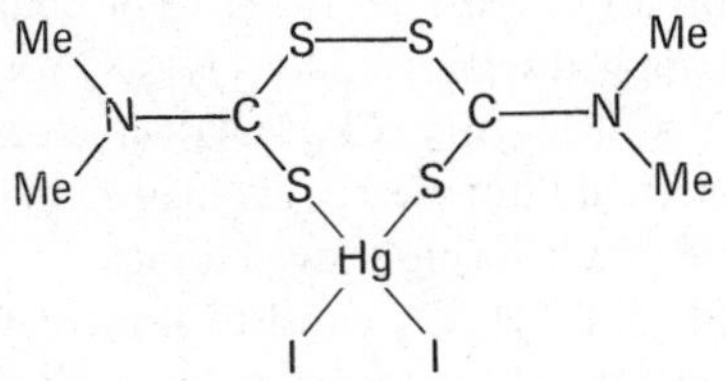

Figure 15.20

Trimethylphosphine sulphide forms $HgCl_2 . (Me_3PS)_2$, a non-electrolyte, the i.r. spectrum of which confirms that the ligand is coordinated through the sulphur.[371] Triphenylphosphine sulphide is reported to form 1:1 complexes, $HgX_2 . (Ph_3PS)$ (X = Cl, Br, I)[376,377] the Ph_3AsS analogues are isomorphous but appear to have a different structure to the Ph_3AsO complexes (q.v.).[376] The report[378] of $HgBr_2 . (Ph_3PS)_2$ has been challenged,[379] and further work is needed to clarify this point. Tris(dimethylamino)phosphine sulphide $(Me_2N)_3PS$ forms only a 1:1 complex with mercury(II) bromide.[421] Tetramethyldiphosphine disulphide forms[380] $HgX_2 . Me_2P(S)P(S)Me_2$ (X = Cl, Br, I) the i.r. spectra of which suggest that the ligand has a *cis* structure and is chelating to the mercury.[381] The diphosphinothioyl ligands (figure 15.21) form $HgBr_2 . L$ in which L functions as a bidentate S_2 donor.[382] A number of other mercury(II) complexes of diphosphine disulphides, and related ligands have been prepared.[383] Trimethylstibine sulphide forms HgL_2X_2 (X = Br, I), probably tetrahedral monomers, and $HgLI_2$, which appears to be sulphur bridged.[384] Mercuric chloride and Me_3SbS produce Me_3SbCl_2 and HgS and no adduct could be isolated.[384]

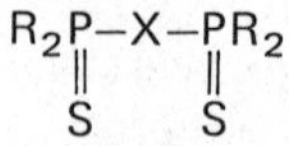

$$R_2P-X-PR_2'$$
$$\underset{S}{\|} \quad \underset{S}{\|}$$

X = CH_2	R = Ph,	R' = Ph
X = CH_2	R = Ph,	R' = Me
X = O	R = Me,	R' = Me

Figure 15.21

Mercuric–thiourea complexes have been much investigated both in solution and in the solid state.[385–393] It has been established that under appropriate conditions $HgCl_2 . nSC(NH_2)_2$ (n = 1, 2, 3, 4, and 2/3) can be isolated,[394] as can $HgBr_2 . nSC(NH_2)_2$ (n = 1, 2, 4) and $HgI_2 . nSC(NH_2)_2$ (n = 1, 2). The literature contains reports of 1:1 and 2:1 adducts with $Hg(CN)_2$,[395–397] and 1:1 and 4:1 complexes with $Hg(SCN)_2$.[391,395] The $Hg[SC(NH_2)_3]_xY_2$ (x = 2, 3, 4; Y = ClO_4, BF_4, CF_3COO, $HCOO$) have been prepared and the exchange with ^{35}S-thiourea examined.[393] An i.r. spectral study of $Hg[SC(NH_2)_2]_2Y_2$ (Y = ClO_4, BF_4, CF_3COO) produced evidence of anion coordination in the solid state, and it was proposed that the structure in the solid state may be trigonal $\{Hg[SC(NH_2)_2]_2-Y\}Y$.[392] The perchlorate and tetrafluoroborate complexes are 1:2 electrolytes in methanol, but the trifluoroacetate species seems to be considerably associated even in solution. A series of complexes $\{Hg[SC(NH_2)_2]_2X\}Y$ were subsequently obtained (X = Cl, Br, I; Y = ClO_4, BF_4, CF_3COO) which are 1:1 electrolytes in methanol, but the i.r. spectra do not clearly distinguish between an essentially three-coordinate structure or a halogen-bridged dimer.[392]

The structure of $Hg[SC(NH_2)_2]_4Cl_2$ consists of mercury in a distorted tetrahedral S_4 environment, the chlorines being ionic.[398] A second form of this complex has been isolated and its crystal structure determined. Once again the mercury is in a highly distorted tetrahedral environment, with Hg–S = 0.251, 0.255, 0.253, 0.261 nm.[399] The 3:1 complex Hg[SC(NH$_2$)$_2$]$_3$Cl$_2$ contains[400] distorted trigonal bipyramidal mercury with axial Hg–Cl = 0.224, 0.236 nm and equatorial thiourea, Hg–S = 0.237, 0.261, 0.310 nm. Trigonal planar mercury is present[358,359,401] in the 2:1 complex Hg[SC(NH$_2$)$_2$]$_2$Cl$_2$ (S$_2$Cl donor set), Hg–S = 0.242 nm, Hg–Cl = 0.257 nm, the other chlorine being normal to the plane and bridging two mercury atoms at 0.322 nm. In HgBr$_2$[SC(NH$_2$)$_2$]$_2$ the mercury is very distorted octahedrally, coordinated to terminal thiourea and four bridging bromines.[402] The iodo analogue has a different structure,[403] with the mercury coordinated to I$_2$S$_2$ donors in a distorted tetrahedral arrangement, Hg–I = 0.284 nm, Hg–S = 0.246 nm, with two further iodines at 0.386 nm completing a very distorted octahedron. In Hg[SC(NH$_2$)$_2$]$_2$(SCN)$_2$ the mercury is hexa-coordinated to four thiourea molecules at 0.328 nm and two –SCN groups at 0.242 nm.[404] A minor byproduct from the preparation of Hg[SC(NH$_2$)$_2$]$_2$Cl$_2$ was shown by X-ray studies to be HgCl$_2$. 2/3SC(NH$_2$)$_2$, formulated as {HgCl(L)]Cl}$_n$. n/2HgCl$_2$, and consisting of Cl–Hg–S units (Cl–Hg–S = 155°) with the other chlorines bridging the mercury atoms into infinite chains (figure 15.22), and with discrete HgCl$_2$ molecules occupying interstices between the chains.[394] The structure of the 1:1 adduct, HgCl$_2$. SC(NH$_2$)$_2$, is unknown, but based on its i.r. spectrum it is likely that a polymer of the type shown in figure 15.22 is present.[405]

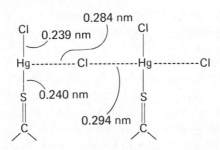

Figure 15.22

The different structures observed in the mercury–thiourea complexes are a further illustration of the richness of the structural chemistry of Hg(II), and exemplify the great care that must be taken in inferring structures from spectroscopic data alone.

Complexes of several substituted thioureas have been prepared but in general the structures have not been elucidated in detail. The reported complexes include those of SC(NMe$_2$)$_2$[406] SCNH(CH$_2$)$_2$NH,[407] *o*- or *p*-tolylthiourea,[408] allyl-thiourea,[408,409] N-methylthiourea,[410] N-acetylthiourea,[411,412] di-N-phenyl-thiourea,[412] and 1-phenyl-3(2-pyridyl)-2-thiourea.[413] It seems reasonable to assume that all, except possibly the last ligand, are S-bonded to mercury.

The mercuric complex of O-ethylthioacetatothioacetate (figure 15.23), HgL$_2$, appears to contain a tetrahedral HgS$_4$ moiety, on the basis of its n.m.r. spectrum,

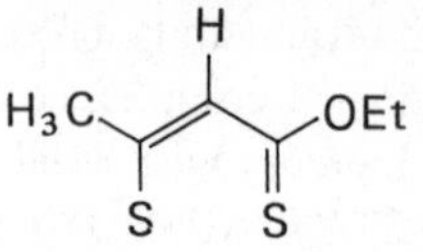

Figure 15.23

which is similar to that of the zinc analogue, which as been studied by X-ray methods.[414]

Dithioxamide (LH) forms $Hg(LH)X_2$ (X = Cl, Br) which, unlike most metal complexes of this ligand (which are S,N bonded), contain the ligand bonded as a bidentate disulphur donor.[415] The N,N'-dimethyl and dicyclohexyl analogues[416] are also S,N coordinated in $Hg(Me_2LH)X_2$ (X = Cl, Br) $Hg(Me_2LH)_3(ClO_4)_2$ and $Hg(Cy_2LH)_2(ClO_4)_2$. Thiobenzamide[417] and o-methoxythiocinnamamide[418] form HgL_2Cl_2 complexes containing neutral ligands bonded to the mercury via the sulphur atoms.

Mercury(II) forms stable complexes with anionic sulphur ligands (see reference 419 for a recent review of the donor properties of anionic sulphur ligands). Mercury(II) sulphite does not seem to be stable enough to isolate, but the complex $Na_2Hg(SO_3)_2 \cdot H_2O$ has been prepared and its crystal structure determined.[420,231] It contains discrete $Hg(SO_3)_2{}^{2-}$ groups with S-bonded SO_3 units, Hg–S = 0.2402, 0.2411 nm. Mercury(II) thiosulphate complexes obtained in solution also appear to be S-bonded.[422]

Mercuric dithioacid complexes have been known for many years.[423] Mercury ethylxanthate, $Hg(S_2COEt)_2$ has been studied by X-ray crystallographic techniques;[424] it contains mercury in a distorted tetrahedral environment, Hg–S = 0.234, 0.249, 0.276, 0.294 nm. Mercury(II) dithiocarbamates, $Hg(S_2CNR_2)_2$ (R = Et, $-(CH_2)_4-$),[423,425-428] are formed from HgX_2 and NaS_2CNR_2. The structures of two forms of mercury(II)-N,N'-diethyldithiocarbamate have been determined.[423,424] One form consists of isolated dimeric units, $Hg_2(S_2CNEt_2)_4$, which have a similar structure to that of the zinc and cadmium analogues, and contain penta-coordinate mercury with Hg–S = 0.2418, 0.2520, 0.2663, 0.2698, 0.3137 nm (figure 15.24). The second form consists of $Hg(S_2CNEt_2)_2$ units with planar HgS_4 coordination, Hg–S = 0.2398, 0.2965 nm, with weak intermolecular interactions, Hg–S = 0.3292 nm, forming infinite chains (figure 15.25). A third complex, $Hg_3Cl_2(S_2CNEt_2)_4$, is also known.[427] S-alkyl-N,N'-dialkyldithiocarbamate esters function as neutral ligands to Hg(II). Brinkhoff and Dautzenberg[429] prepared a series of $HgX_2[R'(R_2dtc)]$ complexes (X = Cl, Br, I; R' = Me, Et, $PhCH_2$; R = Me, Et), which on the basis of i.r., Raman, and n.m.r. spectra are formulated as three-coordinated mercury complexes (X_2S donor sets). The structure of one of these complexes, containing the methylpyrrolidine-1-carbodithioate ligand, has been established by an X-ray study[430] and reveals an irregular four-coordination about the mercury (figure 15.26), with Hg–Cl(bridge) = 0.257, 0.278 nm, Hg–Cl(terminal) = 0.237 nm, and Hg–S = 0.242 nm. The methylated sulphur is further than 0.38 nm from the mercury and is thus not

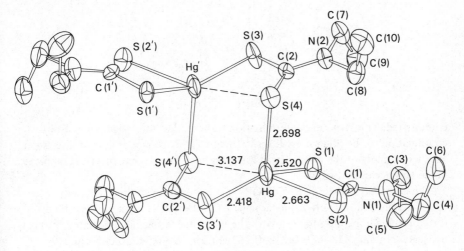

Figure 15.24. $Hg_2(S_2CNEt_2)_4$

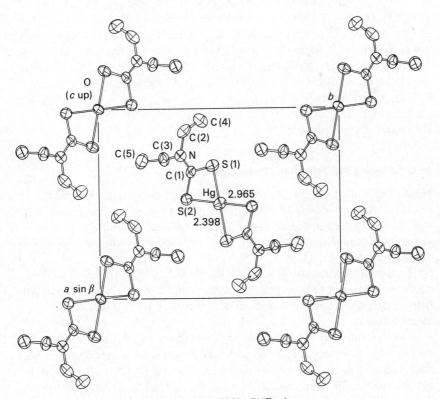

Figure 15.25. $Hg(S_2CNEt_2)_2$

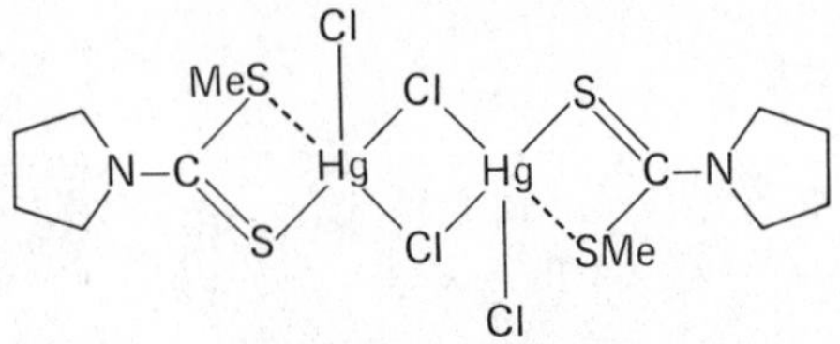

Figure 15.26

coordinated. This structure determination suggests that the assignment of three-coordination to the other complexes[429] may well be in error and, if so, this is yet another example of the difficulty of assigning structures to mercury(II) complexes in the absence of X-ray data.

Thiocarbamate esters, $R'RNC(S)OR''$, have been little studied. The only reported complex of O-ethylthiocarbamate (L), HgL_2Cl_2, has a monomeric distorted tetrahedral configuration, with the individual molecules associated through hydrogen bonding, $Hg-Cl = 0.258, 0.262$ nm, $Hg-S = 0.243, 0.245$ nm.[431]

Mercury(II) dithiophosphinates, $Hg(S_2PR_2)_2$, can be made by reaction of a mercuric halide with the sodium or ammonium dithiophosphinate (R = Me, Ph, Et, Pr^n),[432,433] or by reaction of mercury metal with $(R_2PS_2)_2$ (R = F, CF_3).[433,434] The $Hg[S_2P(OEt)_2]_2$ contains mercury coordinated to four sulphurs at approximately 0.256 nm,[424] while $Hg[S_2P(OPr^i)_2]_2$, unlike the zinc and cadmium analogues which are dimeric, is polymeric with mercury surrounded by five sulphurs, produced by one chelating $S_2P(OPr^i)_2$ ligand, $Hg-S = 0.2388, 0.2391$ nm, and three other sulphur atoms of different ligands at 0.2748 nm (bridge), 0.2888, 0.3408 nm.[435] In benzene solution the complex is monomeric. The species $Hg[SSeP(OPh)_2]_2$ has been mentioned briefly,[436] and the dimethyldithioarsinate complex, $Hg(Me_2AsS_2)_2$, prepared.[437]

15.6 Selenium and Tellurium Donor Ligands

Mercury selenide, HgSe, and mercury telluride, HgTe, are readily formed by direct combination of the elements. Both compounds are polymorphic the common structure being the zinc-blende type.[1] Both compounds are insoluble in water and are semiconductors, HgTe in particular also shows thermoelectric and infrared photoconductivity effects.

Much effort has been devoted to the study of HgE/HgX_2 (E = Se, Te; X = halide) systems, and to HgS/HgE.[1] Since most of the work has been concerned with the structures and physical properties of these systems it will not be discussed here.

Selenium and tellurium donor ligands constitute a rather neglected area of coordination chemistry, and even for the few reported mercuric complexes very little data are available. Diethylselenide forms dimeric $[HgLX]_2$ (X = Cl, Br, I) species; and the following have also been prepared $[HgCl_2 . TeBu_2]_2$,[344] $[HgCl_2 . TeEt_2]$,[438] $[HgX_2 . TeMe_2]$,[440] and $[HgX_2 L']$ [L' = Ph_2Te, o-, m-, and p-tolyl$_2$Te,

$(ROC_6H_4)_2Te$, *p*-, and *m*-xylyl$_2$Te].[441] Aynsley *et al.*[442] have prepared 1 : 1 mercuric chloride complexes of the bidentate selenoethers $MeSe(CH_2)_n SeMe$ (*n* = 2, 3). A structure (figure 15.27) has been proposed for the $(HgCl_2)_2$. L

Cl_2HgSe

$SeHgCl_2$

Figure 15.27

adduct with 1,4-diselenane (L);[443] and figure 15.28 illustrates an adduct with a cyclic telluride.[444]

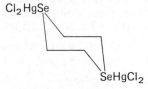

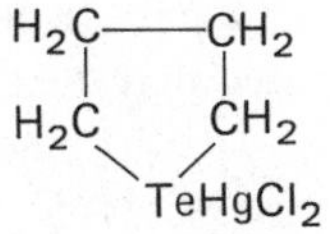

Figure 15.28

Triphenylphosphine selenide forms HgX_2 . Ph_3PSe (X = Cl, Br, I) complexes,[445,446] and a crystal-structure investigation[447] has shown $HgCl_2$. Ph_3PSe to be dimeric with chlorine bridges, Hg—Cl = 0.260, 0.278 nm, terminal chlorines Hg—Cl = 0.233 nm, and Hg—Se = 0.253 nm, the mercury being in a distorted tetrahedral environment.

Selenourea (Su) forms $HgCl_2$.(Su), HgX_2(Su)$_2$ (X = Cl, Br, I),[448] and $Hg(Su)_3(NO_3)_2$[449] complexes. On the basis of i.r. spectral data $HgCl_2$.(Su) appears to be a chlorine-bridged dimer, while $HgBr_2$.(Su)$_2$ probably contains trigonal planar $[Hg(Su)_2Br]^+$ ions, but the nature of HgX_2 .(Su)$_2$ (X = Cl, I) is unclear.[448]

Bis(trifluoromethylseleno)mercury, $(CF_3Se)_2Hg$, is obtained on shaking metallic mercury with $CF_3SeSeCF_3$.[450] Reaction with HgX_2 (X = Cl, Br, I) in methanol produces $[CF_3SeHgX]$, while alkali-metal and tetraalkylammonium halides produce complex anions of types $[Hg(CF_3SE)_2X_2]^{2-}$ and $[Hg(CF_3Se)_2]_2X^-$; 1 : 1 and 1 : 2 adducts can be isolated with triphenylphosphine.[450]

15.7 Mercury(II)–Nitrogen Compounds

The chemistry of compounds containing Hg—N bonds is very extensive and complex. In addition to coordination complexes of nitrogen ligands there are compounds of types $Hg(NR_2)_2$ and $[Hg_{n/2}(NH_{4-n})X]_x$. The latter group, to which the substances described in the literature as 'infusible white precipitates' and Millon's base belong, were for many years of doubtful constitution, and it is only with the application of modern spectroscopic techniques that the structures and bonding have been elucidated.[451,452]

15.7.1 Amides

Compounds containing mercury coordinated to two nitrogen atoms with electro-negative substitutents are known. Mercury bis(bistrifluoromethyl)amide, $Hg[N(CF_3)_2]_2$, is prepared from HgF_2 and perfluoro-2-azapropene, $CF_3N=CF_2$, or cyanogen chloride.[453,454] A second compound, $Hg[(CF_3)_2NNCF_3]_2$ if formed from $(CF_3)_2NN=CCl_2$ and HgF_2.[455] Both are stable monomeric species, which instantly decompose to HgO on contact with moisture. The Hg–N bonds are readily cleaved by halogens and suitable halides, and they are valuable precursors to N-halogenoamines and $(CF_3)_2N-$ compounds.[456] Two related compounds are $Hg[F_5SN(CF_3)_2]_2$[457] and $Hg[N(SiMe_3)_2]_2$.[458] The latter has been investigated by ^1H n.m.r. and i.r. spectroscopy, and the presence of a linear N–Hg–N system with trigonal planar nitrogen was demonstrated.[458]

Mercury(II) acetamide,[459] formed by dissolving HgO in acetamide, is relatively less reactive and has been variously thought to contain Hg–N, Hg–O or Hg–O,N bonds. An X-ray analysis[460] suggested that the coordination is via the nitrogen,

with the acetamide in the tautomeric $Hg(N=C\overset{\displaystyle CH_3}{\underset{\displaystyle OH}{\diagup\diagdown}})_2$ form, but Brown and

Robin[461] showed that in solution the structure consists of the normal acetamide form, $Hg[NHC(O)CH_3]_2$, and proposed that the solid state structure is similar. Irrespective of the doubt about the tautomeric form of the 'acetamide' ligand, the compound contains linear N–Hg–N (Hg–N = 0.206 nm) units linked into ribbons by H-bonds.[461] Diacetylhydrazine displaces acetamide from $Hg[NHCOCH_3]_2$ to form $Hg[N_2(COCH_3)_2]_n$, which is thought to have a chain structure,[462] and $Hg[N_2(COCF_3)_2]$ formed from $CF_3CONHNHCOCF_3$ and $Hg(OAc)_2$ is probably similar.[463] Mercury amidosulphates, $[M(O_3SN)]Hg$ (M = Na, K, etc.), are formed by dissolving HgO or $HgCl_2$ in the alkali-metal amidosulphate in alkaline solution[464,465] and are thought to have the structure shown in figure 15.29. On heating this type of compound in the presence of moisture, $Hg[O_3SNHg]_2$ is produced and this can also be obtained directly by heating HgO with $(NH_4)_2SO_4$.[465] On heating *in vacuo* the amidosulphatomercurates decompose into bis(amidobisulphato)mercurates, $M_4\{[(O_3S)_2N]_2Hg\}$ (M = Na, K, $\frac{1}{2}$Ba, $\frac{1}{2}$Sr),[466] also obtained directly from $Hg(NO_3)_2$ and the appropriate metal

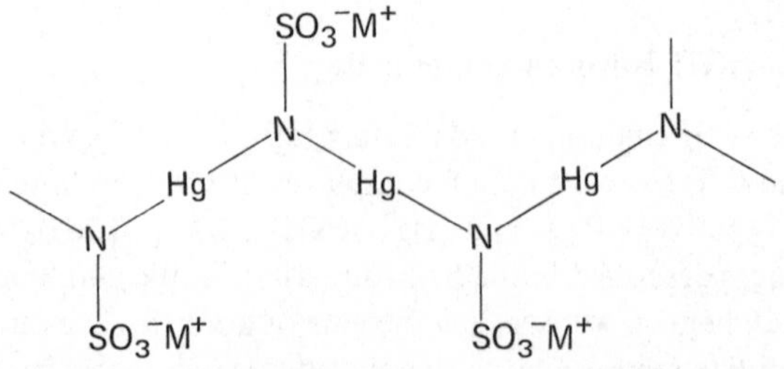

Figure 15.29

imidosulphate, $HN(SO_3M)_2$.[467] The $M_4\{[(O_3S)_2N]_2Hg\}$ also decompose on heating to approximately 150 °C in the presence of water to yield $Hg[O_3SNHg]_2$.[468]

Sulphamide, $SO_2(NH_2)_2$, reacts with mercury acetamide or acetate in aqueous solution to yield complex mixtures,[469] which may contain $\{HN-SO_2-HN-Hg\}_n$ or $\{HN-SO(NH)-O-Hg\}_n$ units. Sulphimide[470] derivatives, $Hg_3[(NSO_2)_3]_2$, and hydroxylamidosulphuric acid[469] derivatives. $[H_3O][(O_2SNO)Hg]$, possibly contain Hg−O as well as Hg−N bonds. Mercury acetate reacts with S_7NH[471] and $(SNH)_4$,[472] to form $Hg(NS_7)_2$ and $Hg(NS)_2$ or $Hg_5(NS)_8$; no structural data seem to be available on these compounds. Mercuric fluoride is converted by $OCFNSF_2$ into $Hg(NSF_2)_2$,[473] which decomposes on heating *in vacuo* into NSF and HgF_2, and is converted by halogens into $XNSF_2$ (X = F, Cl, Br, I). The structure[474] of $Hg(NSF_2)_2$ consists of linear N−Hg−N with Hg−N = 0.205 nm, and contains the anomalously short N−S distance of 0.144 nm. Mercury bis(sulphinylamide) and mercury fluorosulphurylisocyanate have been prepared.[439,475]

The reaction of mercury(II) salts with ammonia in aqueous solution in the presence of the corresponding ammonium salt leads to diamminemercury(II) complexes, for example $Hg(NH_3)_2Cl_2$−'fusible white precipitate'. At higher pH and lower ammonium-salt concentration, amido complexes, for example $HgNH_2Cl$−'infusible white precipitate' result. Complete replacement of the hydrogen in ammonia produces the derivatives of Millon's base, $Hg_2NOH.2H_2O$, usually prepared by dissolving HgO in aqueous ammonia. Since all three types of compound can be obtained from HgX_2 and NH_3 in aqueous solution, depending on the conditions, and mixtures are obtained if the conditions are not closely controlled, the confusion surrounding mercury–ammonia compounds many years ago is easily appreciated. It is not possible to review the mercury–nitrogen compounds of these types in detail without producing a work of inordinate length. Thus, the following sections aim to outline the preparations, properties, and structures of the more important types of compound, rather than provide comprehensive coverage (see also reference 451). Although the $[Hg(NH_3)_2]^{2+}$ compounds can be considered as normal coordination complexes it is also possible to regard them as mercury-substituted ammonium ions, which allows them to be treated as the first members of the $_\infty^{n-1}[NH_{4-n}Hg_{n/2}]X$ series.[452]

Diamminemercury(II) complexes are produced from aqueous solution only in the presence of high concentrations of the corresponding NH_4X, and are better obtained by reaction with ammonia in a non-polar solvent.[476−479] The importance of the ammonium salt is due to the reaction

$$Hg(NH_3)_2X_2 \rightleftharpoons HgNH_2Cl + NH_4Cl$$

The structure[480,481] of $Hg(NH_3)_2Cl_2$ consists of linear $H_3\overset{+}{N}-Hg-\overset{+}{N}H_3$ groups inserted into a simple cubic lattice (figure 15.30), each mercury being effectively six-coordinate; $Hg(NH_3)_2X_2$ (X = Br, I) are similar.[482,483] The $Hg(NH_3)_2X_2$ (X = Cl, Br) are stable compounds, but the iodide loses ammonia on standing.[484] Water or alkalis convert them into $HgNH_2X$ and Hg_2NX.

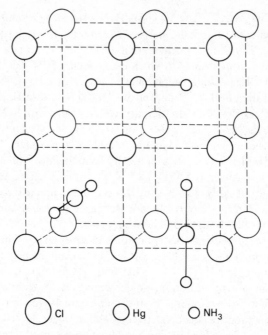

Figure 15.30. $Hg(NH_3)_2Cl_2$

The amidohalides (mercurioammonium halides), $HgNH_2X$ (X = Cl, Br) are orthorhombic (Cl) or cubic (Br), the latter having a very similar structure to $Hg(NH_3)_2Br_2$, with which it forms solid solutions,[482,485-487] When small amounts of $HgBr_2$ are incorporated into $HgNH_2Br$ a rhombic structure results.[486] The cubic $HgNH_2Br$ contains infinite chains of $+H_2N-Hg-NH_2-Hg+_n$, which thread through the cubic lattice with no particular orientation (figure 15.31), but in the rhombic $HgNH_2X$ the $+H_2N-Hg-NH_2-Hg+_n$ chains are parallel (figure 15.32).

The reaction of HgO, $HgCl_2$ and aqueous NH_3 in the appropriate ratio produces Hg_2NHCl_2 of unknown structure.[484] A similar reaction employing $HgBr_2$ produces Hg_2NHBr_2,[488] originally obtained from $HgNH_2Br$ and aqueous $HgBr_2$.[489] The structure[490,491] consists of $HgBr_3{}^-$ layers alternating with $[Hg_3(NH)_2]_n{}^{2n+}$ and Br^-. The cation has the same structure as $[Hg_3O_2]_n{}^{2n+}$, and there are two types of coordination to mercury, six-coordinate octahedral cations and penta-coordinate trigonal bipyramidal anions. The structure can only accommodate anions of the size of Br^-.

Millon's base, $Hg_2NOH \cdot xH_2O$, originally obtained by Millon in 1845 by dissolving yellow HgO in aqueous ammonia, forms salts with numerous anions, the structures of which are anion dependent. The essential unit consists of three-dimensional polymeric $[NHg_2]_n{}^+$ moieties, each nitrogen being tetrahedrally coordinated to four mercury atoms, and each mercury is coordinated to two nitrogens (Hg–N = 0.204–0.209 nm).[486,492-496] With weakly coordinating

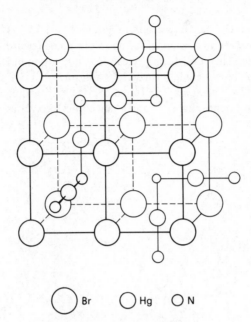

○ Br   ○ Hg   ○ N

Figure 15.31. Cubic H$_2$ NHgBr

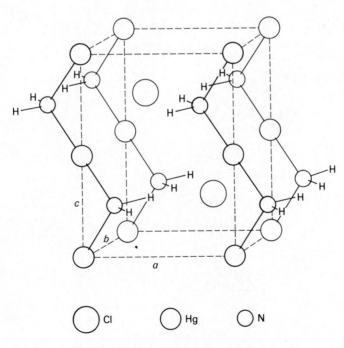

○ Cl   ○ Hg   ○ N

Figure 15.32. Rhombic H$_2$ NHgX

anions (NO_3^-, ClO_4^-) a cubic structure results (figure 15.33).[493] If the anion is halide or hydroxide an hexagonal form is more stable (figure 15.34) although it is possible to prepare cubic forms by careful substitution in $Hg_2N . NO_3$. Individual salts of Millon's base include the azide,[497] nitrate,[476,491,495,498] fluoride,[499] chloride,[476,498,500] bromide,[479,495] iodide,[492,495,501] sulphate[493,502], phosphate,[493,502] pyrophosphate,[503] and a range of carboxylates.[504] The substance

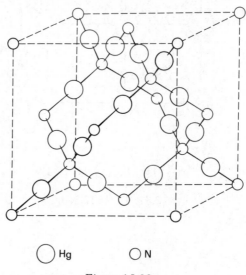

$\bigcirc$ Hg $\bigcirc$ N

Figure 15.33

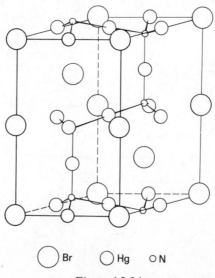

$\bigcirc$ Br $\bigcirc$ Hg $\circ$ N

Figure 15.34

formulated as $HgNH_2F$[505] is probably $Hg_2NF \cdot NH_4F$.[499] Millon's base itself can be dehydrated to a number of compounds containing Hg_2N moieties, the possible structures of which have been discussed by Weber.[506-508] The structural analogy between Millon's salts and the silicate structures has been discussed.[452] The presence of a very heavy atom such as mercury means that the positions of the light nitrogen atoms can only be determined with low accuracy by X-ray diffraction due to the great differences in scattering power, and much additional information on the nature of the above compounds has been obtained by the application of n.m.r.,[509] i.r., and Raman spectroscopy.[510-513]

Finally, mention should be made of the reaction of alkaline potassium iodomercurate with ammonia, which forms the basis of Nessler's method for the detection of ammonia (see reference 451).

Hydrazine forms compounds of formulae $[Hg(N_2H_4)_2]X_2$ and $Hg(N_2H_4)X_2$, which are thought to contain $[H_4N_2HgN_2H_4]^{2+}$ and infinite $\{NH_2-NH_2-Hg\}_n$ chains, respectively, similar to the ammonia analogues.[514] Similarly, formally analogous to $HgNH_2Cl$, there is $[Hg_2(N_2H_2)]Cl_2$, which has a layer structure with the chloride ions between the layers.[514,515]

Several compounds of unknown structures, which are formally alkylamine analogues of the mercury–ammonia compounds above, have been prepared, viz: $Hg(NHR)X$,[516] Hg_2NRX,[488] and $(RN)_2Hg_3X_2$.[452] Organomercurioammonium complexes $_{\infty}^{n-1}[NH_{4-n}HgMe_n]^+$, $N(HgMe)_3$,[517] and $[NH_{4-(m+n)}Me_m(HgMe)_n]^+$ can also be obtained.[452,518-520]

The action of ammonia on mercury(I) salts produces mercury metal and $HgNH_2X$ or $Hg_2NX \cdot H_2O$.[492,521]

15.7.2 Amines

In addition to the $Hg(NH_3)_2X_2$ compounds discussed above, numerous other ammonia adducts have been reported, but very little structural or spectroscopic data have been forthcoming on them. The mercury(II) halides are reported to form the following adducts

$HgF_2 \cdot nNH_3$	$n = 2, 4, 5$	(reference 522)
$HgCl_2 \cdot nNH_3$	$n = 3/2, 2, 4, 8, 9.5, 12$	(references 478, 480, 483)
$HgBr_2 \cdot nNH_3$	$n = 2, 8$	(references 478, 482, 483, 523)
$HgI_2 \cdot nNH_3$	$n = 4/3, 2, 6, 12$	(references 478, 523)

Mercury oxosalts form bis and tetrakis ammine adducts, $Hg(NH_3)_xY_2$ (Y = NO_3,[476,524] ClO_4,[524] $\frac{1}{2}SO_4$[477,524]). The i.r. spectrum of $[Hg(NH_3)_4](I_3)_2$ has been recorded.[525] The hydrazine complexes of HgX_2 have been described above; the corresponding $Hg(N_2H_4)(CN)_2$[526] $Hg(N_2H_4)(NO_3)_2$,[526] and $Hg(N_2H_4)(SO_4)$[527] have not been studied recently.

Primary amines form 2:1, and more rarely 1:1 and 4:1 complexes with mercury(II) halides: $MeNH_2$,[528] $EtNH_2$,[376,527,529,530] Pr^nNH_2,[476,530]

Bu^nNH_2,[530] Bu^iNH_2.[530] The cyanide complexes $Hg(RNH_2)_2(CN)_2$ (R = Me, Et, Pr^n, Bu^n) are assigned an octahedral CN-bridged structure on the basis of i.r. spectra.[531] Aniline complexes of mercury(II) have been known for many years. Mercury(II) chloride forms a 2:1 complex, obtained as colourless needles from the constituents in ethanol.[532,533] Similar complexes are formed by $HgSO_4$, $Hg(NO_3)_2$, $HgBr_2$, HgI_2, $Hg(SCN)_2$,[534] and $Hg(CN)_2$. Substituted anilines, o- and p-toluidine, form 1:1 and 2:1 complexes with $HgCl_2$; the former are chloride bridged, the latter monomeric tetrahedral moieties.[532,535,536] Only 2:1 complexes are formed with m-toluidine,[532,535,536] 2-chloro-6-methylaniline,[537] 5-chloro-2-methylaniline,[542] m- and p-anisidine,[532] and 2,3-, 2,4-, 2,5- and 2,6-dimethylaniline.[538] Mercury(II) thiocyanate does not form complexes with 2-, 3-, 4-methyl- or 3-chloro-aniline.[539] Benzylamine, α- and β-phenylethylamine, and β-naphthylamine also form 2:1 complexes.[532,540,541]

Little is known about the coordinating ability of secondary and tertiary monoamines towards Hg(II), the few old reports suggest that 1:1 complexes are formed, but clearly re-investigation would be desirable.

Ethylenediamine (en) complexes have been investigated much more fully. Tris complexes are formed by mercuric sulphate and nitrate, which contain hexacoordinate $HgN_6{}^{2+}$ units in the solid state and in concentrated D_2O solutions,[542] although in dilute solutions dissociation to $Hg(en)_2{}^{2+}$ occurs. Bis complexes are known with nitrate, perchlorate,[543] iodide,[544] and sulphate.[545] The monoethylenediamine complexes are of two distinct types; with $Hg(CN)_2$ and $Hg(SCN)_2$ the infrared spectra suggest mononuclear complexes with chelating en.[546] The corresponding $Hg(en)X_2$ (X = Cl, Br) contain bridging en ligands on the basis of i.r. and Raman spectra.[545,547-550] The structures of the 1:1 complexes with $HgSO_4$,[550] HgI_2,[530] $Hg(NO_3)_2$ and HgC_2O_4[550] have not been established. 1,3-diaminopropane, 1,4-diaminobutane, and 1,6-diaminohexane form 2:1 complexes with mercuric perchlorate.[543] Complexes with $p\,p'$-diaminodiphenylmethane[551] and p-phenylenediamine[552] have been briefly mentioned.[552]

N,N-diethylethylenediamine, $Et_2NCH_2CH_2NH_2$,[553] o-phenylene-bis(dimethylamine),[554,555] and N,N,N',N'-tetramethylethylenediamine[529,556-560] form 1:1 complexes with mercury(II) halides and pseudohalides that are probably tetrahedral monomers. Diethylenetriamine, $H_2NCH_2CH_2NHCN_2CH_2NH_2$, forms fivecoordinate $HgLX_2$ (X = Cl, Br, I, CN) and six-coordinate octahedral $[HgL_2]Y_2$ (Y = ClO_4, SCN).[559] However, N,N-dimethylpropylenediamine, $Me_2NCH_2CH_2$-CH_2NH_2, and dipropylenetriamine, $H_2NCH_2CH_2CH_2NHCH_2CH_2CH_2NH_2$, form complexes of stoichiometry $Hg_3L_2Cl_6$, irrespective of the ratio of the reactants.[529] Triethylenetetramine forms an $HgLCl_2$ complex,[529] while the hexamethyl analogue, N,N,N',N',N''',N'''-hexamethyl-3,6-diazaoctane-1,8-diamine, yields octahedral $HgLX_2$ (X = Cl, Br) and tetrahedral Hg_2LI_4.[560] Tris(2-aminoethyl)-amine and its hexamethyl analogue produce five-coordinate $[HgLX]^+$ (X = Cl, Br, I, SCN), isolated as the $Hg_2X_6{}^{2-}$ (X = Br, I), $HgX_4{}^{2-}$ (X = Cl, Br) or $BPh_4{}^-$ salts.[561]

1,2-dimethylhydrazine forms a tetrahedral $HgCl_2$ complex,[562] while tetramethyl-2-tetrazine yields HgLX$_2$ (X = Cl, Br), probably with the structure shown in figure 15.35.[563]

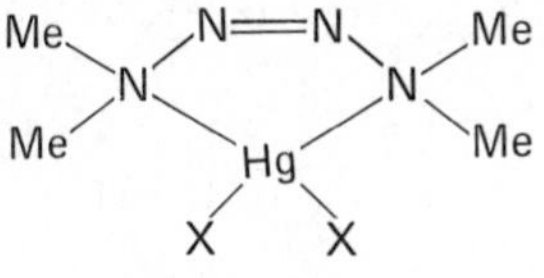

Figure 15.35

15.7.3 Heterocycles

Pyridine forms mercuric chloride complexes in ratios HgCl$_2$. py of 1:1, 1:2, 2:3 and, probably, 3:2; mercuric bromide behaves similarly, while for HgI$_2$ 1:1 and 2:1 complexes have been reported.[476,564-567] The i.r. spectra[564,568] of the 1:1 and 2:1 compounds have been interpreted in terms of polymeric halide-bridged structures. The Hg(py)$_2$Cl$_2$ complex has been X-rayed[569] and shown to be six-coordinate HgCl$_4$N$_2$ with Hg–Cl = 0.234, 0.325 nm, Hg–N = 0.260 nm. Mercury(II) perchlorate forms complexes Hg(py)$_n$(ClO$_4$)$_2$ (n = 2, 4, 6), which exhibit i.r. spectra characteristic of ionic perchlorate groups.[569-571] Thermal decomposition of Hg(py)$_6$(ClO$_4$)$_2$ produces the tetrakis and bis complexes successively. Mercury(II) nitrate appears to give only a bis adduct,[572] which is probably tetrahedral, and loses pyridine on gentle heating.[573] Mercuric cyanide[574] and thiocyanate[575] give 1:2 pyridine adducts only. Substituted pyridines, 2-picoline, 3-picoline, 4-picoline, 2,4-dimethylpyridine, produce 1:1 and 2:1 complexes with HgCl$_2$ and HgBr$_2$,[564] but the more sterically hindered 2,6-dimethylpyridine (2,6-lutidine) forms only 1:1 complexes.[564,576] Mercuric cyanide forms 1:2 complexes with 2-, 3-, and 4-picoline and 1:1 complexes with 2,4- and 2,6-dimethylpyridine; all appear to be CN-bridged polymers on the basis of i.r. spectra.[574,577] The i.r. spectra of the 1:1 Hg(SCN)$_2$. L complexes of 2-methyl-, 2,6-dimethyl-, and 3-ethylpyridine suggest a structure of the type shown in figure 15.36, while 2-ethylpyridine forms tetrahedral 1:2 complexes with terminal Hg–SCN linkages.[575,577] The only cyanate complex, Hg(2-pic)(NCO)$_2$,

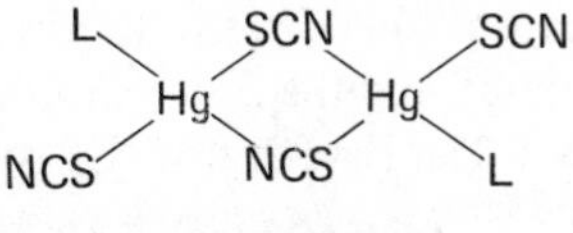

Figure 15.36

is of unknown structure.[577] The complexes of 2-, 3-, and 4-cyanopyridine all contain the ligand bonded only through the ring nitrogen. The stoichiometries show unusual variations: HgL$_n$Cl$_2$ (n = 1,2 for 3-CNpy, 4-CNpy; n = 2 for 2-CNpy), HgL$_n$Br$_2$ (n = 1 for 2-CNpy, 3-CNpy; n = 2 for 4-CNpy). With HgL$_n$(CN)$_2$ (n = 2 for 2-CNpy, 3-CNpy; n = 1 for 4-CNpy) are formed tetrahedral monomers (the former), but probably CN-bridged polymers for the latter complexes. Only 3-CNpy complexed with Hg(SCN)$_2$ to yield a 1:1 complex, probably an SCN-bridged dimer.[558]

The $HgCl_2$–collidine complex (collidine = 2,4,6-trimethylpyridine) contains $HgNCl_2$ units, with further weak interaction with 2-chlorines, providing an essentially trigonal-bipyramidal coordination about the mercury (figure 15.37).[578]

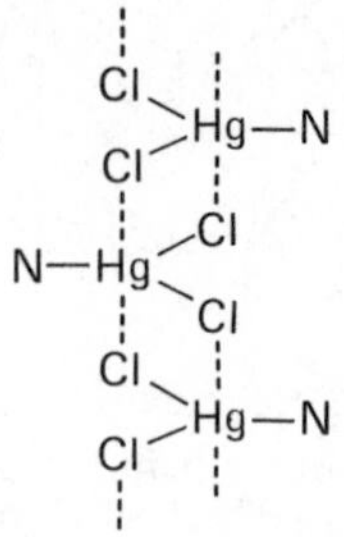

Figure 15.37

Picolinic, nicotinic and isonicotinic acids behave as neutral N-donors towards $HgCl_2$,[579] as do nicotinamide and isonicotinamide, while picolinamide behaves as a bidentate ligand (figure 15.38).[580] 2-picolylamine ($C_5H_4NCH_2NH_2$) forms $HgLX_2$ (X = Cl, Br, I) and $[HgL_2]Y_2$ (Y = NO_3, ClO_4), which are presumably four-coordinate HgN_2X_2 and HgN_4 moieties, respectively.[581]

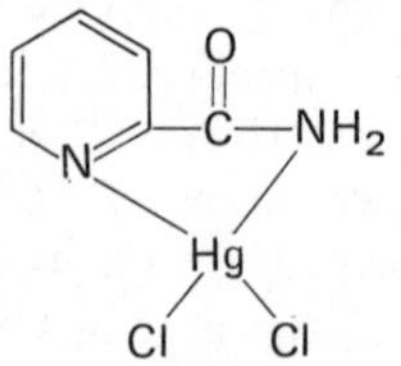

Figure 15.38

Quinoline (Q) forms 1:1 and 2:1 complexes with all three mercury halides.[566,572,582] On the basis of their i.r. spectra the 1:1 complexes are thought to have halide-bridged structures, and the 2:1 complexes to have an analogous structure to $Hg(py)_2X_2$ (q.v.). The $Hg_2(Q)_3(SCN)_4$ complex is probably an SCN-bridged octahedral polymer.[582] Other complexes are $Hg(Q)_2(NO_3)_2$[572] and $(HgSO_4)_2 \cdot Q$.[583] The isoquinoline and 4-methylquinoline complexes are probably of similar constitution.[572,582] Mercuric cyanide was reported to form polymeric octahedral $HgL(CN)_2$ with both quinoline and isoquinoline,[582] but a more recent report[584] has proposed that these complexes (and the 6-methylquinoline, 3-methylisoquinoline, and β-naphthylquinoline analogues) are three-coordinate monomers.

Little recent work has been devoted to mercury(II) complexes of pyridazines, pyrimidines, or pyrazines. Pyridazine forms a 1:1 complex with $HgCl_2$,[581] while the i.r. spectra of the 1:1 $HgLX_2$ (L = pyrazine, pyrimidine, and methylpyrazine;

X = Cl, Br, I) suggest that the ligands bond through both ring nitrogens and hence bridge HgX$_2$ units, producing chain polymers.[564,585,586]

Aziridine (figure 15.39a) forms HgLX$_2$ (X = Cl, Br) and HgL$_2$(NO$_3$)$_2$ of unknown structure,[587] while piperidine (figure 15.39b) forms HgL$_2$X$_2$ (X = Cl, Br, I).[588] It would seem that a considerable number of the complexes reported many years ago with piperidine and related ligands are in need of re-investigation.[588] Piperazine, L, and 1,4-dimethylpiperazine, L', form HgLCl$_2$ and Hg$_2$L'Cl$_4$.[589] Grandberg *et al.* have described a number of substituted pyrazole complexes.[590]

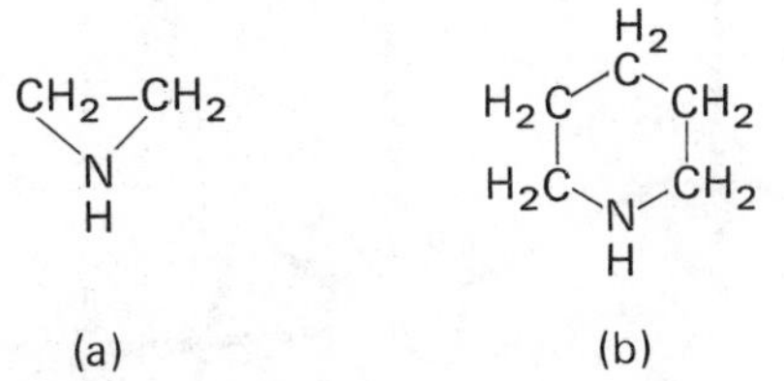

(a) (b)

Figure 15.39

2,2'-bipyridyl and 1,10-phenanthroline form HgLX$_2$ (X = Cl, Br, I), which are non-electrolytes in nitrobenzene,[591] and were originally suggested to be tetrahedral monomers.[592] However, an X-ray study of Hg(bipy)Br$_2$ established[593] the presence of five-coordinate mercury with a distorted square-pyramidal structure (N$_2$Br$_2$ in the basal plane and apical Br). Penta-coordination is achieved by dimerisation via bromide bridges, Hg–N = 0.237, 0.240 nm. With weakly co-ordinating anions HgL$_2$Y$_2$ (L = bipy, phen; Y = NO$_3$, ClO$_4$) and Hg(phen)$_3$(ClO$_4$)$_2$ are formed.[591,594,595]

Mercuric thiocyanate forms 1 : 1 and 1 : 2 complexes with 1,10-phenanthroline and a 1 : 1 complex with 2,2'-bipyridyl.[592] The i.r. spectra suggest that only terminal S-bonded thiocyanate groups are present and hence that all the complexes are monomeric. This has been confirmed in the case of Hg(phen)$_2$(SCN)$_2$ by an X-ray study, which revealed a *cis*-octahedral structure (figure 15.40), Hg–N = 0.242–0.252 nm, Hg–S = 0.2692, 0.2582 nm.[596] Mercuric cyanide forms only 1 : 1 complexes with these two ligands; only one type of CN group is revealed by the i.r. spectra, suggesting monomeric tetrahedral compounds.[592]

The mercuric thiocyanate complex of 4,4'-bipyridyl, HgL(SCN)$_2$, contains polymeric four-coordinate mercury with bridging bipyridyl ligands.[597] The tridentate ligand 2,2',6',2''-terpyridine forms HgLX$_2$ (X = Cl, Br, I) complexes, which are assigned penta-coordinate structures in view of the isomorphism of the chloro complex with the zinc analogue, the latter having been X-rayed.[598] The Hg(terpy)$_3$(ClO$_4$) is a HgN$_6^{2+}$ complex, but it is not definitely established whether Hg(terpy)(NO$_3$)$_2$, the i.r. spectrum of which reveals coordinated nitrate, is five- or six-coordinate.[598] Nicotine (C$_{10}$H$_{14}$N$_2$) forms HgL$_2$X$_2$ (X = Cl, Br, I) species.[599]

Tris(1,8-naphthyridine)perchloratomercury(II) perchlorate contains irregular seven-coordinate mercury, consisting of six nitrogens from three asymmetrically

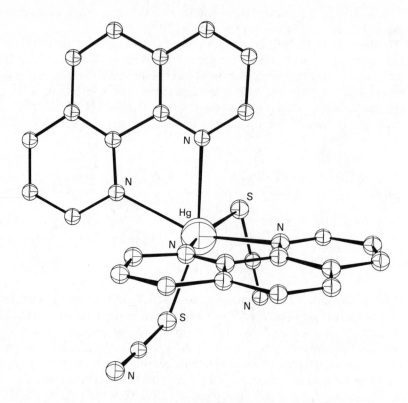

Figure 15.40. Hg(phen)$_2$(SCN)$_2$

coordinated napy ligands and one oxygen from a perchlorate group (Hg–N = 0.264, 0.230, 0.284, 0.214, 0.287, 0.220 nm; Hg–O = 0.293 nm) (figure 15.41).[600] Cadmium, by comparison, yields the eight-coordinate [Cd(napy)$_4$](ClO$_4$)$_2$.

Nelson and co-workers[601] have examined a series of potentially ter- and tetra-dentate ligands (figure 15.42) and isolated HgLX$_2$ (L = dpma, X = Cl, Br, I; L = mdmpa, X = Cl, Br, I; L = Me$_3$dmpa, X = I; L = dbpma, X = Br; L = tmpa, X = Br, I; L = Me$_2$tpma, X = Cl, Br, I; L = Me$_3$tpma, X = Cl, Br, I). The potentially ter-dentate ligands coordinate all three nitrogen atoms to form penta-coordinate complexes,[601] but four nitrogens in the second type of ligand are only coordinated in the Hg(Me$_2$tpma)Cl$_2$ complex; for tpma and Me$_3$tpma, and in Hg(Me$_2$tpma)X$_2$ (X = Br, I) only three nitrogens coordinate producing penta-coordination. Comparison with the zinc and cadmium analogues suggests that for this type of ligand the tendency to favour six- over five-coordination is Cd > Zn ≈ Hg.[601]

The older literature contains numerous reports of mercury(II) complexes of other heterocyclic nitrogen donors. Much of this work needs to be repeated and the complexes subjected to spectroscopic investigation before useful conclusions can be drawn.

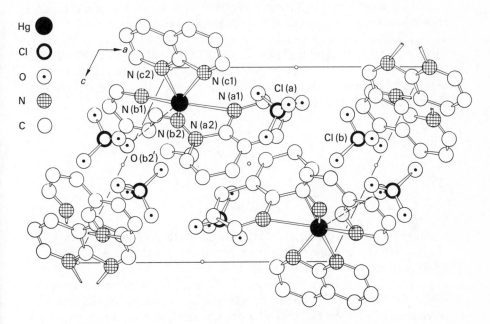

Figure 15.41. Unit cell of Hg(1,8-naphthyridine)$_3$ClO$_4$ClO$_4$

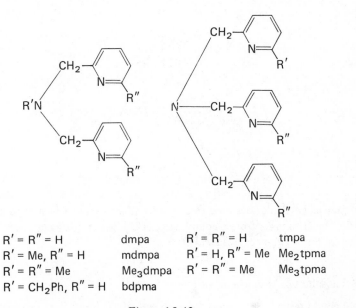

R' = R" = H   dmpa   R' = R" = H   tmpa
R' = Me, R" = H   mdmpa   R' = H, R" = Me   Me$_2$tpma
R' = R" = Me   Me$_3$dmpa   R' = R" = Me   Me$_3$tpma
R' = CH$_2$Ph, R" = H   bdpma

Figure 15.42

Azo and diazo compounds complex with the mercuric ion. Diazoaminobenzene (1,3-diphenyltriazine) behaves as a neutral ligand towards $HgCl_2$ and $HgBr_2$ producing bis complexes (figure 15.43), but with $Hg(NO_3)_2$ or $Hg(CH_3COO)_2$ the amine proton is lost to give $Hg(PhN=N-NPh)_2$.[603,604] Other ligands of this type that are reported to yield mercury(II) complexes are *m*- and *p*-nitrodiazoamino-benzene,[605] benzene diazonium bromide[606] and iodide,[607] and *p*-dimethylamino-benzene diazonium chloride.[608]

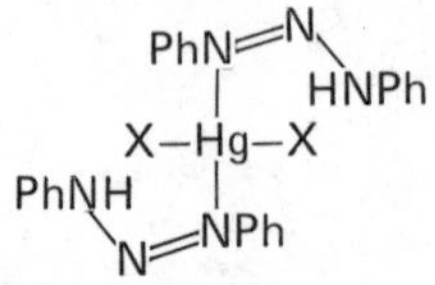

Figure 15.43

Acetonitrile forms a $(HgBr_2)_3 \cdot CH_3CN$ complex, the structure of which consists of $HgBr_6$ and $HgBr_6N$ (seven-coordinate) environments.[609] Nitrogen-donor Schiff bases have been examined by Bahr and Kretzer,[610] who prepared 1:1 mercuric halide complexes with diacetyldianil, diacetyl-di-*p*-tolil, and diacetyl-di-*o*-tolil; and Bähr *et al.*[611,612] prepared similar complexes with pyridine-2-aldehyde-alkylimines (alkyl = Me, Et, Pr^n, Bu^n, n-amyl, n-hexyl, n-heptyl) and with pyridine-2-aldehydeanil. Schiff-base complexes derived from benzalaniline have also been described.[613]

Mercury(II) forms a 1:1 complex in solution with tetraphenylporphin.[614] However, mercury(II) complexes of a series of porphyrins of types shown in figure 15.44

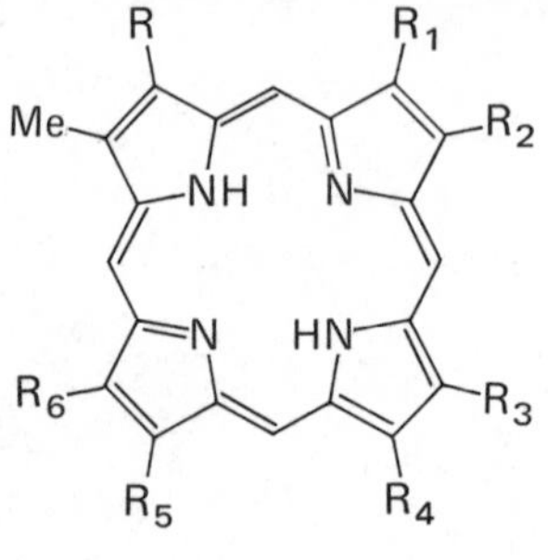

Figure 15.44

have been isolated and suggested to contain a 'double sandwich' structure (figure 15.45). The ^1H n.m.r. of these complexes can be used to distinguish between the aetio- and capro-porphyrin-type isomers.[615]

15.8 Phosphorus, Arsenic and Antimony Ligands

Monodentate tertiary phosphines form mercury(II) complexes in the ratios

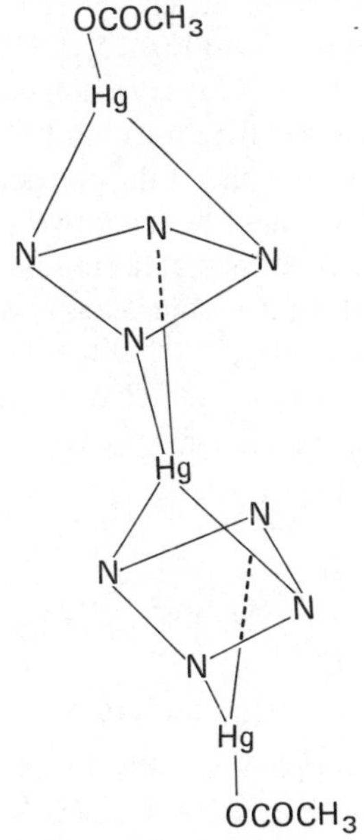

Figure 15.45

1:1, 2:1, and more rarely 3:2, 2:3, 3:1 and 4:1. Much less work has been reported with tertiary arsines and few stibine complexes are known.

Bis(trifluoromethyl)phosphine, $(CF_3)_2PH$, reacts with divinylmercury to produce $Hg[P(CF_3)_2]_2$. The product is spontaneously inflammable in air, decomposes on heating into $P_2(CF_3)_4$, and reacts with $(CF_3)_2AsH$ to produce $Hg[As(CF_3)_2]_2$.[616] The formation of $Hg(PBu^t_2)_2$ is readily achieved from Bu^t_2PH and $HgBu^t_2$, but other secondary phosphines produce mercury and P_2R_4.[617] The $Hg(PBu^t_2)_2$ is remarkably stable towards oxidation and thermal decomposition.

Phosphine, PH_3, produces[618] a compound $P(HgCl)_3$ of unknown structure. Goggin *et al.*[619] found that mercury(II) halides react with one equivalent of $AgNO_3 . PMe_3$ to form $[Hg(PMe_3)X]NO_3$ (X = Cl, Br, I, CN), which are converted to the corresponding $[Hg(PMe_3)X]BF_4$ by fluoroboric acid, and reacts with more $AgNO_3 . PMe_3$ to form $[Hg(PMe_3)_2]^{2+}$. Trimethylarsine similarly yields $[Hg(AsMe_3)X]^+$ and $[Hg(AsMe_3)_2]^{2+}$.[618] A detailed i.r., Raman, and ^1H n.m.r. study of those complexes was reported. Under similar conditions $SbMe_3$ produced only metallic mercury. The HgX_2/PMe_3 system has been further studied by Schmidbauer[620] who found that under appropriate conditions $HgX_2 . nPMe_3$ (n = 1,2,3, X = Cl, Br, I; n = 1,2, X = CN; n = 1,2,4; X = SCN; n = 4, X = NO_3) could be isolated. Raman, i.r., and ^{31}P n.m.r. spectra showed the 1:1 halo-complexes to be halide-bridged dimers and the $HgX_2(PMe_3)$ (X = CN or SCN) to be oligomers.[620] The 1:2, 1:3, and 1:4 complexes contain $Hg(PMe_3)_2^{2+}$,

$Hg(PMe_3)_3{}^{2+}$, and $Hg(PMe_3)_4{}^{2+}$, respectively.[620] The triethylphosphine complexes reported to date are of the types $[Hg(PEt_3)_2X_2]$,[621,622] $[Hg(PEt_3)X_2]_2$ (X = Cl, Br, I)[623] and $Hg_3(PEt_3)_2I_6$.[623] An X-ray crystallographic study of $Hg_2Br_4(PEt_3)_2$, although of low accuracy, showed the presence of dimeric halogen-bridged units (figure 15.46).[623] This is the only one of the possible isomeric forms definitely established, although recently it has been tentatively suggested on the basis of n.m.r. evidence that other isomers may occur in solution.[624] The 2 : 1 compounds are thought to be tetrahedral. Similar complexes have been obtained with PPr_3,[623] $PBu^n{}_3$,[623-626] $P(n$-amyl$)_3$,[623] $PPhEt_2$,[622,624] $PPhBu_2$,[624,625] PPh_2Bu,[624,625] $PPhMe_2$,[622] $P(octyl)_3$.[626] The majority of these complexes have been examined by ^{31}P n.m.r. spectrometry and the $^{199}Hg-^{31}P$ n.m.r. coupling constants evaluated.[624-626]

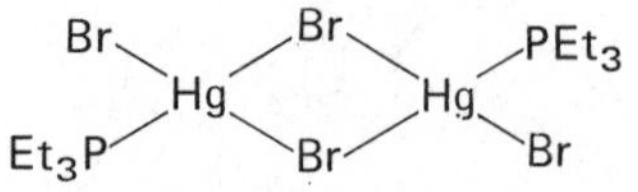

Figure 15.46

The bulky tricyclohexylphosphine affords similar $[Hg(PCy_3)X_2]_2$ and $[Hg(PCy_3)_2X_2]$ with X = halide, but $Hg(PCy_3)(SCN)_2$ appears to be a three-coordinate monomer.[627] A series of substituted allylphosphine[628] and dialkyl(4-dimethylaminophenyl)phosphine[344,622] complexes have been examined. The ligand siphos, $(Me_3SiCH_2)_3P$, forms $[HgCl_2 \cdot (siphos)]_2$ and $Hg_3I_6(siphos)_3$, probably[630] of the structure shown in figure 15.47.

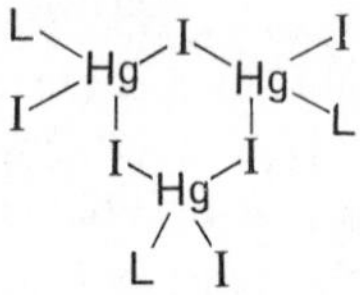

Figure 15.47

Alkylarsines complex readily with mercuric halides, but since they are not so suitable for n.m.r. investigations, they have been studied much less than the phosphine analogues. The complexes are generally of 1 : 1 and 2 : 1 types and are expected to have analogous structures to the corresponding phosphine compounds. The ligands that have been studied are $AsMe_3$,[619,637] $AsEt_3$,[623,629,632] $AsPr_3$,[623] $AsBu_3$,[623,633] $AsPh_2Pr$,[634] $AsPh_2Me$,[532] $AsMe_2$ (p-$C_6H_4NMe_2$),[622] $AsPhMe_2$, and a number of mixed alkylarsines $R_2R'As$.[634,635]

Alkylstibines do not seem to have been complexed with mercuric salts. Diphenyl-methylstibine yields only the halide-bridged $[HgLX_2]_2$ type of complex and not the HgL_2X_2 type (X = Cl, Br, I).[636] McAuliffe and co-workers[637] have recently isolated $HgLX_2$ and HgL_2X_2 complexes with a series of phenyldialkylstibines.

Triphenylphosphine complexes $[Hg(PPh_3)X_2]_2$ and $[Hg(PPh_3)_2X_2]$ have been extensively studied by vibrational spectroscopy,[344,623,629,638,639] which confirmed that the structures were analogous to those of the alkylphosphines. Similar tri-phenylarsine complexes are known,[344,623,640] but triphenylstibine forms only 1 : 1 compounds with mercury(II) halides.[641] Mercury thiocyanate and EPh_3 (E = P, As) yield $Hg(EPh_3)_2(SCN)_2$[642] assigned tetrahedral monomeric structures, which was subsequently confirmed by an X-ray[643] of $Hg(PPh_3)_2(SCN)_2$ Hg–S = 0.2565, 0.2577 nm, Hg–P = 0.2487, 0.2484 nm. Under different conditions $Hg(EPh_3)$-$(SCN)_2$ were prepared and their i.r. spectra interpreted in terms of SCN-bridged dimers,[642] but an X-ray[643] of $Hg(AsPh_3)(SCN)_2$ revealed an essentially three-coordinate structure Hg–As = 0.260 nm, Hg–S = 0.253, 0.255 nm, with two long contacts Hg . . . NCS at 0.280 and 0.270 nm. Recently $Hg(SbPh_3)(SCN)_2$ has been prepared.[644] Mercuric cyanide forms only a 1 : 2 complex with PPh_3. Davis *et al.*[642] prepared $Hg(EPh_3)Y_2$ and $Hg(EPh_3)_2Y_2$ (Y = NO_3, ClO_4), which appeared to contain coordinated anions. In $Hg(PPh_3)(NO_3)_2$ the mercury is approximately tetrahedrally coordinated to one PPh_3, one unidentate NO_3 (Hg–O = 0.219 nm) and two bridging bidentate NO_3 groups (Hg–O = 0.243, 0.256 nm).[645] Tri-(*p*-tolyl)-phosphine[267] and tri(*o*-tolyl)stibine[641] form 1 : 1 complexes with mercury(II) halides. The latter are formed in cold benzene solutions, but under reflux the stibine is chlorinated to R_3SbCl_2 by $HgCl_2$, which is reduced to Hg_2Cl_2. In refluxing THF the products are RHgCl and R_2SbCl.[641] Triarylbismuthines R_3Bi (R = Ph, *o*-tolyl, *p*-tolyl) are chlorinated to R_2BiCl by $HgCl_2$.[641]

Methyl iodide converts $Hg(PPh_3)_2X_2$ into $(PPh_3Me)_2HgX_2I_2$,[629] further exchange leading ultimately to HgI_3^- or HgI_4^{2-}. Tertiary arsine complexes behave similarly.[632]

The olefin phosphine and arsine ligands $Ph_2PCH_2CH_2CH_2CH=CH_2$, $PhE(CH_2CH_2CH_2CH=CH_2)_2$ (E = P, As) and $Me_2AsCH_2CH_2CH_2CH=CH_2$ behave only as monodentate group VB donors.[646]

Mixed cadmium–mercury complexes of PPr_3–$(PPr_3)_2CdHgI_4$– were obtained[647] from $(PPr_3)_3CdI_2$ and HgI_2, or from $(PPr_3)_3(CdI_2)_2$ and $(PPr_3)_2HgI_2$. It was assumed that the structure was $[I(PPr_3)CdI_2Hg(PPr_3)I]$. A $PdHg(AsPr_3)_2Br_4$ was obtained similarly.[647] Much more recently Baker *et al.*[648] prepared $Pt(PPhMe_2)_2X_2(HgX_2)$ (X = Cl, Br), and showed that the structure (X = Cl) was as in figure 15.48a Hg–Cl = 0.231 nm (terminal) and

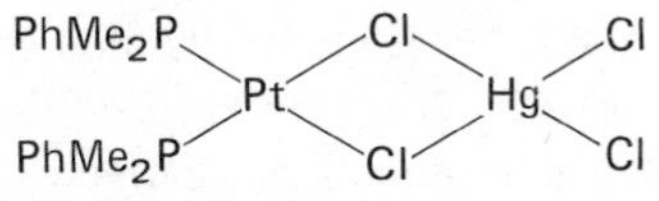

Figure 15.48a

0.284 nm (bridging). The long Hg–Cl in the bridge suggests that the mercury is only weakly coordinated, and this is supported by the evidence of considerable dissociation in solution. Brookes and Shaw[649] isolated similar complexes with both palladium and platinum and a range of phosphines, including a $Pt(PMe_3)_2Cl_2(HgCl_2)_2$ formulated as in figure 15.48b. Subsequently these studies were extended to Rh(III) and Ir(III) phosphine and arsine complexes.[650]

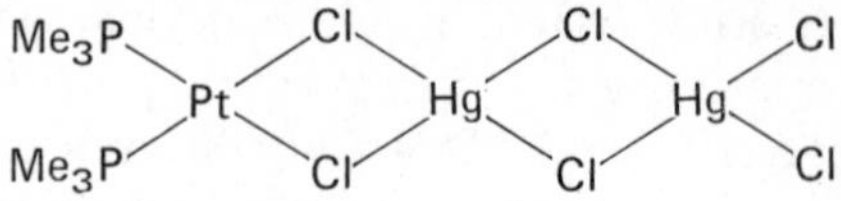

Figure 15.48b

It would be interesting to establish the structure of the Cd–Hg complex of Mann and Purdie[647] since, unlike the cases of the group VIII–Hg compounds, the ER_3 ligands may not reside only on the cadmium. Cobalt(II) and nickel(II) analogues, $MHg(PEt_3)_2Y_4$, appear to contain the PEt_3 only bonded to mercury.[651]

Diphosphines usually produce only 1 : 1 compounds with mercuric halides— $HgLX_2$ L = $Ph_2PCH_2PPh_2$,[652] $Et_2PCH_2CH_2PEt_2$,[653] $Ph_2PC\equiv CPPh_2$[654,655] $Ph_2PCH_2CH_2PPh_2$,[344,656,657] $Ph_2P(CH_2)_4PPh_2$,[656] but an $[HgX_2]_2L$ stoichiometry is also obtainable with $Ph_2P(CH_2)_nPPh_2$ (n = 2,4).[656] The acetylenic diphosphine, $Ph_2PC\equiv CPPh_2$,[655] which incidently does not form stable zinc or cadmium complexes, produces ligand-bridged polymers with mercury(II). Contrary to earlier studies, Strommen[657] assigned both bridging and terminal $\nu(Hg-X)$ in the i.r. and Raman spectra of $(HgX_2)_2(Ph_2PCH_2CH_2PPh_2)$, and only bridging $\nu(Hg-X)$ in the 1 : 1 complexes (X = Cl, Br). The iodocomplex $HgI_2(Ph_2PCH_2CH_2PPh_2)$, however, seems to be a discrete monomer.[657] The structures of $Hg_2L(SCN)_4$ and $HgL(CN)_2$ (L = $Ph_2PCH_2CH_2PPh_2$) are less certain.[653] The ligand 1,1'-ferrocenebis(diphenylphosphine) (L) forms $L(HgX_2)_n$ (X = Cl, Br, I, n = 1,2) which contain Hg–P and not Hg–arene coordination[658] (figure 15.49). $[HgL_2]Y_2$ (Y = BF_4, PF_6) containing $HgP_4{}^{2+}$ moieties are

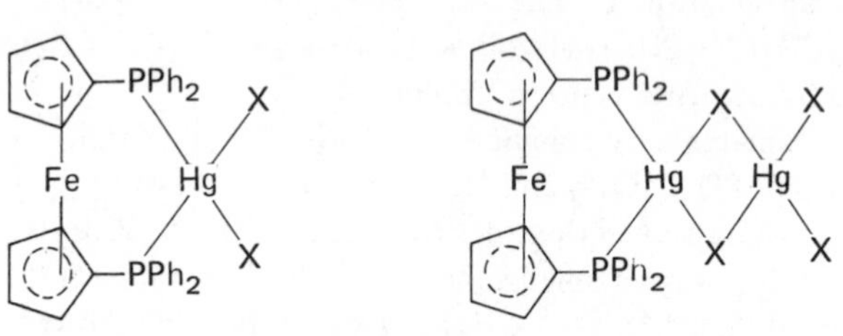

Figure 15.49

formed from $Hg(CN)_2$, HY and L.[658] The tetraalkyldiphosphines R_2PPR_2 (R = Me, Et) behave as bidentate P_2 donors in $Hg(R_2PPR_2)Br_2$ and $Hg(Me_2-PPMe_2)_2Br_2$.[659,660] o-phenylenebis(dimethylarsine) (das) forms $Hg(das)X_2$, which are tetrahedral monomers in the solid state, but exhibit some evidence of isomerism $2Hg(das)X_2 \rightleftharpoons [Hg(das)_2][HgX_4]$ in solution.[550,661,662] The $[Hg(das)_2]^{2+}$ ion is present in the diperchlorate complex,[550,661] while $Hg(das)_2(SCN)_2$ may be six-coordinate.[644] The ligand 1,2-bis(diphenylarsino)ethane forms $Hg_2L(SCN)_4$, analogous to the corresponding phosphine complex.[644] The fluoroalicyclic ligands $Ph_2PC=C(PPh_2)CF_2CF_2$ and $Me_2AsC=C(AsMe_2)CF_2CF_2$ form 1 : 1 complexes with $HgCl_2$, which are coordinated through the group VB element.[663] The only distibine complex of mercury is $Ph_2SbCH_2SbPh_2 . HgI_2$, an insoluble yellow solid probably with a polymeric structure.[664] The tripod tetradentates tris(*o*-diphenylphosphinophenyl)phosphine (QP) and the arsenic analogue (QAS) exhibit unusual bidentate behaviour in [HgLX$_2$], established by an X-ray of Hg(QAS)-Br$_2$[665] (figure 15.50). The [Hg(QP)Cl]$^+$ ion appears to be five-coordinate, and the [HgL](ClO$_4$)$_2$ (L = QP, QAS) may be trigonal pyramids since i.r. spectral examination showed that only ionic ClO$_4$$^-$ ions were present.[665]

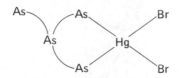

Figure 15.50

Amine-phosphines form 1 : 1 and 2 : 1 mercury(II) iodide complexes, which are presumed to be structurally the same as the alkylphosphine analogues. Complexes have been isolated with P(NMe$_2$)$_3$,[666] P(NR$_2$)(NR$'_2$)Ph[667] and P(NR$_2$)$_2$-Ph[666,667] (R = Me, Et, Pr). The green Hg[(R$_2$N)$_2$PPhMe]I$_3$[667] and yellow Hg[Me$_2$NPMe$_3$]I$_3$[668] were formulated as phosphonium triiodomercurates(II), but it seems possible that Hg–P or Hg–N coordination could be present. Seidal[669–672] investigated the reaction of mercury(II) iodide with (pip)$_n$-(C$_6$H$_{11}$)$_{4-n}$P$_2$ and (pip)$_2$Ph$_2$P$_2$ (pip = piperidine). While (pip)$_2$Ph$_2$P$_2$ and (pip)$_2$(C$_6$H$_{11}$)$_2$P$_2$ gave unstable complexes, (pip)(C$_6$H$_{11}$)$_3$P$_2$ underwent P–P bond fission to (C$_6$H$_{11}$)$_2$P–HgI, which could be alkylated with MeI to (C$_6$H$_{11}$)$_2$MePHgI$_2$.

Phosphites P(OR)$_3$ (R = Me, Et, Pr^i) form 1 : 1 and, rarely, 1 : 2 [P(OR)$_3$: Hg] complexes.[673,674] Phosphonates form [HgLX]$_2$ (X = Cl, Br, I, SCN, CN, OAc, L = OP(OR)$_2$$^-$ and [HgL$_2$].[675–678] The structure[679] of [HgCl(EtO)$_2$PO] revealed two different coordination geometries about the mercury, a very distorted octahedron, and a distorted trigonal bipyramid. The mercury forms two strong bonds to a chlorine (0.236 nm) and the phosphorus (0.240 nm) resulting in digonal coordination, with weak oxygen contacts at 0.253–0.291 nm. The bis(phosphonate) Hg[OP(OMe)$_2$]$_2$ also contains[680] approximately linear P–Hg–P (166°) Hg–P = 0.241 nm with weak Hg–O contacts at approximately 0.254 nm.

15.9 Group IIIB and IVB Compounds

Decaborane(14) B$_{10}$H$_{14}$ reacts with HgCl$_2$ in diethylether to form B$_{10}$H$_{14-n}$-(HgCl)$_n$ (n = 7.8), but in THF or with HgBr$_2$, reduction to Hg(I) occurs.[681] Alkylmercury halides and B$_{10}$H$_{13}$MgX react in ethereal solution to produce (MgX$_2$)$_x$. {Mg[Hg(B$_{10}$H$_{12}$)$_2$]} (solvent)$_y$, which are hydrolysed in water to [Hg(B$_{10}$H$_{12}$)$_2$]$^{2-}$.[681,682] The latter, formally a mercury(II) complex of the dodecahydro-nidodecaborate(2–) ion has been isolated as Cs$^+$, Me$_4$N$^+$, and

Ph_4P^+ salts. The $[Me_4N]_2[Hg(B_{10}H_{12})_2]$ salt has also been prepared from $HgCl_2$ and $[Me_4N][B_{10}H_{13}]$ in THF.[683] The ^{11}B n.m.r. spectrum of the anion suggests that it is isostructural with the zinc analogue (figure 15.51).

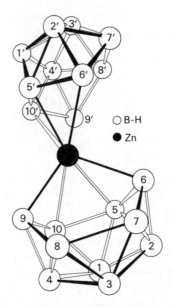

Figure 15.51

Organomercury compounds are discussed in part 3 and it is merely necessary to note here that organomercury compounds are very numerous, of considerable importance, and often of high stability. The properties of the Hg–C bond in relation to other Hg–element bonds has been discussed.[2] While organomercurials have been known for over one hundred years, other Hg–group-IVB element bonds have a much more recent history. In spite of this, Hg–Si and Hg–Ge compounds have been the subject of a considerable amount of work. Suitable synthetic routes include the reactions of the appropriate R_3EH with R'_2Hg, and of R_3EX with sodium amalgam (R,R' alkyl or aryl, X = Cl, Br, E = Si, Ge).[2,684–686] The products are highly reactive, the mercury–E bond being cleaved by oxygen [to $(R_3E)_2O$] by heat (to R_3EER_3), by water (to R_3EOH) and by most metals [to $(R_3E)_nM$]. Transition-metal–mercury bonds and transition-metal–E bonds are also formed under appropriate conditions, for example

$$Pt[Ph_2P(CH_2)_2PPh_2]Cl_2 + (Me_3E)_2Hg \longrightarrow Pt[Ph_2P(CH_2)_2PPh_2](EMe_3)_2$$

$$+ Pt[Ph_2P(CH_2)_2PPh_2](EMe_3)Cl \quad \text{(reference 686)}$$

$$Ir(CO)(PEt_3)_2Cl + Hg(EMe_3)_2 \longrightarrow (Et_3P)_2Ir(HgEMe_3)(EMe_3)_2$$

$$\text{(reference 687)}$$

Reactions of $Hg(ER_3)_2$ with numerous organic compounds have been studied, and references to this work can be found in Seyferth's reviews.[688]

Mercury–tin bonds are considerably less stable, the $(R_3Sn)_2Hg$ (R = Me, Et, Pr^n) formed from $Bu_2{}^tHg$ and R_3SnH decompose above $-10\,^\circ C$ into mercury and R_6Sn_2,[689] although $[(Me_3SiCH_2)_3Sn]_2Hg$ is more stable.[690] Mercury–lead bonded compounds appear to be unknown, the claim for $(Ph_3Pb)_2Hg$ having been withdrawn.[691]

15.10 Mixed Donor Ligands

This section deals with ligands with two or more different donor atoms.

Ethanolamine forms $HgLX_2$ (X = Cl, Br, I), which are polymeric with bridging N,O coordinated ligands, and HgL_2X_2 (X = Cl, Br) in which the ligands are bonded through the amino group only.[692,693] Morpholine (figure 15.52a) forms 2:1

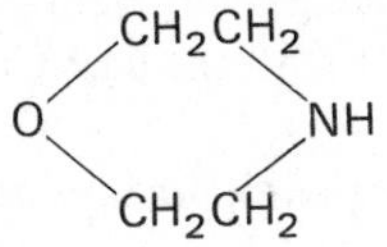

Figure 15.52a

complexes HgL_2X_2 (X = Cl, Br, I) in which the ligand is bonded only through the nitrogen.[694-696] The 1:1 dimorpholinoethane (figure 15.52b) compounds appear

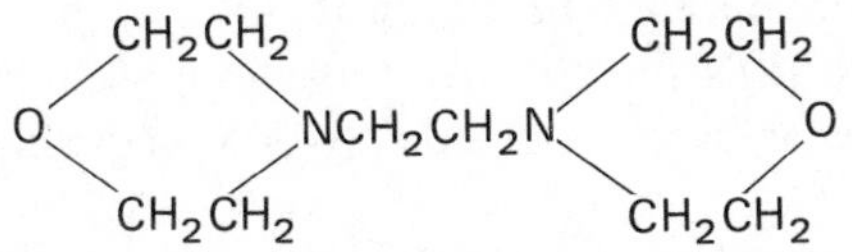

Figure 15.52b

to be similar.[695,697] The mercuric chloride complex of 2-imidazolidinone contains an octahedral mercury atom ($N_2O_2Cl_2$ donor set), the units being linked into infinite chains by imidazolidine ligands O-coordinated to one mercury, and N-coordinated to the next mercury, (Hg–Cl = 0.231 nm, Hg–O = 0.267 nm, Hg–N = 0.295 nm).[698] Substituted sulphonamides bond to mercury(II) through the sulphonyl and $-NH$ groups,[699] and the cyano-ethylated sulphonamides via the $-SO_2$ and $-CN$ groups. Several mercury compounds of 2-amino-1,3,4-oxadiazoles have been prepared but the coordination site is unclear.[700]

Thiosalicylic acid forms a series of compounds of the $[HgL_2]^{2-}$ ion (figure 15.53) with alkali, alkaline earth, or lanthanide cations.[701] Prochazkova *et al.*[702] prepared $HgL.xH_2O$, $Na_2HgL_2.xH_2O$, and H_2LHgX_2 complexes of three thiocarboxylic acids $H_2L–HO_2CCH_2SCH_2CO_2H^*$, $HO_2CCH_2SCH_2CH_2SCH_2$-CO_2H, $S(CH_2CH_2SCH_2CO_2H)_2$. The 1:1 complexes contain Hg–S- and Hg–O_2C-coordination, but the 2:1 complexes, except for those of (*) are bonded

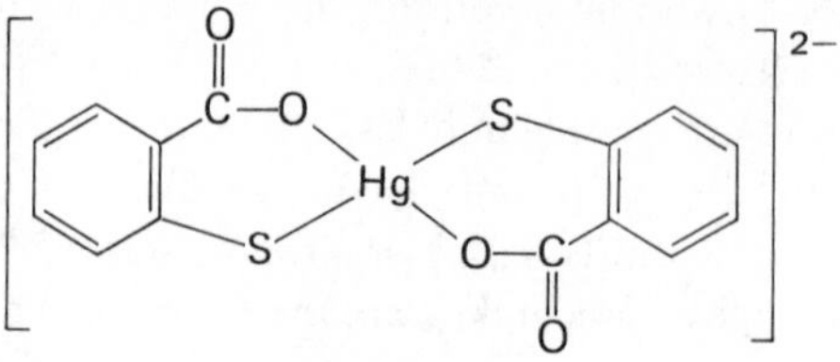

Figure 15.53

only through the sulphur. Mercury forms HgL_2 with the monothioketone $C_6H_5C(S)CH_2C(O)CF_3$ (LH).[703] Phenylisothiocyanate and mercury(II) methoxide yields $Hg[S(OMe)C=NPh]_2$, which contains two Hg–S bonds (Hg–S = 0.233 nm), with weak contacts to the methoxy oxygens at 0.286, 0.297 nm.[704] The complexes of o-methoxyphenyldimethylarsine[705] and o-di(R)arsinobenzoic acids[706] (R = Me, Ph) of type $HgLX_2$ probably contain O,As coordination. o-di(p-tolyl)arsino-benzoic acid forms similar complexes with $HgBr_2$ and HgI_2, but with $HgCl_2$, the product is a chloro-bridged dimer with the ligand bonded only through the arsenic.[706]

8-methylthioquinoline is N,S coordinated in the tetrahedral $HgLCl_2$.[707] The thiol analogue 8-mercaptoquinoline (thio-oxine) (LH) is bound as a neutral ligand in $Hg(LH)Cl_2$, $Hg(LH)_2Cl_2$ and $Hg(LH)_4X_2$ (X = Cl, Br, I), and as an anionic N,S donor in HgL_2 and $HgLCl$.[708-711] 2,3-dimercaptoquinoxaline is S-bonded in the HgL_2Cl_2 compound.[712]

Thiopicoline acid amide $C_5H_4NC(S)NH_2$ (LH) forms $Hg(LH)I_2$ and $[Hg(LH)_2]$-$(ClO_4)_2$, which are presumably four-coordinate with N,S-bonded ligands.[713] The N-substituted $C_5H_4NC(S)NHR$ (L'H) deprotonates to yield HgL'_2.[714,715] A number of thioamide complexes with either S- or N,S-coordinated ligands have been described.[716,717]

Sulphur–nitrogen coordination to Hg(II) is also present in the $HgLCl_2$ and $[HgL_2](ClO_4)_2$ complexes of 2-(2-methylthioethyl)pyridine,[718] and in the HgL'_2 derived from pyridine-2-thiol (L'H).[719] The latter ligand and the pyridine-4-thiol and 2-methylpyridine-6-thiol analogues bond as neutral S-donors to HgX_2.[719] Thiosemicarbazide $H_2NC(S)NHNH_2$,[720,721] and thiocarbohydrazide $H_2NNHC(S)$-$NHNH_2$[722] are N,S coordinated in the HgL_2Cl_2 compounds. Dithizone (diphenyl-thiocarbazone) PhN=NCSNHNHPh forms HgL_2 (figure 15.54).[723,724]

The dehydrodithiazone (anhydro-5-mercapto-2,3-diphenyltetrazolium hydro-xide) complex, $HgLCl_2$, has been shown by an X-ray study[725] to be trigonal

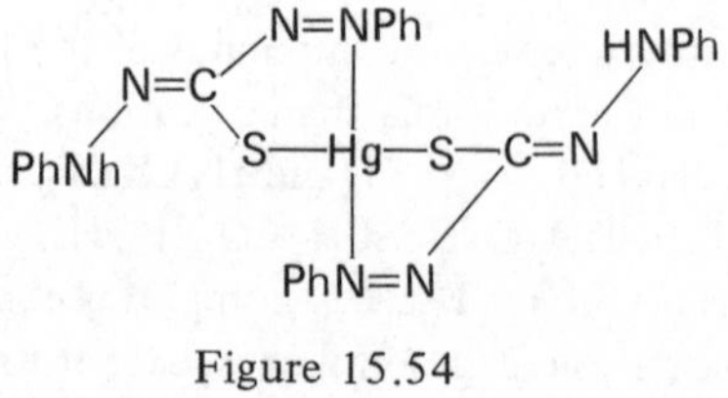

Figure 15.54

bipyramidal (S_3Cl_2), with the units linked into chains by bridging ligands. Further examples of N,S-coordinated ligands are found in the complexes of guianylthiourea,[726] 3-hydroxy-1-mercaptopyrimidine,[727] and quinoxaline(1H,3H)-2,4-dithione.[728]

The ligands thiomorpholin-3-one (tm), thiazolidine-2-thione (ttz) and thiomorpholinethione (ttm) (figure 15.55) provide a means of comparing the competition between N,S and O donors for a metal ion. The 2:1 complexes of these

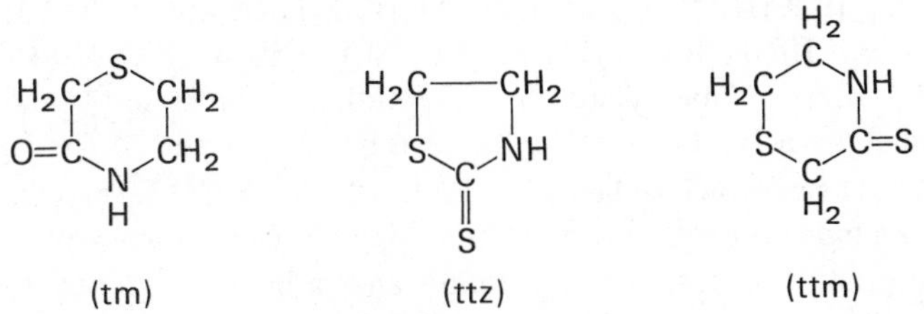

(tm)   (ttz)   (ttm)

Figure 15.55

ligands with HgX$_2$, are on the basis of i.r. spectral evidence, S-bonded (tm) and N-bonded (ttz, ttm).[759,730] On similar grounds Hg(tm)$_2$(SCN)$_2$ and Hg(tm)$_2$(CN)$_2$ contain S-bonded tm, but Hg(SCN)$_2$ is reduced to the metal by ttz.[731]

The Hg(ttz)$_4$(BF$_4$)$_2$ (N-coordinated) and Hg(tm)$_2$(BF$_4$)$_2$ (S-coordinated) are also known.[731] The diphosphinopyridine (figure 15.56) is P,P coordinated in

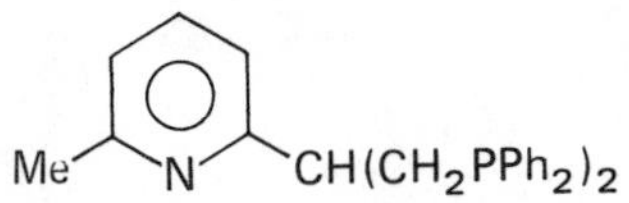

Figure 15.56

HgLX$_2$ (X = Cl, I),[732] but the coordination site of the phosphinoaminopyridine (figure 15.57) in HgL$_2$Cl$_2$ is uncertain.[733]

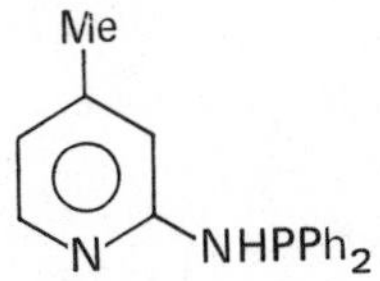

Figure 15.57

Houk and Dobson[734] prepared HgLX$_2$ complexes from 1-diethylamino-2-diphenylphosphinoethane (NP), bis(diethylaminoethyl)phenylphosphine, 1,3-bis(diethylamino)-2-diphenylphosphinopropane (TNP), and bis(diphenylphosphino-ethyl)ethylamine: all appear to be tetrahedral. 2-cyanoethyldiphenylphosphine (PCN) and bis(2-cyanoethyl)phenylphosphine form 1:1 adducts, which are dimeric halide-bridged entities, with the ligands bonded only through the phosphorus. The ligands NP, TNP, and PCN are also capable of forming 2:1 adducts, which are probably tetrahedral with coordination via the phosphorus only.[734]

Mercury(II) complexes of amino acids have received considerable attention.[735]

Cysteine (LH) reacts rapidly with $HgCl_2$ to form HgL_2, which contains sulphydryl coordination only.[736-738] There is considerable uncertainty about other complexes of this amino acid, the reported compounds include $[ClHgSCH_2CH-(NH_3)COOH]_2[HgCl_4]$,[739] $Hg_2[SCH_2CH(NH_2)CO_2]_2$, $Hg_2[SCH_2CH(NH_2)CO_2]_2 \cdot HCl$, $Hg_3[SCH_2CH(NH_2)CO_2]_2Cl_2$ and $Hg_3[SCH_2CH(NH_3)CO_2H]_2Cl_5 \cdot 2H_2O$.[740-742] Cysteine methyl ester is N,S coordinated in HgL_2,[743] and evidence for the formation of $Hg_2L_3^+$ and $Hg_2L_4^{2+}$ has been reported.[744] The $Hg(LH)_2Cl_2 \cdot 1\frac{1}{2}H_2O$[744] has been shown to be $HgL_2 \cdot HCL \cdot HNO_3 \cdot H_2O$[745]. Penicillamine can act either as a bidentate ligand through the sulphydryl and amine groups or as a terdentate ligand through sulphydryl, amino, and carboxylato groups, to Hg(II).[735] Glutathione forms 2:1 (S-coordinated), 2:2 (S,N) and 2:2 (S,N,O) complexes with Hg^{2+}.[746,747] Methionine, ethionine, and S-methylcysteine do not apparently bond through the sulphur to mercury(II), and the solid complexes are thought to be polymeric with amino and bidentate bridging carboxylate groups.[748,749] In acidic solution n.m.r. measurements suggest Hg—S bonding.[741,750,751] N-phenylsulphonylglycine,[752] N-(p-tolylsulphonyl)-α and -β-alanine,[753] and N-benzenesulphonyl-α-alanine[754] complexes have been reported. The mercuric chloride–histidine complex has been X-rayed[755] and found to contain an $HgCl_3O$ unit.

16 Mercury–Metal Bonds

Mercury forms a unique series of compounds in which it is bonded to a second mercury atom (Hg_2^{2+} p. 51), and many transition metals form heteronuclear M–M bonds to mercury.[12,756,757]

The most widely used synthetic route is the reaction of HgX_2 (X = halide or CN) with a carbonylmetalate anion, for example

$$Hg(CN)_2 + 2Mn(CO)_5^- \longrightarrow Hg[Mn(CO)_5]_2 \quad \text{(reference 758)}$$

Mercury halides will oxidatively add to some transition metal complexes

$$trans \ Ir(CO)Cl(PPh_3)_2 + HgCl_2 \longrightarrow Ir(CO)(PPh_3)_2Cl_2(HgCl)$$
$$\text{(reference 759)}$$

or react by oxidative elimination

$$Pt(PPh_3)_4 + HgCl_2 \longrightarrow Pt(PPh_3)_2(HgCl)Cl + (2PPh_3)$$
$$\text{(reference 760)}$$

In other cases Lewis-acid–base adduct formation results

$$Co(\pi\text{-}C_5H_5)(CO)_2 + HgCl_2 \longrightarrow Co(\pi\text{-}C_5H_5)(CO)_2 \cdot HgCl_2$$
$$\text{(references 761 and 762)}$$

although if this reaction is conducted with excess $HgCl_2$, a salt $[Co(\pi\text{-}C_5H_5)(CO)_2\text{-}HgCl]Cl \cdot 2HgCl_2$ is formed.[763]

Mercury–metal-bonded species will sometimes react further with HgX_2

$$\{Hg[Fe(CO)_2NO(AsEt_2Ph)]_2\} + HgBr_2 \longrightarrow 2[BrHgFe(CO)_2(NO)(AsEt_2Ph)]$$
$$\text{(reference 764)}$$

A number of examples of the reaction of metallic mercury with M–X bonds are known

$$(\pi\text{-}C_5H_5)Fe(CO)_2Br + Hg \xrightarrow{h\nu} (\pi\text{-}C_5H_5)Fe(CO)_2HgBr$$
$$\text{(reference 765)}$$

Many Hg—M-bonded complexes undergo substitution by Lewis bases at M, without Hg—M bond fission

$$Hg[Fe(CO)_3NO]_2 + 2PPh_3 \longrightarrow Hg[Fe(CO)_2NO(PPh_3)]_2$$

(reference 766)

although in other cases disproportionation occurs

$$Hg[Co(CO)_4]_2 + P(NMe_2)_3 \longrightarrow Hg + [Co(CO)_3P(NMe_2)_3][Co(CO)_4]$$

(reference 767)

In solution in polar solvents ionic dissociation may occur

$$Hg[Fe(CO)_3NO]_2 \xrightarrow[DMSO]{} Hg^{2+} + 2[Fe(CO)_3NO]^-$$ (reference 764)

or reversible dissociation into the constituents

$$X_2HgCo(CO)_2(\pi\text{-}C_5H_5) \rightleftharpoons (\pi\text{-}C_5H_5)Co(CO)_2 + HgX_2$$

(reference 761)

Thermal decomposition with elimination of elemental mercury is sometimes observed

$$Hg[Fe(CO)_2(\pi\text{-}C_5H_5)]_2 \xrightarrow{80\text{-}90\,^\circ C} Hg + [Fe(\pi\text{-}C_5H_5)(CO)_2]_2$$

(reference 768)

Zinc will displace mercury from several of the compounds, as will cyanide ions, while Hg—M bonds are usually cleaved by acids.

$$Zn + Hg[Mn(CO)_5]_2 \longrightarrow Zn[Mn(CO)_5]_2 + Hg$$ (reference 769)

$$Hg[Co(CO)_4]_2 + xs\ CN^- \longrightarrow Hg(CN)_4{}^{2-} + 2Co(CO)_4{}^-$$

(reference 770)

Table 16.1 lists some examples of Hg—M bonded species. The detailed chemistry of many of these compounds is yet to be investigated, but it is apparent that much of it is concerned with reaction at M rather than Hg, and hence lies outside the scope of this review. The remainder of this chapter will deal with structural data on the Hg—M compounds. Aspects of their chemistry are covered in other reviews.[756,757,771]

X-ray data on Hg—M compounds are given in table 16.2. All the known examples of L_nM—Hg—ML_n for which data are available appear to be essentially linear, and the effect of changes in L on the Hg—M bond is small. Thus the Hg—Co bond lengths in $Hg[Co(CO)_4]_2$ and $Hg[Co(CO)_3PEt_3]_2$ are the same.[804,805] In the X—Hg—M compounds, approximately linear arrangements are found,[808,809] reminiscent of HgX_2. In contrast, the Lewis donor–acceptor interaction in $(\pi\text{-}C_5H_5)Co(CO)_2 \cdot HgCl_2$ (figure 16.1)[762] has a considerably longer Hg—Co bond and Hg—Cl (approximately 0.247 nm) is also significantly longer than in $HgCl_2$ (0.225 nm). The ionic $[(\pi\text{-}C_5H_5)(CO)_2CoHgCl]Cl.2HgCl_2$ contains a much shorter

Table 16.1. Examples of Mercury–Metal Bonded Compounds†

Compound	Remarks	Reference
RHgTa(CO)$_6$	prep. RHgX + Ta(CO)$_6^-$	772
[(π-C$_5$H$_5$)(CO)$_3$M]$_2$Hg	M = Cr, Mo, W	773, 774
[(π-C$_5$H$_5$)(CO)$_3$MHgX]	M = Cr, Mo, W	774, 775, 776
Hg[M(CO)$_5$]$_2$	M = Mn, Re, also (CO)$_5$Mn–Hg–Re(CO)$_5$	758, 777, 778
XHgMn(CO)$_5$	prep. from HgX$_2$ + Hg[Mn(CO)$_5$]$_2$	779
Hg[Mn(CO)$_{5-n}$(ER$_3$)n]$_2$	n = 1,2, ER$_3$ = wide range of phosphine and arsine ligands	780–782
Fe(CO)$_4$(HgCl)$_2$	HgFe(CO)$_4$ + HCl *cis* configuration	783–785
[(π-C$_5$H$_5$)Fe(CO)$_2$]$_2$Hg	prep. [Fe(CO)$_2$(π-C$_5$H$_5$)]$^-$ + Hg(CN)$_2$	773
[(π-C$_5$H$_5$)Fe(CO)$_2$]- Hg[Co(CO)$_4$]	prep. [(π-C$_5$H$_5$)Fe(CO)$_2$Co(CO)$_4$] + Hg	786
Hg[Fe(CO)$_3$NO]$_2$		787
Hg[Fe(CO)$_2$NO(ER$_3$)]$_2$	ER$_3$ = wide range of phosphines, arsines or stibines	764, 766, 788
[M(CO)$_3$(PPh$_3$)$_2$(HgX)]HgX$_3$	M = Ru,Os prep. M(CO)$_3$(PPh$_3$)$_2$ + HgX$_2$	789
[(π-C$_5$H$_5$)$_2$RuHgX$_2$]		790
Hg[Co(CO)$_4$]$_2$	prep. Co$_2$(CO)$_8$ + Hg or Co(CO)$_4^-$ + HgX$_2$	770, 791, 792
Hg[Co(CO)$_3$ER$_3$]$_2$	ER$_3$ = phosphines, arsines, diphosphines	793–795
Hg[Co(CO)$_4$]$_3^-$	prep. Hg[Co(CO)$_4$]$_2$ + Co(CO)$_4^-$	796, 797
Hg[M(PF$_3$)$_4$]$_2$	M = Rh, Ir, prep. M(PF$_3$)$_4$Cl + PF$_3$ + K/Hg	798
RhX$_2$(HgY)(AsMePh$_2$)$_3$	X = Cl, Br, Y = Cl, Br, I, OAc	759
[Rh(CO)$_2$(PPh$_3$)$_2$]$_2$Hg		799
[Ir(CO)$_3$PPh$_3$]$_2$Hg		799, 800
Pt(PPh$_3$)$_2$(HgX)X	Pt(PPh$_3$)n + HgX$_2$	760

† This list is intended to illustrate the types of compounds formed and is not comprehensive.

Table 16.2. Mercury–Metal Bond Length Data

Compound	Hg–M (nm)	Remarks	Reference
[ClHg–Mo(CO)$_3$bipy]	0.270(7)	7-coord. Mo figure 16.2	801
Hg–[Mn(CO)$_5$]$_2$	0.261(2)	linear Mn–Hg–Mn	802, 803
[(π-C$_5$H$_5$)(CO)$_2$Co–HgCl$_2$]	0.2578	figure 16.3	762
[(π-C$_6$H$_5$)(CO)$_2$CoHgCl] Cl	0.2504	also 2HgCl$_2$ adduct	763
Hg–[Co(CO)$_3$(PEt$_3$)]$_2$	0.2499(5)	Hg$\stackrel{\frown}{-}$Co$-$C = 86° (av.)	804
Hg–[Co(CO)$_4$]$_2$	0.2499(7)	Co$\stackrel{\frown}{-}$Hg$-$Co = 178.4° 0.3	805
Hg–[Fe(CO)$_2$(NO)(PEt$_3$)]$_2$	0.2534(2)	approx. linear P–Fe–Hg–Fe–P	806
[(π-C$_5$H$_5$)(CO)$_2$FeHgCo(CO)$_4$]	0.249 (Hg–Fe) 0.256 (Hg–Co)	Fe$\stackrel{\frown}{-}$Hg$\stackrel{\frown}{-}$Co = 176°	807
(BrHg)$_2$Fe(CO)$_4$	0.244, 0.259	Hg$\stackrel{\frown}{-}$Fe$-$Hg = 75°	808
(pyClHg)$_2$Fe(CO)$_4$	0.2552(8)	Hg$\stackrel{\frown}{-}$Fe$-$Hg = 77°	809
Os(HgCl)Cl$_2$(NO)(PPh$_3$)$_2$	0.2577(5)	Os$\stackrel{\frown}{-}$Hg$-$Cl = 177(1)°	810

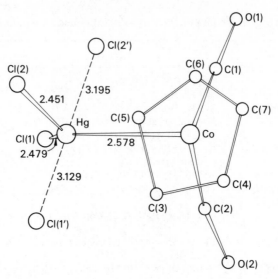

Figure 16.1. $[(\pi\text{-}C_5H_5)(CO_2CoHgCl_2]$

Hg–Co bond (0.2504 nm). It is probable that $(EPh_3)Fe(CO)_3 \cdot HgX_2$ are of the donor–acceptor type, and that $[Mo(CO)_3(PPh_3)_2 \cdot 2HgX_2]^{789}$ are ionic $[Mo(CO)_3\text{-}(PPh_2)_2HgX]HgX_3$. An unusual complex is $(pyHgCl)_2Fe(CO)_4$,[809,811] which contains six-coordinate iron (*cis* $Hg_2(CO)_4$), with each mercury coordinated to one pyridine, the iron, and two chlorines; the molecules being linked into chains by Hg–Cl bridges. The Hg . . . Hg separation is 0.317 nm, and coupled with Hg–Fe–Hg = 77° (there seems to be no crystallographic reason why it should not be 90°) may suggest Hg . . . Hg interaction.

Few Hg–M bonds containing early transition metals in positive oxidation states are known, but the characterisation of the seven-coordinate $[(HgCl)MoCl(CO)_3\text{-}bipy]^{801}$ (figure 16.2) suggests that further study would be worthwhile.

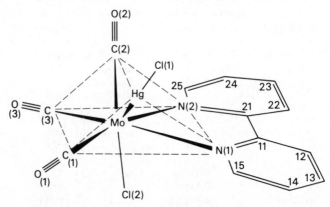

Figure 16.2. $[Mo(HgCl)Cl(CO)_3(bipy)]$

17 Mercury(II) in Solution

This review has concentrated on those mercury(II) compounds that can be isolated in the solid state. Extensive studies on the solution behaviour of Hg^{2+} have been reported. Mercury(II) is a typical class B ion, and this is reflected in the formation-constant data obtained with various ligands.[2,5] Stability constant data are tabulated in the Chemical Society Special Publication.[5] Attention is also drawn to the discussion of mercury–nitrogen base ligands in the papers of Worth and Davidson.[48,812] The thermodynamics of the formation of several mercury(II) complexes in solution has been discussed by Graddon et al.[813-815]

Acknowledgement

We are grateful to the following for permission to reproduce figures: The Chemical Society (figures 13.1, 13.2, 15.8, 15.41, 15.51, 16.1; The Australian Chemical Society (figure 16.2); The American Chemical Society (figures 15.30, 15.32, 15.33); The Canadian Chemical Society (figures 14.2, 15.40); Acta Chemica Scandinavica (figure 15.6); Acta Crystallographia (figures 15.24, 15.25); Zeitschrift fur Anorganische und Allgemeine Chemie (figures 15.31, 15.34).

References

1. *Gmelin's Handbuch der Anorganische Chemie 8th Auflage* 'Quecksilber'. Teil B. Lfg. 1, 2, 3, 4. Verlag. Chemie. GMBH (1965–69).
2. H. L. Roberts, *Adv. inorg. Chem. Radiochem.*, 11 (1968) p. 309.
3. C. G. S. Phillips and R. J. P. Williams, *Inorganic Chemistry*. (Oxford University Press, 1965) vol. 2.
4. L. G. Hepler and G. Olofsson, *Chem. Revs.* (1975) p. 585.
5. Chemical Society Special Publication 17. Stability Constants of Metal-Ion Complexes, London, 1964.
6. R. J. Gillespie and J. Passmore, *Adv. Inorg. Chem. Radiochem.*, 17 (1975) p. 49.
7. M. K. Cooper and D. M. Foster, *J. chem. Soc. (A)* (1968) p. 2968.
8. L. A. Woodward, *Phil. Mag.*, 18 (1934) p. 823.
9. H. Stammreich and T. T. Sans, *J. mol. Struct.*, 1 (1967) p. 55.
10. K. Brodersen and N. Hacke, *Chem. Ber.*, 107 (1974) p. 3260.
11. J. Limmer, N. Hacke and K. Brodersen, *Chem. Ber.*, 106 (1973) p. 2185.
12. D. Breitinger, K. Brodersen and J. Limmer, *Chem. Ber.*, 103 (1970) p. 2388.
13. T. G. Spiro, *Prog. inorg. Chem.*, 11 (1970) p. 1.
14. E. Dorm, *J. C. S. Chem. Comm.* (1971) p. 466.
15. R. J. Havinghurst, *J. Am. chem. Soc.*, 48 (1926) p. 2113.
16. D. Grdenic, *J. chem. Soc.* (1956) p. 1312.
17. E. Dorm, *Acta Chem. Scand.*, 21 (1967) p. 2834.
18. E. Dorm, *Acta Chem. Scand.*, 23 (1969) p. 1607.
19. E. Johansson, *Acta Chem. Scand.*, 20 (1966) p. 533.
20. B. Lindh, *Acta Chem. Scand.*, 21 (1967) p. 2743.
21. H. Puff, G. Lorbacher and R. Skrabs, *Z. Krist.*, 122 (1965) p. 156.
22. R. C. Elder, J. Halpern and J. S. Pond, *J. Am. chem. Soc.*, 89 (1967) p. 6877.
23. D. L. Kepert and D. Taylor, *Aust. J. Chem.*, 27 (1974) p. 1199.
24. D. L. Kepert, D. Taylor and A. H. White, *Inorg. Chem.*, 11 (1972) p. 1639.
25. D. L. Kepert, D. Taylor and A. H. White, *J. C. S. Dalton* (1973) p. 1658.
26. D. L. Kepert, D. Taylor and A. H. White, *J. C. S. Dalton* (1973) p. 392.
27. D. L. Kepert, D. Taylor and A. H. White, *J. C. S. Dalton* (1973) p. 893.
28. J. C. Dewan, D. L. Kepert and A. H. White, *J. C. S. Dalton* (1975) p. 490.
29. K. Brodersen and L. Kunkel, *Chem. Ber.*, 91 (1958) p. 2698.
30. K. Brodersen, N. Hacke and G. Liehr, *Z. anorg. allg. Chem.*, 409 (1974) p. 1.
31. E. Dorm, *Acta Chem. Scand.*, 25 (1971) p. 1655.
32. B. Kamener and B. Kaitner, *Acta Cryst.*, B29 (1973) p. 1666.

33. Ref. 1, Lief 2, pp. 403, 424, 722, 820.
34. D. Grdenić and C. Djordjevic, *J. chem. Soc.* (1956) p. 1316.
35. E. Dorm, *Chem. Comm. Univ. Stockholm*, 1970, No. 3; *Chem. Abs.*, 75 (1971) p. 102345g.
36. E. Dorm and B. Lindh, *Acta Chem. Scand.*, 21 (1967) p. 1661.
37. K. Huttner and S. Knappe, *Z. anorg. allg. Chem.*, 190 (1930) p. 27.
38. S. K. Deb and A. D. Yoffe, *Proc. Roy. Soc. (London)*, A256 (1960) p. 528.
39. E. Sonderback, *Acta Chem. Scand.*, 11 (1957) p. 1622.
40. H. J. Emeléus and A. A. Woolf, *J. chem. Soc.* (1950) p. 164.
41. E. Newbury, *Trans. Electrochem. Soc.*, 69 (1936) p. 611.
42. G. Brauer, *Handbook of Preparative Inorganic Chemistry*, vol. 2 (Academic Press, New York, 1963).
43. O. G. Sheintis, *Zh. Prikl. Khim.*, 13 (1940) p. 1101; *Chem. Abs.* (1941) p. 2087.
44. W. D. Larson and W. J. Tomsicek, *J. Am. chem. Soc.*, 63 (1941) p. 3329.
45. R. A. Potts and A. L. Allred, *Inorg. Chem.*, 5 (1966) p. 1066.
46. G. Anderegg, *Helv. Chim. Acta*, 42 (1959) p. 344.
47. T. Yamane and N. Davidson, *J. Am. chem. Soc.*, 82 (1960) p. 2123.
48. T. H. Wirth and N. Davidson, *J. Am. chem. Soc.*, 86 (1964) p. 4314.
49. A. Fubahashi, M. Kawano, N. Tashiro, and A. Ouchi, *J. inorg. nucl. Chem.*, 34 (1972) p. 2960.
50. D. Grdenić, *Quart. Rev.*, 19 (1965) p. 303.
51. J. W. Bouknight, R. Layton and J. E. Lewis, *Inorg. Nucl. Chem. Letts.*, 3 (1967) p. 103.
52. D. Breitinger and K. Brodersen, *Angew. Chem. Int. Ed.*, 9 (1970) p. 357.
53. W. N. Lipscomb, *Anal. Chem.*, 25 (1953) p. 737.
54. M. Goehring and G. Zirker, *Z. anorg. Allg. Chem.*, 285 (1956) p. 70.
55. G. Tosi and G. Mamantov, *Inorg. Nucl. Chem. Letts.*, 6 (1970) p. 843.
56. G. Tosi, K. W. Fung, G. M. Begun and G. Mamantov, *Inorg. Chem.*, 10 (1971) p. 2285.
57. G. Mamantov, Report 1974 ORO-3518; *Chem. Abs.*, 72 (1975) p. 36598h.
58. R. D. Ellison, H. A. Levy and K. W. Fung, *Inorg. Chem.*, 11 (1972) p. 833.
59. C. G. Davies, P. A. W. Dean, R. J. Gillespie and P. K. Ummat, *Chem. Comm.* (1971) p. 782.
60. B. D. Cutforth, C. G. Davies, P. A. W. Dean, R. J. Gillespie, P. Ireland and P. K. Ummat, *Inorg. Chem.*, 12 (1973) p. 1343.
61. B. D. Cutforth, R. J. Gillespie and P. R. Ireland, *J. C. S. Chem. Comm.* (1973) p. 723.
62. R. J. Gillespie and P. K. Ummat, *J. C. S. Chem. Comm.* (1971) p. 1168.
63. J. D. Brown, B. D. Cutforth, C. G. Davies, R. J. Gillespie, P. Ireland and J. E. Vekris, *Can. J. Chem.*, 52 (1974) p. 791.
64. R. J. Booth, H. C. Starkie and M. C. R. Symons, *J. chem. Soc. (A)* (1971) p. 3198.
65. O. Ruff and G. Bahlan, *Ber. Deut. Chem. Ges.*, 51 (1918) p. 1752.
66. J. H. Simons, *Fluorine Chemistry*, vol. V (Academic Press, New York, 1954).
67. F. Ebert and H. Woitinek, *Z. anorg. allg. Chem.*, 210 (1933) p. 269.
68. R. Dotzer and A. Meuwsen, *Z. anorg. allg. Chem.*, 308 (1961) p. 79.
69. D. Grdenić and M. Sikirica, *Inorg. Chem.*, 12 (1973) p. 544.
70. J. E. Roberts and G. H. Cady, *J. Am. chem. Soc.*, 82 (1960) p. 353; J. M. Schreeve and W. Scholten, *J. Am. chem. Soc.*, 83 (1961) p. 4521.
71. H. Braekben and W. Scholten, *Z. Krist.*, 89 (1934) p. 448.

72. D. Grdenić, *Arkiv. Kemi*, 22 (1950) p. 14.
73. H. J. Verweel and J. M. Bijvoet, *Z. Krist.*, 77 (1931) p. 122.
74. G. A. Jeffrey and M. Vlasse, *Inorg. Chem.*, 6 (1967) p. 396.
75. R. F. Rolsten, *Iodide-Metals and Metal-Iodides* (Wiley and Sons Inc., New York, 1961) p. 215.
76. D. Schwartzenbach, *Z. Krist.*, 128 (1969) p. 97.
77. G. Jander, *Die Chemie in Wasserähnlichen Lösungmitteln* (Springer, Berlin, 1949).
78. R. P. Rastogi and B. L. Dubey, *J. Am. chem. Soc.*, 89 (1967) p. 200.
79. R. P. Rastogi and B. L. Dubey, *J. inorg. nucl. Chem.*, 31 (1967) p. 1530.
80. R. P. Rastogi, B. L. Dubey and N. D. Agrawal, *J. inorg. nucl. Chem.*, 37 (1975) p. 1167.
81. Y. Marcus, *Acta Chem. Scand.*, 11 (1957) pp. 610, 811.
82. I. Eliezer, *J. phys. Chem.*, 68 (1964) p. 2722.
83. R. L. Ammlung and T. B. Brill, *Inorg. Chim. Acta*, 11 (1974) p. 201.
84. I. R. Beattie and J. R. Hordar, *J. chem. Soc.* (1970) p. 2433.
85. G. J. Janz and D. W. James, *J. chem. Phys.*, 38 (1963) p. 902.
86. K. Aurivillius, *Acta Chem. Scand.*, 8 (1954) p. 523.
87. S. Ščavničar, *Acta Cryst.*, 8 (1955) p. 379.
88. K. Aurivillius, *Arkiv. Kemi*, 23 (1964) p. 505; K. Aurivillius and C. Stalhandske, *Acta Cryst.*, B30 (1974) p. 1907.
89. S. Ščavničar and D. Grdenić, *Acta Cryst.*, 8 (1955) p. 275.
90. K. Aurivillius, *Arkiv. Kemi*, 22 (1964) p. 517; 22 (1964) p. 537.
91. A. Weiss, G. Nagorsen and A. Weiss, *Z. Naturforsch*, 9B (1954) p. 81.
92. K. Aurivillius, *Arkiv. Kemi*, 23 (1964) p. 469; 28 (1967) p. 279.
93. K. Aurivillius, *Acta Chem. Scand.*, 18 (1964) p. 1305.
94. R. Hoppe and R. Homann, *Z. anorg. allg. Chem.*, 369 (1969) p. 212.
95. C. Hebeker, *Naturwiss.*, 60 (1973) p. 154.
96. G. B. Deacon, *Rev. pure appl. Chem.*, 13 (1963) p. 189.
97. Chem. Soc. Special Publicn., 17II (1958).
98. L. G. Sillen, *Acta Chem. Scand*, 3 (1949) p. 505.
99. G. Ellendt and K. Cruse, *Z. phy. Chem.*, 201 (1952) p. 130.
100. T. R. Griffiths and M. C. R. Symons, *Trans. Farad. Soc.*, 56 (1960) p. 1752.
101. G. B. Deacon and B. O. West, *J. chem. Soc.* (1961) p. 3929.
102. N. Gallo, V. S. Bianco and S. Doronzo, *J. inorg. nucl. Chem.*, 33 (1971) p. 3950.
103. F. Gaizer and G. Johansson, *Acta Chem. Scand.*, 22 (1968) p. 3013.
104. M. A. Hooper and D. W. James, *Aust. J. Chem.*, 24 (1971) p. 1345.
105. G. B. Deacon and B. O. West, *Aust. J. Chem.*, 16 (1963) p. 579.
106. Gmelin, Teil B. Lief 4.
107. I. N. Belyaev and K. E. Mironov, *Zh. Obsch. Khim.*, 22 (1952) pp. 1484, 1490.
108. I. N. Belyaev, A. E. Shurginov and N. S. Kudryashov, *Zh. Neorg. Khim.*, 17 (1972) p. 2812.
109. M. Stan and F. Zalaru, *Ann. Univ. Bucaresti Chem.*, 21 (1972) p. 53; *Chem. Abs.*, 79 (1972) p. 38082x.
110. M. Stan, F. Zalaru and G. S. Ionescu, *Ann. Univ. Bucaresti Chem.*, 20 (1971) p. 45; *Chem. Abs.*, 78 (1972) p. 66392w.
111. M. L. Deliwaule, *Bull. Soc. Chim. France* (1955) p. 1294.
112. J. A. Rolfe, D. E. Sheppard and L. A. Woodward, *Trans. Farad. Soc.*, 50 (1954) p. 1275.
113. J. T. R. Dunsmuir and A. P. Lane, *J. inorg. nucl. Chem.*, 33 (1971) p. 4361.

114. R. M. Barr and M. Goldstein, *J. C. S. Dalton* (1974) p. 1180.
115. M. A. Hooper and D. W. James, *Aust. J. Chem.*, 24 (1971) p. 1331.
116. D. E. Scaife, *Aust. J. Chem.*, 24 (1971) p. 1753.
117. G. B. Deacon, J. H. S. Green and W. Kynaston, *Aust. J. Chem.*, 19 (1966) p. 1603.
118. P. Day and R. H. Seal, *J. C. S. Dalton* (1972) p. 2054.
119. T. R. Griffiths, *Anal. Chem.*, 35 (1963) p. 1077.
120. C. H. MacGillavry, J. H. Wilde and J. M. Bijvoet, *Z. Krist.*, 100 (1939) p. 212.
121. K. Aurivillius and C. Stalhandske, *Acta Chem. Scand.*, 27 (1973) p. 1086.
122. H. C. McMurdle, J. De Groot, M. Morris and H. E. Swanson, *J. Res. Nat. Bur. Stand.*, Sect. A, 73 (1969) p. 621; *Chem. Abs.*, 72 (1969) p. 17989.
123. E. J. Harmsen, *Z. Krist.*, 100 (1938) p. 208.
124. R. M. Barr and M. Goldstein, *Inorg. nucl. Chem. Lett.*, 10 (1974) p. 33.
125. J. A. D. Jeffries, G. A. Sim, R. H. Burnell, W. I. Taylor, R. E. Corbett, J. Murray and B. J. Sweetman, *Proc. chem. Soc.* (1963) p. 171.
126. E. Fatuzzo, R. Nitsche, H. Roetschi and S. Zingg, *Phys. Rev.*, 125 (1962) p. 514.
127. D. V. Ninkovic, *Bull. Inst. Sci. Boris Kidrič (Belgrade),* 7 (1957) p. 81; *Chem. Abs.*, 52 (1958) p. 2493f.
128. R. G. Wyckoff, *Crystal Structures*, Vols. I–V, 2nd edn. (Wiley and Sons, Inc., New York, 1963–5).
129. V. M. Padmanabham and V. S. Yadava, *Acta Cryst.*, B25 (1969) p. 647.
130. J. G. White, *Acta Cryst.*, 16 (1963) p. 397.
131. K. Brodersen, G. Thiele and G. Gorz, *Z. anorg. allg. Chem.*, 401 (1973) p. 217.
132. V. A. Gerken and V. I. Pakhomov, *Zh. Strukt. Khim.*, 10 (1969) p. 753.
133. L. Nyqvist and G. Johansson, *Acta Chem. Scand.*, 25 (1971) p. 1615.
134. V. I. Pakhomov and P. M. Fedorov, *Kristallografiya*, 17 (1972) p. 942; *Chem. Abs.*, 77 (1972) p. 169898z.
135. R. H. Fenn, J. W. H. Oldham and D. C. Phillips, *Nature*, 198 (1963) p. 381.
136. K. Kohler, D. Breitinger and G. Thiele, *Angew. Chem.*, 86 (1974) p. 863.
137. K. Damm and A. Weiss, *Z. Naturforsch.*, 10B (1955) p. 535.
138. J. Huart, *Bull. Soc. Franc. Mineral Crist.*, 88 (1965) p. 65; 89 (1966) p. 23.
139. A. F. Wells, *Structural Inorganic Chemistry*, 3rd edn. (Oxford Univ. Press, 1962) p. 174.
140. J. Gaizer and M. T. Beck, *J. inorg. nucl. Chem.*, 29 (1967) p. 21.
141. D. Breitinger and K. Koehler, *Inorg. nucl. Chem. Lett.*, 8 (1972) p. 957.
142. W. Biltz, *Z. anorg. allg. Chem.*, 170 (1928) p. 161.
143. J. Hroslef, *Acta Chem. Scand.*, 12 (1958) p. 1568.
144. R. C. Seccombe and C. H. L. Kennard, *J. organometallic Chem.*, 18 (1969) p. 243.
145. L. H. Jones, *Spectrochim. Acta*, 19 (1963) p. 1675.
146. L. L. Bircumshaw, F. H. Taylor and D. H. Whiffen, *J. chem. Soc.* (1954) p. 931.
147. W. P. Griffith, *J. chem. Soc.* (1964) p. 4070.
148. R. A. Pennman and L. H. Jones, *J. inorg. nucl. Chem.*, 20 (1961) p. 19.
149. N. Tanaka and T. Murayama, *Z. phys. Chem.*, 11 (1957) p. 366.
150. S. Glasstone, *J. chem. Soc.* (1930) p. 1237.
151. L. D. C. Bok and S. S. Basson, *J. Afr. chem. Inst.*, 19 (1966) p. 62; *Chem. Abs.*, 66 (1967) p. 89259z.
152. R. G. Dickinson, *J. Am. chem. Soc.*, 44 (1922) p. 774.

153. G. Jander and B. Gruttner, *Chem. Ber.*, 81 (1948) p. 114.
154. G. Thiele, R. Bauer and O. Messmer, *Naturwiss*, 61 (1974) p. 215.
155. K. G. Ashurst, N. P. Finkelstein and L. A. Gould; *J. chem. Soc. (A)* (1971) p. 1899.
156. F. H. Kruse, *Acta Cryst.*, 16 (1963) p. 105.
157. C. Mahon, *Inorg. Chem.*, 10 (1971) p. 1813.
158. A. Weiss and G. Hofmann, *Z. Naturforsch*, 15B (1960) p. 679.
159. P. T. Beurskens, W. P. J. H. Bosman and J. A. Cras, *J. Cryst. Molec. Struct.*, 2 (1972) p. 183.
160. A. Y. Tsivadze, G. V. Tsintsadze and Y. Y. Kharitonov, *Russ. J. inorg. Chem.*, 17 (1972) p. 1417.
161. K. F. Chew, W. Derbyshire, N. Logan, A. H. Norbury and A. I. P. Sinha, *J. C. S. chem. Comm.* (1970) p. 1708.
162. Houben-Weyl, *Methoden der Organischen Chemie*, Bd. VIII, Georg Thieme, Verlag, Stuttgart, 5 (1952) p. 355.
163. C. D. Garner and B. Hughes, *Adv. inorg. chem. Radiochem.*, 17 (1975) p. 1.
164. T. L. Davis, *The Chemistry of Powder and Explosives*, vol. 2 (New York, 1943) p. 410.
165. W. Beck and E. Schuierer, *Z. anorg. Allg. Chem.*, 347 (1966) p. 304.
166. L. Hackspill and W. Schumacher, *Ann. Acad. Sci. Techn. Varsovic*, 3 (1936) p. 84; *Chem. Abs.*, 32 (1938) p. 4338.
167. W. Beck, W. P. Feldhammer, P. Poellman, E. Schiuerer and K. Fedl, *Chem. Ber.*, 100 (1967) p. 2335.
168. L. Wohler and A. Berthmann, *Ber. Deutsch. Chem. Ges.*, 62 (1929) p. 2748.
169. A. L. Beauchamp and D. Goutier, *Can. J. Chem.*, 50 (1973) p. 977.
170. H. Puff and H. Becker, *Acta Cryst.*, 18 (1965) p. 299.
171. N. Tanaka, K. Ebata and T. Murayama, *Bull. chem. Soc. Japan*, 35 (1962) p. 124.
172. G. J. Nyman and G. S. Alberts, *Anal. Chem.*, 32 (1960) p. 207.
173. A. Larbot and A. L. Beauchamp, *Rev. Chim. Mineral.*, 10 (1973) p. 465.
174. A. Sakhri and A. L. Beauchamp, *Acta Cryst.*, B31 (1975) p. 409.
175. A. Sakhri and A. L. Beauchamp, *Inorg. Chem.*, 14 (1975) p. 740.
176. Z. V. Zvonkova, *Russ. J. Phys. Chem.*, 26 (1952) p. 1798.
177. G. S. Zhdanov and V. V. Sanadze, *Zh. Fiz. Khim.*, 26 (1952) p. 469.
178. J. W. Jeffery and K. M. Rose, *Acta Cryst.*, B24 (1968) p. 653.
179. R. Gronback and J. D. Dunitz, *Helv. Chim. Acta*, 47 (1964) p. 1889.
180. R. Makhifa, L. Pazdernik and R. Rivest, *Can. J. Chem.*, 51 (1973) p. 438.
181. R. Makhifa, L. Pazdernik and R. Rivest, *Can. J. Chem.*, 51 (1973) p. 2987.
182. D. Forster and D. M. L. Goodgame, *J. chem. Soc.* (1965) p. 268; *Inorg. Chem.*, 4 (1965) p. 823; 4 (1965) p. 715.
183. F. A. Cotton, D. M. L. Goodgame, M. Goodgame and A. Sacco, *J. Am. Chem. Soc.*, 83 (1961) p. 4157.
184. A. Korczynski, *Rocz. Chem.*, 36 (1962) p. 1539.
185. A. Y. Tsivadze, Y. Y. Kharitonov and G. V. Tsintsadze, *Russ. J. Inorg. Chem.*, 15 (1970) p. 1094.
186. Z. V. Zvonkova and G. S. Zhdanov, *Zh. Fiz. Khim.*, 26 (1952) p. 586.
187. P. Biscarini, L. Fusina and G. Nivellini, *J. C. S. Dalton* (1974) p. 2140.
188. R. M. Alasaniya, V. V. Skopenko and G. V. Tsintsadze, *Tr. Gruz. Politekh. Inst.*, 7 (1967) p. 21.
189. V. F. Torpova, *Zh. Neorg. Khim.*, 1 (1956) p. 243.
190. T. Murayama and A. Takayanagi, *Bull. Chem. Soc. Japan*, 45 (1972) p. 3459.

191. Y. Y. Kharitonov, and V. V. Skopenko, *Russ. J. inorg. Chem.*, 10 (1965) p. 984.
192. A. Swinarski and A. Lodzinska, *Rocz. Chem.*, 32 (1958) p. 1053.
193. A. Turco, C. Pecile and M. Nicolini, *J. chem. Soc.* (1962) p. 3008.
194. G. V. Tsintsadze, A. E. Shvelashvili and V. V. Skopenko, *Tr. Gruz. Politekh. Inst.*, 1 (1967) p. 29; 4 (1967) p. 53.
195. D. M. Czakis-Sulikowska, *Rozc. Chem.*, 39 (1965) p. 1161.
196. F. D. Miles, *J. chem. Soc.* (1931) p. 2532.
197. F. P. Bowden, *Proc. Roy. Soc. London*, A246 (1958) p. 146.
198. D. Seybold and K. Dehnicke, *Z. anorg. allg. Chem.*, 361 (1968) p. 277.
199. U. Mueller, *Z. anorg. allg. Chem.*, 399 (1973) p. 183.
200. G. R. Levi, *Gazz. Chim. Ital.*, 54 (1924) p. 709.
201. P. Laruelle, *Compt. Rend.*, 241 (1955) p. 802.
202. W. L. Roth, *Acta Cryst.*, 9 (1956) p. 277.
203. K. Aurivillius, *Acta Chem. Scand.*, 8 (1954) p. 523; 10 (1956) p. 852.
204. P. Laruelle, *Ann. Chim. (Paris)*, 5 (1960) p. 1315.
205. K. Aurivillius and I. B. Carlsson, *Acta Chem. Scand.*, 12 (1958) p. 1297.
206. N. G. Vannerberg, *Arkiv. Kemi*, 13 (1959) p. 515.
207. N. G. Vannerberg, *Prog. inorg. Chem.*, 4 (1962) p. 125.
208. A. Weiss, *Zehn Jahre Fonds der Chemischen Industrie* (Weinheim, Verlag Chemie) 1960, p. 157.
209. R. Hoppe and H. J. Röhrborn, *Naturwiss.*, 49 (1962) p. 419.
210. R. Hoppe and H. J. Röhrborn, *Z. anorg. allg. Chem.*, 329 (1964) p. 110.
211. E. R. Allen, J. Cartlidge, M. M. Taylor and C. F. H. Tipper, *J. phys. Chem.*, 63 (1959) p. 1442.
212. Y. A. Ugai, *Zh. Obsch. Khim.*, 24 (1954) p. 1315.
213. A. S. Tichonov, *Zh. Neorg. Khim.*, 3 (1958) p. 296.
214. N. V. Sidgwick, *The Chemical Elements and their Compounds*, vol. I (Oxford, 1950) p. 326.
215. D. Hall and R. V. Holland, *Proc. Chem. Soc.* (1963) p. 204; *Inorg. Chim. Acta*, 3 (1969) p. 235.
216. L. F. Power, K. E. Turner and F. H. Moore, *Inorg. nucl. Chem. Letts.*, 8 (1972) p. 809.
217. D. M. L. Goodgame and M. A. Hitchman, *J. chem. Soc. (A)* (1967) p. 612.
218. A. Ferrari and C. Colla, *Gazz. Chim. Ital.*, 65 (1935) p. 789.
219. G. Jander and K. Brodersen, *Z. anorg. allg. Chem.*, 262 (1950) p. 33.
220. C. C. Addison and N. Logan, *Adv. inorg. Chem. Radiochem.*, 6 (1964) p. 71.
221. J. Bullock and D. G. Tuck, *J. chem. Soc.* (1965) p. 1877.
222. R. Klemmert and H. Haselbeck, *Z. anorg. allg. Chem.*, 334 (1964) p. 27.
223. J. C. Huttner, *Ann. Chim. (Paris)*, 8 (1953) p. 450.
224. M. T. Fournier and M. Capestan, *Bull. Soc. Chem. France* (1972) p. 573.
225. E. Thilo and I. Grunze, *Z. anorg. allg. Chem.*, 290 (1957) p. 209.
226. H. Guérin and B. Boulitrop, *Compt. Rend.*, 230 (1950) p. 447; 232 (1951) p. 65.
227. A. Bonifačič, *Croat Chim. Acta*, 35 (1963) p. 195; *Chem. Abs.*, 60 (1964) p. 1193c.
228. K. Aurivillius and B. Malmros, *Acta Chem. Scand.*, 15 (1961) p. 1932.
229. P. A. Kokkoros and P. J. Rentzeparis, *Z. Krist.*, 119 (1964) p. 234.
230. A. Bonifačič, *Acta Cryst.*, 14 (1961) p. 116.
231. L. K. Templeton, D. H. Templeton and A. Zalkin, *Acta Cryst.*, 17 (1964) p. 933.
232. M. J. Redman and W. W. Harvey, *J. less-common Metals*, 12 (1967) p. 395.

233. M. T. Falqui, *Ric. Sci. Rend. Sez. A*, 3 (1963) p. 627; *Chem. Abs.*, 62 (1965) p. 1143c.
234. O. H. J. Christie, *Acta Cryst.*, 15 (1962) p. 94.
235. L. S. Lilich and B. F. Dzhurinskii, *Zh. Obsch. Khim.*, 26 (1956) p. 1549.
236. A. M. Golub, A. A. Baran and T. I. Tsitsurina, *Ukrain. Khim. Zh.*, 27 (1961) p. 443; *Chem. Abs.*, 56 (1962) p. 8276c.
237. R. P. J. Cooney and J. R. Hall, *Aust. J. Chem.*, 22 (1969) p. 337.
238. G. Johansson, *Acta Chem. Scand.*, 25 (1971) pp. 2787, 2799.
239. B. Ribar, B. Malkovic, M. Sljukic and F. Gabela, *Z. Kristallogr.*, 134 (1971) p. 311.
240. B. Matkovic, B. Ribar, B. Prelesnik and R. Herak, *Inorg. Chem.*, 13 (1974) p. 3006.
241. A. Weiss, S. Lyng and A. Weiss, *Z. Nat.*, 15B (1960) p. 678.
242. G. Bjornlund, *Acta Chem. Scand.*, 25 (1971) p. 1645.
243. R. P. J. Cooney and J. R. Hall, *Aust. J. Chem.*, 25 (1972) p. 1159.
244. G. Nagorsen, S. Lyng, A. Weiss and A. Weiss, *Angew. Chem.*, 74 (1962) p. 119.
245. A. Bonifačič, *Acta Cryst.*, 16A (1963) p. 30.
246. G. Bjornlund, *Acta Chem. Scand.*, 28A (1974) p. 169.
247. G. Johansson, *Acta Chem. Scand.*, 25 (1971) p. 1905.
248. R. L. Carlin, J. Roitman, M. Dankleff and J. O. Edwards, *Inorg. Chem.*, 1 (1962) p. 182.
249. A. J. Pappas, J. F. Villa and H. B. Powell, *Inorg. Chem.*, 8 (1969) p. 550.
250. D. L. Kepert, D. Taylor and A. H. White, *J. C. S. Dalton* (1973) p. 670.
251. S. K. Madan and W. E. Bull, *J. inorg. nucl. Chem.*, 26 (1964) p. 2211.
252. T. B. Brill and D. W. Wertz, *Inorg. Chem.*, 9 (1970) p. 2692.
253. G. Schmauss and H. Specker, *Naturwiss.*, 54 (1967) p. 248.
254. G. Schmauss and H. Specker, *Z. anorg. allg. Chem.*, 363 (1968) p. 113.
255. I. S. Ahuja, *Inorg. nucl. Chem. Letts*, 6 (1970) p. 879.
256. I. S. Ahuja and P. Rastogi, *J. chem. Soc. (A)* (1970) p. 378.
257. G. Sawitzki and H. G. von Schnering, *Chem. Ber.*, 107 (1974) p. 3266.
258. I. S. Ahuja and A. Garg, *J. inorg. nucl. Chem.*, 34 (1972) p. 2074.
259. I. S. Ahuja and R. Singh, *Ind. J. Chem.*, 11 (1973) p. 1070.
260. A. J. Papas, F. A. Osterman and H. B. Powell, *Inorg. Chem.*, 9 (1970) p. 2695.
261. T. B. Brill and Z. Z. Hugas, *J. inorg. nucl. Chem.*, 33 (1971) p. 371.
262. A. T. McPhail and G. A. Sim, *Chem. Comm.* (1966) p. 21.
263. F. Genet, J. C. Leguen and G. Tsoucaris, *C. R. Acad. Sci.*, 262C (1966) p. 989; F. Genet and J. C. Leguen, *Acta Cryst.*, B25 (1969) p. 2029.
264. I. Lindqvist and G. Olofssen, *Acta Chem. Scand.*, 13 (1959) p. 1753.
265. M. J. Frazer, W. Gerrard and R. Twaits, *J. inorg. nucl. Chem.*, 25 (1963) p. 637.
266. M. Negoui and P. Spacu, *Ann. Univ. Bucuresti*, 20 (1971) p. 71; *Chem. Abs.*, 78 (1972) p. 66393x.
267. M. Zachrisson and K. I. Alden, *Acta Chem. Scand.*, 14 (1960) p. 994.
268. F. J. Welch and H. J. Paxton, *J. Polymer Sci.*, A3 (1965) p. 3427.
269. D. J. Phillips and S. Y. Tyree, *J. Am. chem. Soc.*, 83 (1961) p. 1806.
270. C. I. Bränden, *Proc. VIII Int. Conf. Coord. Chem.*, Vienna, 1964, p. 114; *Arkiv. Kemi*, 37 (1964) p. 485.
271. C. I. Bränden, *Acta Chem. Scand.*, 17 (1963) p. 1363.
272. M. D. Joeston and J. F. Forbes, *J. Am. chem. Soc.*, 88 (1966) p. 5465.
273. K. P. Lannert and M. D. Joeston, *Inorg. Chem.*, 7 (1968) p. 2048.

274. G. B. Deacon and J. H. S. Green, *Spectrochim. Acta*, 25A (1969) p. 355.
275. J. Selbin, W. E. Bull and L. H. Holmes, *J. inorg. nucl. Chem.*, 16 (1961) p. 219.
276. P. Biscarini, L. Fusina and G. D. Nivellini, *J. chem. Soc. (A)* (1971) p. 1128.
277. J. Gopalakrishnan and C. C. Patel, *Inorg. Chim. Acta*, 1 (1967) p. 165.
278. S. C. Jain and R. Rivest, *Inorg. Chim. Acta*, 3 (1969) p. 552.
279. P. Biscarini, L. Fusina and G. D. Nivellini, *J. C. S. Dalton* (1972) p. 1003.
280. P. Biscarini, L. Fusina, G. D. Nivellini, A. Mangia and G. Pellizzi, *J. C. S. Dalton* (1973) p. 159.
281. P. Biscarini, L. Fusina, G. D. Nivellini, A. Mangia and G. Pelizzi, *J. C. S. Dalton* (1974) p. 1846.
282. D. W. Meek, D. K. Straub and R. S. Drago, *J. Am. chem. Soc.*, 82 (1960) p. 6013.
283. R. S. Drago and D. W. Meek, *J. phys. Chem.*, 65 (1961) p. 1446.
284. R. S. McEwen, G. A. Sim and C. R. Johnson, *J. C. S. Chem. Comm.* (1967) p. 885.
285. E. S. Gould and J. D. McCullough, *J. Am. chem. Soc.*, 73 (1951) p. 3196.
286. R. Paetzold and P. Vordank, *Z. anorg. allg. Chem.*, 347 (1966) p. 296.
287. K. A. Jensen and V. Krishnan, *Acta Chem. Scand.*, 21 (1967) p. 1988.
288. F. Madaule-Aubry, *Compt. Rend.*, 261 (1965) p. 1283; *Ann. Chim. (Paris)*, 10 (1965) p. 367.
289. F. Madaule-Aubry and H. Gillier-Pandraud, *Compt. Rend.*, 260 (1965) p. 6613.
290. H. Brusset and F. Madaule-Aubry, *Bull. Soc. Chim. France* (1966) p. 3121.
291. J. C. Monier and M. Griffon, *Compt. Rend.*, 250 (1960) p. 4011.
292. M. Ledesert, M. Frey, S. Nakajima and J. C. Monier, *Bull. Soc. Fr. Miner. Crystallog.*, 92 (1969) p. 342.
293. E. P. Tureskaya, N. Y. Turova and A. V. Novoselova, *Izv. Akad. Nauk SSSR. Otdel, Khim. Nauk.* (1968) p. 1667.
294. H. J. Eméléus, J. M. Shreeve and P. M. Spaziante, *J. chem. Soc. (A)* (1969) p. 431.
295. D. P. Babb and J. M. Schreeve, *Intra-Science Chem. Reports* (1971) p. 55.
296. H. Schmidbauer, M. Bergfeld and F. Schindler, *Z. anorg. allg. Chem.*, 363 (1968) p. 73.
297. A. K. Gosh, C. E. Hansing, A. L. Stutz and A. G. MacDiarmid, *J. chem. Soc.* (1962) p. 403.
298. M. Frey, *C. R. Acad. Sci. Ser. C,* 270 (1970) p. 1265.
299. M. Frey, H. Leligny and M. Ledesert, *Bull. Soc. Fr. Miner. Crystallog.*, 94 (1972) p. 467.
300. M. Frey and M. Ledesert, *Acta Cryst. B,* 27 (1971) p. 2119.
301. P. A. Laurent and E. Arsénio, *Bull. Soc. Chim. France* (1958) p. 618.
302. P. A. Laurent and P. Tarte, *Bull. Soc. Chim. France* (1958) p. 1374; (1957) p. 403.
303. K. Brand and I. Turck, *Pharm. Zentralhalle*, 77 (1936) p. 591.
304. H. Reinboldt, A. Luyken and H. Schmittmann, *J. prakt. Chem.*, 149, (1937) p. 30.
305. F. Reiff, P. Pöhls and W. Overbeck, *Z. anorg. allg. Chem.*, 223 (1935) p. 113.
306. M. Frey and J. C. Monier, *Acta Cryst.*, B27 (1971) p. 2487.
307. M. Frey, *C. R. Acad. Sci.*, 270C (1970) p. 413.
308. O. Hassel and J. Hvoslef, *Acta Chem. Scand.*, 8 (1954) p. 1953.
309. G. P. Rossetti and B. P. Susz, *Helv. Chim. Acta*, 49 (1966) p. 1899.
310. P. Groth and O. Hassel, *Acta Chem. Scand.*, 18 (1964) p. 1326.
311. S. T. Yuan and S. K. Madan, *Inorg. Chim. Acta*, 6 (1972) p. 463.

312. S. Prasad and V. N. Garg, *J. Ind. chem. Soc.*, 42 (1965) p. 259.
313. I. Baxter and G. A. Swan, *J. chem. Soc.* (1965) p. 3011.
314. S. Manay and R. S. Tobias, *Inorg. Chem.*, 14 (1975) p. 287.
315. M. L. Tosata, L. Soccarsi, M. Cignitti and L. Paolini, *Tetrahedron*, 29 (1973) p. 1339.
316. L. Paolini and G. B. Marini-Bettolo, *Rend. 1st Super-Sanita*, 23 (1960) p. 77; *Chem. Abs.*, 54 (1960) p. 24496g.
317. L. Paolini, *Gazz. Chim. Ital.*, 87 (1957) p. 408.
318. J. A. Carrabine and M. Sundaralingram, *Biochem.*, 10 (1971) p. 292.
319. D. C. Nonhebel, *J. chem. Soc.* (1963) p. 738.
320. M. Hassanein and I. F. Hewaidy, *Z. anorg. allg. Chem.*, 373 (1970) p. 80.
321. K. Flatau and H. Musso, *Angew. Chem. Int. Ed.*, 9 (1970) p. 379; R. Allmann, K. Flatau and H. Musso, *Chem. Ber.*, 105 (1972) p. 3067; R. Allmann and H. Musso. *Chem. Ber.*, 106 (1973) p. 3001.
322. L. C. Newell, R. S. Maxson and M. H. Filson, *Inorg. Synth.*, 1 (1939) p. 19.
323. G. Brauer, *Handbook of Preparative Inorganic Chemistry*. vol. II (1963).
324. K. Aurivillius, *Acta Chem. Scand.*, 4 (1950) p. 1413.
325. A. F. Wells, *Structural Inorganic Chemistry* (Ox. U.P., 3rd edn.) p. 528.
326. Gmelin, lief 3, p. 988.
327. P. Biscarini and G. D. Nivellini, *J. chem. Soc. (A)* (1969) p. 2206.
328. W. B. Faragher, I. C. Morrel and S. Comay, *J. Am. chem. Soc.*, 51 (1929) p. 2774.
329. D. T. McAllan, T. V. Cullum, R. A. Dean and F. A. Fidler, *J. Am. chem. Soc.*, 73 (1951) p. 3627.
330. R. D. Obolentzev, V. G. Bukharov and N. K. Faizullina, *Chem. Abs.*, 57, (1962) p. 7218e.
331. J. E. Fergusson and K. S. Lok, *Aust. J. Chem.*, 26 (1973) p. 2615.
332. N. S. Faizullina and E. N. Gur'yanov, *Zh. Obsch. Khim.*, 34 (1964) p. 941.
333. M. Vêcêra, I. Gasparic, D. Snobl and M. Juracek, *Coll. Czech. Chem. Comm.*, 21 (1956) p. 1284; 24 (1959) p. 640.
334. B. M. Mikhailov and F. B. Tutorskaya, *Zh. Obsch. Khim.*, 32 (1962) p. 833.
335. C. I. Branden, *Arkiv. Kemi*, 22 (1964) pp. 83, 495, 501.
336. P. Biscarini, L. Fusina and G. D. Nivellini, *J. C. S. Dalton* (1972) p. 1921.
337. T. A. Danilova, I. N. Tits-Skvortsova, I. Nasyrov and D. N. Kutznetsov, *Vest. Mosk. Univ. Ser. II Khim.*, 20 (1965) p. 79; *Chem. Abs.*, 63 (1965) p. 5569a.
338. P. Biscarini, L. Fusina and G. D. Nivellini, *Inorg. Chem.*, 10 (1971) p. 2564.
339. P. Biscarini, L. Fusina and G. D. Nivellini, *J. chem. Soc. (A)* (1971) p. 1128.
340. R. W. Bost and M. W. Conn, *Ind. Eng. Chem.*, 25 (1933) p. 526; 23 (1931) p. 93.
341. H. J. Worth and H. M. Haendler, *J. Am. chem. Soc.*, 64 (1942) p. 1232.
342. G. T. Morgan and W. Ledbury, *J. chem. Soc.*, 121 (1922) p. 2882.
343. D. M. Sweeney, S. I. Mizushima and J. V. Quagliano, *J. Am. chem. Soc.*, 77 (1955) p. 6521.
344. G. E. Coates and D. Ridley, *J. chem. Soc.* (1964) p. 166.
345. M. F. Shostokovskii, E. N. Prilezhaeva and N. I. Uvarova, *Izv. Akad. Nauk. SSSR* (1954) p. 526.
346. D. C. Goodall, *J. chem. Soc. (A)* (1967) p. 1387.
347. D. C. Goodall, *J. chem. Soc. (A)* (1968) p. 887.
348. R. A. Walton, *Inorg. Chem.*, 5, (1966) p. 643.
349. R. S. McEwen and G. A. Sim, *J. chem. Soc. (A)* (1967) p. 271.

350. P. A. Laurent and J. L. Cardoso Pereira, *Bull. Soc. Chim., France*, (1963) p. 1158.
351. J. W. Bouknight and G. McP. Smith, *J. Am. chem. Soc.*, 61 (1939) p. 28.
352. J. B. Schroyer and R. M. Jackman, *J. chem. Ed.*, 24 (1947) p. 146.
353. J. A. W. Dalziel and T. G. Hewitt, *J. chem. Soc. (A)*, (1966) p. 233.
354. J. A. W. Dalziel, T. G. Hewitt and S. D. Ross, *Spectrochim. Acta*, 22 (1966) p. 1267.
355. J. A. W. Dalziel, M. J. Hitch and S. D. Ross, *Spectrochim. Acta*, 25A (1969) p. 1055.
356. W. R. Costello, A. T. McPhail and G. A. Sim, *J. chem. Soc. (A)* (1966) p. 1190.
357. R. S. McEwen and G. A. Sim, *J. chem. Soc. (A)* (1969) p. 1897.
358. K. K. Cheung, R. S. McEwen and G. A. Sim, *Nature*, 205 (1965) p. 383.
359. K. K. Cheung and G. A. Sim, *J. chem. Soc.* (1965), 5988.
360. D. C. Bradley and N. R. Kunchar, *J. chem. Phys.*, 40 (1964) p. 2258.
361. N. R. Kunchar, *Nature*, 204 (1964) p. 468.
362. D. C. Bradley and N. R. Kunchar, *Can. J. Chem.*, 43 (1965) p. 2786.
363. A. J. Downs, E. A. V. Ebsworth and H. J. Emeleus, *J. chem. Soc.* (1961) p. 3187.
364. E. H. Man, D. D. Coffman and E. L. Muetterties, *J. Am. chem. Soc.*, 81 (1959) p. 3575.
365. R. N. Haszeldine and J. M. Kidd, *J. chem. Soc.* (1953) p. 3219.
366. G. A. R. Brandt, H. J. Emeleus and R. N. Haszeldine, *J. chem. Soc.* (1952) p. 2198.
367. W. Reid, W. Merkel and R. Oxenius, *Chem. Ber.*, 103 (1970) p. 32.
368. P. R. Brown and J. O. Edwards, *J. inorg. nucl. Chem.*, 32 (1970) p. 2671.
369. G. McP. Smith and W. L. Semon, *J. Am. chem. Soc.*, 46 (1924) p. 1325.
370. A. Johansson, *Arhiv. Kemi. Min. Geol. Ser. A*, 13 (1939) p. 1.
371. H. C. Brinkhoff, J. A. Cras, J. J. Steggerda and J. Willemse, *Rec. Trav. Chim.*, 88 (1969) p. 633.
372. H. C. Brinkhoff, A. M. Grotens and J. J. Staggerda, *Rec. Trav. Chim.*, 89 (1970) p. 11.
373. A. M. Grotens and F. W. Pijpers, *Rec. Trav. Chim.*, 92 (1973) p. 619.
374. P. T. Beurskins, J. A. Cras, J. H. Noordik and A. M. Spruijt, *J. Cryst. Mol. Struct.*, 1 (1971) p. 93.
375. D. W. Meek and P. E. Nicpon, *J. Am. chem. Soc.*, 87 (1965) p. 4951.
376. M. G. King and G. P. McQuillan, *J. chem. Soc. (A)* (1967) p. 898.
377. J. A. W. Dalziel, A. F. le C. Holding and B. E. Watts, *J. Chem. Soc. (A)* (1967) p. 358.
378. J. Philip and C. Curran, *Abstracts 147th Nat. Melt. Amer. Chem. Soc.*, Philadelphia (April 1964) p. 282.
379. K. C. Malhotra, *Ind. J. Chem.*, 12 (1974) p. 823.
380. M. A. A. Beg and K. S. Hussain, *Chem. Ind.* (1966) p. 1181.
381. M. A. A. Beg and S. H. Khawaja, *Spectrochim. Acta.*, 24 (1968) p. 1031.
382. D. A. Wheatland, C. H. Clapp and R. W. Waldron, *Inorg. Chem.*, 11 (1972) p. 2340.
383. J. S. Harman and C. A. McAuliffe. unpublished work 1972.
384. T. Saito, J. Otera and R. Okawara, *Bull. Chem. Soc. Japan*, 43 (1970) p. 1733.
385. C. J. Nyman and E. P. Parry, *Anal. Chem.*, 30 (1958) p. 1255.
386. I. Aucken and R. S. Drago, *Inorg. Synth.*, 6 (1960) p. 26.
387. N. S. Autonenko and Y. A. Nuger, *Zh. neorg. Khim.*, 11 (1964) p. 1072.
388. M. Nardelli, L. Cavalea and A. Braibanti, *Gazz. Chim. Ital.*, 86 (1956) p. 867.

389. M. Czakis-Sulikowska, *Rooz. Chem.*, 38 (1964) p. 1741.

390. R. Battistruzzi and G. Marcotrigiano, *Gazz. Chim. Ital.*, 104, (1974) p. 617.

391. M. Czakis-Sulikowska and R. Soloniewicz, *Rocz. Chem.*, 37 (1963) p. 1405.

392. G. O. Marcotrigiano, G. Peyronel and R. Battistuzzi, *Inorg. Chim. Acta* 9 (1974) p. 5.

393. G. Peyronel, G. O. Marcotrigiano and R. Battistuzzi, *J. inorg. nucl. Chem.*, 35 (1973) p. 1117.

394. P. D. Brotherton and A. H. White, *J. C. S. Dalton* (1973) p. 2698.

395. A. Rosenheim and V. J. Meyer, *Z. anorg. allg. Chem.*, 41 (1906) p. 13.

396. M. P. Rodriguez, M. Cubero, A. Lopez-Castro and C. Moreno-Bajo, *Nature*, 206 (1965) p. 392.

397. A. N. Sergeeva, L. A. Kiseleva and S. M. Galitskaya, *Zh. neorg. Khim.*, 16 (1971) p. 1782.

398. A. Korezynski, M. Nardelli and M. A. Pellinghelli, *Cryst. Struct. Commun.*, 1 (1972) p. 327; *Rocz. Chem.*, 47 (1973) p. 905.

399. P. D. Brotherton and A. H. White, *J. C. S. Dalton* (1973) p. 2696.

400. A. Korczynski, *Rocz. Chem.*, 42 (1968) p. 1207.

401. P. D. Brotherton, P. C. Healy, C. L. Raston and A. H. White, *J. C. S. Dalton* (1973) p. 334.

402. A. Korczynski, *Zeszytz. Nauk. Politechniki Codzkief*, 19 (1969) p. 85; *Chem. Abs.*, 72 (1970) p. 6926q.

403. A. Korczynski, *Rocz. Chem.*, 42 (1968) p. 392.

404. A. Korczynski, *Rocz. Chem.*, 37 (1963) p. 1647; 40 (1966) p. 547.

405. G. Marcotrigiano and R. Battistuzzi, *Inorg. nucl. chem. Letts.*, 8 (1972) p. 969.

406. M. Nardelli and I. Chierici, *Rec. Sci.*, 29 (1959) p. 1733.

407. M. Nardelli and I. Chierici, *Gazz. Chim. Ital.*, 88 (1958) p. 248.

408. P. C. Rath and B. K. Mohapatra, *J. Ind. chem. Soc.*, 51 (1974) p. 710.

409. M. Gencher and V.St. Krunster, *Natura*, 5 (1972) p. 49; *Chem. Abs.*, 80 (1970) p. 66258g.

410. B. Knobloch and R. Soloniewicz, *Rocz. Chem.*, 47 (1973) p. 1731.

411. S. K. Siddhanta and S. N. Banergee, *J. Ind. chem. Soc.*, 38 (1961) p. 675.

412. S. N. Banergee and A. C. Sukthanker, *J. Ind. chem. Soc.*, 40 (1963) p. 573.

413. P. K. Mandal, *Ind. J. Chem.*, 12 (1974) p. 845.

414. A. R. Hendrickson and R. C. Martin, *Aust. J. Chem.*, 25 (1972) p. 257.

415. G. C. Pallacani, A. C. Fabretti and G. Peyronel, *Inorg. nucl. Chem. Letts.*, 9 (1973) p. 897.

416. A. C. Fabretti, G. C. Pellacani and G. Peyronel, *Gazz. Chem. Ital.*, 103 (1973) p. 1259.

417. F. Kasparek and J. Mollin, *Coll. Czech. Chem. Comm.*, 25 (1960) p. 2919.

418. D. Negiou and V. Mureson, *Ann. Univ. Bucaresti Chim.*, 20 (1971) p. 121; *Chem. Abs.*, 78 (1973) p. 66380r.

419. R. P. Burns, J. Dwyer and C. A. McAuliffe, *Coord. Chem. Rev.*, to be submitted.

420. B. Nyberg and I. Cynkier, *Acta Chem. Scand.*, 26 (1972) p. 4175.

421. W. E. Slinkard and D. W. Meek, *Inorg. Chem.*, 8 (1969) p. 1811.

422. C. J. Nyman and J. Salazor, *Anal. Chem.*, 33 (1961) p. 1467.

423. D. Coucouvanis, *Prog. inorg. Chem.*, 11 (1970) p. 233.

424. Y. Watanabe and H. Hayihara, *Acta Cryst.* A28 Suppl. (1972) p. 589.

425. R. Keller, *Anal. Chim. Acta*, 68 (1974) p. 49.

426. J. M. Cheremisina and S. V. Larionov, *Izv. Akad. Nauk. S.S.S.R. Ser. Chim.* (1971) p. 2150.

427. H. Iwasaki, *Acta Cryst.*, A28 Supp. S (1972) p. 85; *Chem. Lett.* (1972) p. 1105; *Acta Cryst.*, B24 (1973) p. 2115.
428. P. C. Healy and A. H. White, *J. C. S. Dalton* (1973) p. 284.
429. H. C. Brinkhoff and J. M. A. Dautzeuberg, *Rec. Trav. Chim.*, 91 (1972) p. 117.
430. P. D. Brotherton, J. M. Epstein, A. H. White and A. C. Willis, *J. C. S. Dalton* (1974) p. 2341.
431. G. Bandoli, D. A. Clemente, L. Sindelleri and E. Tondello, *Cryst. Struct. Comm.*, 3 (1974) p. 289; *J. C. S. Dalton* (1975) p. 449.
432. W. Kuchen, J. Metten and A. Judat, *Chem. Ber.*, 97 (1964) p. 2306.
433. R. G. Cavell, E. D. Day, W. Byers and P. M. Watkins, *Inorg. Chem.*, 11 (1972) p. 1759.
434. F. N. Tebbe and E. L. Muetterties, *Inorg. Chem.*, 9 (1970) p. 629.
435. S. L. Lawton, *Inorg. Chem.*, 10 (1971) p. 328.
436. L. A. Il'ina, N. I. Zemlyanskii, S. V. Larionov and N. M. Chernaya, *Isz. Akad. Nauk S.S.S.R.*, 3 (1969) p. 198.
437. W. Kucher, M. Foeroter, H. Hertel and B. Hoehm, *Chem. Ber.*, 105 (1972) p. 3310.
438. J. E. Ferguson and K. S. Lok, *Aust. J. Chem.*, 26 (1973) p. 2615.
439. R. E. Noftle and J. Crews, *Inorg. Chem.*, 13 (1974) p. 3031.
440. F. Carr and T. G. Pearson, *J. chem. Soc.* (1938) p. 436.
441. K. Lederer, *Ber. Deut. Chem. Ges.*, 47 (1914) p. 277; 48 (1915) p. 1422; 49 (1916) p. 1071; 49 (1916) p. 334; 50 (1917) p. 238; 52B (1919) p. 1989; 53B (1920) p. 712.
442. E. E. Aynsley, N. N. Greenwood and J. B. Leech, *Chem. Ind.* (1966) p. 379.
443. P. J. Hendra and N. Sadaswan, *J. chem. Soc.* (1965) p. 2063.
444. G. T. Morgan and F. H. Burstall, *J. chem. Soc.* (1931) p. 180.
445. P. Nicpon and D. W. Meek, *Chem. Comm.* (1966) p. 398.
446. M. G. King and G. P. McQuillan, *J. chem. Soc. (A)* (1967) p. 898.
447. D. Glasser, L. Ingram, M. G. King and G. P. McQuillan, *J. chem. Soc. (A)* (1969) p. 2501.
448. G. B. Aitken, J. L. Duncan and G. P. McQuillan, *J. C. S. Dalton* (1972) p. 2103.
449. D. R. Goddard, B. D. Lodam, S. O. Ajayi and M. J. Campbell, *J. chem. Soc. (A)* (1969) p. 506.
450. H. J. Close and E. A. V. Ebsworth, *J. chem. Soc.* (1965) p. 940.
451. Gmelin, Lief 1, pp. 128–270.
452. D. Breitinger and K. Brodersen, *Angew Chem. Int. Ed.*, 9 (1970) p. 357.
453. J. A. Young, S. N. Tsoukales and R. D. Dresdner, *J. Am. chem. Soc.*, 80 (1958) p. 3604.
454. H. J. Emeléus and G. L. Hurst, *J. chem. Soc.* (1964) p. 396.
455. R. C. Dobbie and M. J. Emeléus, *J. chem. Soc. (A)* (1966) p. 933.
456. H. G. Ag and Y. C. Sin, *Adv. inorg. Chem. Radiochem.*, 16 (1974) p. 1.
457. C. W. Tullock, D. D. Coffman and E. L. Muetterties, *J. Am. chem. Soc.*, 86 (1964) p. 357.
458. H. Burger, W. Sawodny and U. Wannagat, *J. organometallic Chem.*, 3 (1965) p. 113.
459. W. Schoeller and W. Schrauth, *Chem. Ber.*, 42 (1909) p. 786.
460. B. Kamenar and D. Grdenic, *Inorg. Chim. Acta*, 3 (1969) p. 25.
461. D. B. Brown and M. B. Robin, *Inorg. Chim. Acta*, 3 (1969) p. 644.
462. K. Brodersen and L. Kunkel, *Z. anorg. allg. Chem.*, 298 (1959) p. 34.

463. J. A. Young, W. S. Durrell and R. D. Dresdner, *J. Am. chem. Soc.*, 84 (1962) p. 2105.
464. K. A. Hoffmann, E. Biesalski and E. Söderlund, *Ber. Deut. Chem. Ges.*, 45 (1912) p. 1731.
465. B. Picaud and M. Capestan, *Bull. Soc. Chim. France* (1966) p. 3984.
466. P. Picaud and M. Capestan, *C. R. Acad. Sci.*, C264 (1967) p. 1118.
467. E. Divers and T. Haga, *J. chem. Soc.*, 61 (1892) p. 943; 69 (1896) p. 1620.
468. M. Bolte and M. Capestan, *Bull. Soc. Chim. France* (1966) p. 3981.
469. K. Brodersen, L. Stumpp and G. Krauss, *Chem. Ber.*, 93 (1960) p. 375.
470. G. Heinze and A. Meuwsen, *Z. anorg. allg. Chem.*, 275 (1954) p. 49.
471. A. Meuwsen and F. Schlossnagel, *Z. anorg. allg. Chem.*, 271 (1953) p. 226.
472. A. Meuwsen and M. Lösel, *Z. anorg. allg. Chem.*, 271 (1953) p. 217.
473. O. Glemser, R. Mews and H. W. Roesky, *Chem. Ber.*, 102 (1969) p. 1523.
474. B. Krebs, E. Meyer-Hussein, O. Glemser and R. Mews, *Chem. Comm.* (1968) p. 1578.
475. W. Verbeek and W. Sundermeyer, *Angew. Chem.*, 81 (1969) p. 330.
476. D. Strömholm, *Z. anorg. allg. Chem.*, 57 (1908) p. 72.
477. E. C. Franklin, *J. Am. chem. Soc.*, 29 (1907) p. 35.
478. W. Biltz and C. Mau, *Z. Anorg. allg. Chem.*, 148 (1925) p. 170.
479. G. Jander and K. Brodersen, *Z. anorg. allg. Chem.*, 265 (1951) p. 117.
480. C. H. MacGillavry and J. M. Bijvoet, *Z. Krist.*, 94 (1936) p. 231.
481. W. N. Lipscomb, *Anal. Chem.*, 25 (1953) p. 737.
482. W. Rudorf and K. Brodersen, *Z. anorg. allg. Chem.*, 270 (1952) p. 145.
483. K. R. Manolov, *Zh. Neorg. Khim.*, 9 (1964) p. 207.
484. W. Peters, *Z. anorg. allg. Chem.*, 77 (1912) p. 137.
485. W. N. Lipscomb, *Acta Cryst.*, 4 (1951) p. 266.
486. W. N. Lipscomb, *Acta Cryst.*, 5 (1952) p. 604.
487. K. Brodersen and W. Rudorf, *Z. anorg. allg. Chem.*, 275 (1954) p. 141.
488. A. Meuwsen and G. Weiss, *Z. anorg. allg. Chem.*, 289 (1957) p. 5.
489. R. Widman, *Z. anorg. allg. Chem.*, 68 (1910) p. 1.
490. K. Brodersen and W. Rüdorff, *Z. Nat.*, 98 (1955) p. 164.
491. K. Brodersen, *Acta Cryst.*, 8 (1955) p. 723.
492. S. D. Arora, W. N. Lipscomb and M. C. Sneed, *J. Am. chem. Soc.*, 73 (1951) p. 1015.
493. E. Hayek and P. Inama, *Monatsch*, 96 (1965) p. 1454.
494. K. Brodersen and W. Rudorf, *Angew. Chem.*, 64 (1952) p. 617.
495. W. Rudorf and K. Brodersen, *Z. anorg. allg. Chem.*, 274 (1953) p. 323.
496. L. Nijssen and W. N. Lipscomb, *Acta Cryst.*, 7 (1954) p. 103.
497. W. Strecker and E. Schwinn, *J. prakt. Chem.*, 156 (1939) p. 205.
498. K. A. Hoffman and E. C. Marburg, *Z. anorg. allg. Chem.*, 23 (1900) p. 126.
499. K. Brodersen and W. Rudorf, *Z. anorg. allg. Chem.*, 287 (1956) p. 24.
500. M. Francois, *Bull. Soc. Chim. France* 47 (1930) p. 825.
501. E. C. Franklin, *J. Am. chem. Soc.*, 27 (1905) p. 820.
502. R. Airoldi, *Ann. Chim. Rome*, 48 (1958) p. 491.
503. M. T. Fournier and M. Capestan, *Bull. Soc. Chim. France* (1972) p. 577.
504. W. Cuisa and G. Adamo, *Ann. Chim. Rome*, 43 (1953) p. 827.
505. A. Meuwsen and R. Dötzer, *Angew. Chem.*, 67 (1955) p. 616.
506. R. Weber, *Z. Natusforsch*, 9B (1954) p. 612; *Liebigs Ann. Chem.*, 158 (1958) p. 161.
507. R. Weber, *Naturwissenschaften*, 44 (1957) p. 465; 43 (1956) p. 107.
508. R. Weber, *Z. anorg. allg. Chem.*, 338 (1965) p. 100.

509. C. M. Deeley and R. E. Richards, *J. chem. Soc.* (1954) p. 3697.
510. K. Brodersen and H. J. Becher, *Chem. Ber.*, 89 (1956) p. 1487.
511. M. G. Miles, J. H. Patterson, C. W. Hobbs, M. J. Hopper, J. Overend and R. S. Tobias, *Inorg. Chem.*, 7 (1968) p. 1721.
512. S. Mizushima, I. Nakagawa and D. M. Sweeney, *J. chem. Phys.*, 25 (1956) p. 1006.
513. I. Nakagawa and T. Shimanouchi, *Spectrochim. Acta.*, 22 (1966) p. 759.
514. K. Brodersen, *Z. anorg. allg. Chem.*, 290 (1957) p. 24.
515. K. Brodersen, *Z. anorg. allg. Chem.*, 285 (1956) p. 5.
516. M. Raffo and A. Scarella, *Gazz. Chim. Ital.*, 45 (1915) p. 123.
517. W. Thiel, F. Weller, J. Lorbeth and K. Dehnicke, *Z. anorg. Allg. Chem.*, 381 (1971) p. 57.
518. K. Brodersen, *Chem. Ber.*, 90 (1957) p. 2703.
519. D. Breitinger and N. Q. Dao, *J. organometallic Chem.*, 15 (1968) p. 21.
520. N. Q. Dao and D. Breitinger, *Spectrochim. Acta*, A27 (1971) 905.
521. L. Nijssen and W. N. Lipscomb, *J. Am. chem. Soc.*, 74 (1952) p. 2113.
522. W. Biltz and E. Rahlfs, *Z. anorg. allg. Chem.*, 366 (1927) p. 351.
523. W. Biltz, K. A. Klatte and E. Rahlfs, *Z. anorg. allg. Chem.*, 166 (1972) p. 339.
524. E. Weitz, K. Blasberg and E. Wernicke, *Z. anorg. allg. Chem.*, 188 (1930) p. 344.
525. J. Breckenridge, W. Warzecha and T. Surles, *Inorg. nucl. Chem. Lett.*, 9 (1973) p. 437.
526. H. Franzen and H. L. Lucking, *Z. anorg. allg. Chem.*, 70 (1911) p. 145.
527. K. A. Hofmann and E. C. Marburg, *Ann. Chem.*, 305 (1899) p. 191.
528. M. Francois, *Compt. Rend.*, 142 (1906) p. 1199.
529. J. Dwyer, W. Levason and C. A. McAuliffe, *J. inorg. nucl. Chem.*, 38 (1976) 1919.
530. M. Straumanis and A. Cirulis, *Z. anorg. allg. Chem.*, 230 (1937) p. 65.
531. K. N. Mikhalerich and S. M. Galitskaya, *Ukr. Khim. Zh.*, 38 (1972) p. 713; *Chem. Abs.*, 77 (1972) p. 120411w.
532. C. H. Misra, S. S. Paramar and S. N. Shukla, *J. inorg. nucl. Chem.*, 28 (1966) p. 147.
533. M. S. Barvinok, I. S. Bukhareva and Y. S. Varshavskii, *Zh. neorg. Khim.*, 10 (1965) p. 2293.
534. H. Grossmann and F. Hünseler, *Z. anorg. allg. Chem.*, 46 (1905) p. 361.
535. I. S. Ahuja and P. Rastogi, *J. inorg. nucl. Chem.*, 32 (1970) p. 2085.
536. I. S. Maslennikova, *Zh. fiz. Khim.*, 46 (1972) p. 168.
537. C. H. Misra, S. S. Parmar and S. N. Shukla, *Can. J. Chem.*, 45 (1967) p. 2459.
538. C. H. Misra, S. S. Parmar and S. N. Shukla, *J. inorg. nucl. Chem.*, 29 (1967) p. 2584.
539. I. S. Ahuja and A. Garg, *J. inorg. nucl. Chem.*, 34 (1972) p. 1924.
540. M. S. Barvinok and I. S. Bukhareva, *Zh. fiz. Khim.*, 39 (1965) p. 1006.
541. I. S. Ahuja and R. Singh, *J. inorg. nucl. Chem.*, 35 (1973) p. 302.
542. K. Krishnan and R. A. Plane, *Inorg. Chem.*, 5 (1966) p. 852.
543. P. Pfeiffer, E. Schmitz and A. Böehm, *Z. anorg. allg. Chem.*, 270 (1952) p. 287.
544. J. Peacock, F. C. Schmidt, R. E. Davies and W. B. Schaap, *J. Am. chem. Soc.*, 77 (1955) p. 5829.
545. T. D. O'Brien, *J. Am. chem. Soc.*, 70 (1948) p. 2771.
546. I. S. Ahuja and R. Singh, *Inorg. nucl. chem. Lett.*, 9 (1973) p. 289.
547. K. Brodersen, *Z. anorg. allg. Chem.*, 298 (1959) p. 142.
548. T. Iwamoto and D. F. Schriver, *Inorg. Chem.*, 10 (1971) p. 2428.
549. G. Newman and D. B. Powell, *J. chem. Soc.* (1961) p. 477.

550. G. J. Sutton, *Aust. J. Chem.*, 12 (1959) p. 637.
551. C. G. Macarovici and A. Donatui, *Studia Univ. Babes-Bolyai* Ser. 1, (1960) p. 97; *Chem. Abs.*, 58 (1963) p. 7590f.
552. M. S. Barvinok and I. S. Bukhareva, *Zh. neorg. Khim.*, 10 (1965) p. 861.
553. I. S. Ahuja and R. Singh, *Inorg. nucl. chem. Lett.*, 10 (1974) p. 421.
554. L. Sindellari, *Ann. Chim. (Rome)*, 56 (1966) p. 386; *Chem. Abs.*, 65 (1966) p. 1746b.
555. R. Graziani, G. Bombieri and E. Forsellini, *Ric. Sci.*, 36 (1966) p. 855; *Chem. Abs.*, 66 (1967) p. 50012p.
556. K. G. Caulton, *Inorg. nucl. chem. Letts*, 9 (1973) p. 533.
557. N. A. Bell, *Thermochim. Acta*, 6 (1973) p. 275.
558. I. S. Ahuja and R. S. Singh, *J. inorg. nucl. Chem.*, 36 (1974) p. 1505.
559. G. Cova, D. Galizzioli, D. Guisto and F. Morazzoni, *Inorg. Chim. Acta*, 6 (1972) p. 343.
560. A. Christini and G. Ponticelli, *J. C. S. Dalton* (1972) p. 2603.
561. M. Ciampolini, A. Christini, A. Diaz and G. Ponticelli, *Inorg. Chim. Acta*, 7 (1973) p. 549.
562. P. Barz and H. P. Fritz, *Z. Naturforsch.*, B27 (1972) p. 1131.
563. W. E. Bull, J. A. Seaton and L. F. Audrieth, *J. Am. chem. Soc.*, 80 (1958) p. 2516.
564. I. S. Ahuja and P. Rastogi, *J. chem. Soc. (A)* (1953) p. 1493.
565. R. S. McBride, *J. phys. Chem.*, 14 (1909) p. 189.
566. D. R. Glasson and S. J. Gregg, *J. chem. Soc.* (1953) p. 1493.
567. S. Prasad and P. O. Sharma, *J. Ind. Chem. Soc.*, 35 (1958) p. 565.
568. R. J. H. Clark and C. S. Williams, *Chem. Ind.* (1964) p. 1317; *Inorg. Chem.*, 4 (1965) p. 350.
569. D. Grdenic and I. Krstanovic, *Arkiv Kemi*, 27 (1955) p. 143; E. R. Allen, J. Cartlidge, M. M. Taylor and C. F. M. Tipper, *J. phys. Chem.*, 63 (1959) p. 1442.
570. L I. Chudinova, *Zh. neorg. Khim.*, 14 (1969) p. 2974; *Zh. Prickl. Khim.*, 42 (1969) p. 189; *Chem. Abs.*, 70 (1969) p. 83746s.
571. L. I. Chudinova and R. M. Klykova, *Zh. neorg. Khim.*, 18 (1973) p. 47.
572. R. H. Wiley, J. C. Hartman and E. L. de Young, *J. Am. chem. Soc.*, 74 (1952) p. 3452.
573. O. G. Strode and J. E. House, *Thermochim. Acta*, 3 (1972) p. 461.
574. I. S. Ahuja and A. Garg, *J. inorg. nucl. Chem.*, 34 (1972) p. 2681.
575. I. S. Ahuja and A. Garg, *J. inorg. nucl. Chem.*, 34 (1972) p. 1929.
576. D. Cook, *Can. J. Chem.*, 42 (1964) p. 2523.
577. S. S. Chauhan and P. C. Sinha, *J. Ind. Chem. Soc.*, 40 (1963) p. 769.
578. S. Kulpe, *Z. Chem.*, 5 (1965) p. 306; *Z. anorg. allg. Chem.*, 349 (1967) p. 314.
579. M. K. Alyariva, R. S. Ryspaeva and A. L. Kats, *Zh. neorg. Khim.*, 17 (1972) p. 957.
580. M. K. Alyariva, R. S. Ryspaeva and A. L. Kats, *Zh. neorg. Khim.*, 17 (1972) p. 1916.
581. I. S. Ahuja and A. Garg, *Inorg. Chim. Acta*, 6 (1972) p. 453.
582. T. L. Chang and W. C. Yu, *Z. anorg. allg. Chem.*, 243 (1939) p. 14.
583. I. S. Ahuja and K. S. Rao, *J. inorg. nucl. Chem.*, 37 (1975) p. 586.
584. J. R. Ferraro, W. Wozriak and G. Roch, *Ric. Sci.*, 38 (1968) p. 433.
585. J. R. Ferraro, W. Wozriak and G. Roch, *Chem. Abs.*, 70 (1969) p. 52631w.
586. H. D. Stidham and J. A. Chandler, *J. inorg. nucl. Chem.*, 27 (1965) p. 397.
587. T. B. Jackson and J. O. Edwards, *J. Am. chem. Soc.*, 83 (1961) p. 355.

588. S. F. Babak and I. A. Kondrashov, *Zh. neorg. Khim.*, 10 (1965) p. 1642.
589. P. J. Hendra and D. B. Powell, *J. chem. Soc.* (1960) p. 5105.
590. J. J. Grendberg, A. N. Kost and N. N. Zheltikova, *Zh. Obsch. Khim.*, 30 (1960) p. 2931.
591. G. J. Sutton, *Aust. J. Chem.*, 12 (1959) p. 637.
592. S. C. Jain and R. Rivest, *Inorg. Chim. Acta*, 4 (1970) p. 291.
593. D. C. Craig, Y. Farhangi, D. P. Graddon and N. C. Stephenson, *Cryst. Struct. Comms.*, 3 (1974) p. 155; *Chem. Abs.*, 80 (1974) p. 101148m.
594. P. P. Feiffer, E. Schmitz, F. Dominick, A. Fritzen and B. Werdelmann, *Z. anorg. allg. Chem.*, 264 (1951) p. 188.
595. A. A. Schilt and R. C. Taylor, *J. inorg. nucl. Chem.*, 9 (1959) p. 211.
596. A. L. Beauchamp, B. Saperas and R. Rivest, *Can. J. Chem.*, 49 (1971) p. 3579; 52 (1974) p. 2923.
597. I. S. Ahuja and R. Singh, *Ind. J. Chem.*, 12 (1974) p. 107.
598. J. E. Douglas and C. J. Wilkins, *Inorg. Chim. Acta*, 3 (1969) p. 635.
599. I. A. Kondrashov and S. F. Babak, *Zh. Obsch. Khim.*, 28 (1958) p. 1105.
600. J. M. Epstein, J. C. Dewan, D. L. Kepert and A. H. White, *J. C. S. Dalton* (1974) p. 1949.
601. D. P. Madden, H. M. da Mota and S. M. Nelson, *J. chem. Soc. (A)* (1970) p. 790.
602. L. A. Kazitsyna, N. B. Kupletskaya, V. A. Ptitsyna, M. N. Bockhareva and O. A. Rentov, *Zh. Organ. Khim.*, 2 (1966) p. 565.
603. C. M. Harris, B. F. Hoskins and R. L. Martin, *J. chem. Soc.* (1959) p. 3728.
604. C. M. Knowles and G. W. Watt, *J. Am. chem. Soc.*, 64 (1942) p. 935.
605. A. Mangini and I. Defudicibus, *Gazz. Chim. Ital.*, 63 (1933) p. 601.
606. A. N. Nesmayanov and T. P. Tolstafa, *Dokl. Akad. Nank. SSSR*, 128 (1959) p. 726.
607. A. N. Nesmayanov, *Z. anorg. allg. Chem.*, 178 (1929) p. 300.
608. A. F. Gremillion, H. B. Jonassen and R. J. O'Connor, *J. Am. chem. Soc.*, 81 (1959) p. 6134.
609. H. Leligny, M. Frey and J. C. Monier, *Acta Cryst.*, B28 (1972) p. 2104.
610. G. Bähr and H. Kretzger, *Z. anorg. allg. Chem.*, 267 (1951) p. 161.
611. G. Bähr and H. Döge, *Z. anorg. allg. Chem.*, 292 (1957) p. 119.
612. G. Bähr and H. Trämlitz, *Z. anorg. allg. Chem.*, 282 (1955) p. 3.
613. L. A. Kazitsynu, A. A. Nilson, N. B. Kupletskaya and O. A. Rentov, *Vestn. Mosk. Univ. Ser. II Khim.*, 21 (1966) p. 95; *Chem. Abs.*, 65 (1966) p. 11742e.
614. J. R. Miller and G. D. Dorough, *J. Am. chem. Soc.*, 74 (1952) p. 3977.
615. M. F. Hudson and K. M. Smith, *J. C. S. Chem. Comm.* (1973) p. 515.
616. J. Grobe and R. Demuth, *Angew. Chem. Int. Ed.*, 11 (1972) p. 1097.
617. M. Baudler and A. Zarkadas, *Chem. Ber.*, 105 (1972) p. 3844.
618. D. V. Sokols'ki, Y. A. Dorfman, I. A. Kazantsera and G. S. Uteyenova, *Zh. fiz. Khim.*, 44 (1970) p. 2263.
619. P. L. Goggin, R. J. Goodfellow, S. R. Haddock and J. G. Eary, *J. C. S. Dalton* (1972) p. 647.
620. H. Schmidbauer and K. H. Raethlein, *Chem. Ber.*, 106 (1973) p. 2491.
621. G. E. Coates and A. Lauder, *J. chem. Soc.* (1965) p. 1857.
622. R. C. Cass, G. E. Coates and R. G. Hayter, *J. chem. Soc.* (1955) p. 4007.
623. R. C. Evans, F. G. Mann, H. S. Peiser and D. Purdie, *J. chem. Soc.* (1940) p. 1207.
624. S. O. Grim, P. J. Lui and R. L. Keiter, *Inorg. Chem.*, 13 (1974) p. 342.
625. R. L. Keiter and S. O. Grim, *Chem. Comm.* (1968) p. 521.
626. A. Yamasaki and E. Fluck, *Z. anorg. allg. Chem.*, 396 (1973) p. 297.

627. F. G. Moers and J. P. Langhout, *Rec. Trav. Chim.*, 92 (1973) p. 996.
628. W. J. Jones, W. C. Davies, S. T. Bowden, C. Edwards, V. E. Davies and L. H. Thomas, *J. chem. Soc.* (1947) p. 1446.
629. G. B. Deacon and B. O. West, *J. inorg. nucl. Chem.*, 24 (1962) p. 169.
630. A. T. T. Hsieh, J. D. Ruddick and G. Wilkinson, *J. C. S. Dalton* (1972) p. 1966.
631. F. Challenger, C. Higginbottom and L. Ellis, *J. chem. Soc.* (1933) p. 95.
632. M. M. Baig, W. R. Cullen and D. S. Dawson, *Can. J. Chem.*, 40 (1962) p. 46.
633. W. J. C. Dyke, G. Davies and W. J. Jones, *J. chem. Soc.* (1931) p. 185.
634. W. J. Jones, W. J. C. Dyke, G. Davies, D. C. Griffiths and J. H. E. Webb, *J. chem. Soc.* (1932) p. 2284.
635. F. Challenger and L. Ellis, *J. chem. Soc.* (1935) p. 396.
636. K. Brodersen, R. Palmer and D. Breitinger, *Chem. Ber.*, 104 (1971) p. 360.
637. C. A. McAuliffe, I. Niven and R. V. Parish, unpublished work.
638. G. B. Deacon and J. H. S. Green, *Chem. Ind.* (1965) p. 1031. *Spectrochim. Acta*, 24A (1968) p. 845.
639. G. B. Deacon, J. H. S. Green and D. J. Harrison, *Spectrochim. Acta*, 24A (1968) p. 1921.
640. G. B. Deacon and J. H. S. Green, *Spectrochim. Acta*, 25A (1969) p. 355.
641. G. Deganello, G. Dolcetti, M. Guistiniani and U. Belluco, *J. C. S. (A)* (1969) p. 2138.
642. A. R. Davis, C. J. Murphy and R. A. Plane, *Inorg. Chem.*, 9 (1970) p. 423.
643. R. C. Makhija, A. L. Beauchamp and R. Rivest, *J. C. S. Dalton* (1973) p. 2447; R. C. Makhija, A. L. Beauchamp and R. Rivest, *J. C. S. Chem. Comm.* (1972) p. 1043.
644. S. C. Jain, *J. inorg. nucl. Chem.*, 35 (1973) p. 413.
645. S. H. Whitlow, *Can. J. Chem.*, 52 (1974) p. 198.
646. H. W. Kouwenhoven, J. Lewis and R. S. Nyholm, *Proc. Chem. Soc.* (1961) p. 220; M. A. Bennett, H. W. Kouwenhoven, J. Lewis and R. S. Nyholm, *J. chem. Soc.* (1964) p. 4570.
647. F. G. Mann and D. Purdie, *J. chem. Soc.* (1940) p. 1230.
648. R. W. Baker, M. J. Braithwaite and R. S. Nyholm, *J. C. S. Dalton* (1972) p. 1924.
649. P. R. Brookes and B. L. Shaw, *J. C. S. Dalton* (1973) p. 783.
650. P. R. Brookes and B. L. Shaw, *J. C. S. Dalton* (1974) p. 1702.
651. M. J. Braithwaite and R. S. Nyholm, *J. inorg. nucl. Chem.*, 35 (1973) p. 2237.
652. K. Isslieb, H. P. Abicht and H. Winkelmann, *Z. anorg. allg. Chem.*, 388 (1972) p. 89.
653. C. E. Wymore and J. C. Bailar, *J. inorg. nucl. Chem.*, 14 (1960) p. 42.
654. A. J. Carty and A. Efraty, *Chem. Comm.* (1968) p. 1559.
655. W. A. Anderson, A. J. Carty and A. Efraty, *Can. J. Chem.*, 47 (1969) p. 3361.
656. S. S. Sandhu, S. S. Sandhu and M. P. Gupta, *Z. anorg. allg. Chem.*, 377 (1970) p. 348.
657. D. P. Strommen, *J. inorg. nucl. Chem.*, 37 (1975) p. 487.
658. K. R. Mann, W. H. Morrison Jr. and D. N. Hendrickson, *Inorg. Chem.*, 13 (1974) p. 1180.
659. K. Isslieb, U. Giesder and H. Hartung, *Z. anorg. allg. Chem.*, 390 (1972) p. 239.
660. K. Isslieb and U. Giesder, *Z. anorg. allg. Chem.*, 379 (1970) p. 9.
661. J. Lewis, R. S. Nyholm and D. J. Phillips, *J. chem. Soc.* (1962) p. 2177.
662. G. B. Deacon and J. H. S. Green, *Spectrochim. Acta*, A24 (1968) p. 959.

663. W. R. Cullen, P. S. Dhaliwal and C. J. Stewart, *Inorg. Chem.*, 6 (1967) p. 2256.
664. W. Levason and C. A. McAuliffe, *J. Coord. Chem.*, 4 (1974) p. 47.
665. G. Dyer, D. C. Goodall, R. H. B. Mais, H. M. Powell and L. M. Venanzi, *J. C. S. (A)* (1966) p. 1110.
666. H. Nöth and H. Vetter, *Chem. Ber.*, 96 (1963) p. 1479.
667. G. Ewart, D. S. Payne, A. L. Porte and A. P. Lane, *J. chem . Soc.* (1962) p. 3984.
668. A. B. Burg and P. F. Slota, *J. Am. chem. Soc.*, 80 (1958) p. 1107.
669. W. Seidal, *Z. Chem.*, 3 (1963) p. 429.
670. W. Seidal, *Z. anorg. allg. Chem.*, 335 (1965) p. 316.
671. W. Seidal, *Z. anorg. allg. Chem.*, 341 (1965) p. 70.
672. W. Seidal, *Z. anorg. allg. Chem.*, 342 (1966) p. 165.
673. E. A. Arbuzov and V. M. Zoroastrova, *Izvest. Akad. Nauk. SSR.* (1952) p. 826, *Chem. Abs.*, 47 (1953) p. 9900a.
674. S. I. Shupack and B. Wagner, *Chem. Comm.* (1966) p. 547.
675. F. K. Butcher, B. E. Deuters, W. Gerrard, E. F. Mooney, R. A. Rothenbury and H. A. Willis, *Spectrochim. Acta*, 20 (1964) p. 759.
676. L. M. Bauer and M. H. B. Stiddard, *J. organometallic Chem.*, 13 (1968) p. 235.
677. R. B. Fox and D. L. Venetzky, *J. Am. chem. Soc.*, 75 (1953) p. 3967.
678. S. F. Spangeberg and H. H. Sisler, *Inorg. Chem.*, 8 (1969) p. 1004.
679. J. Bennett, A. Pidcock, C. R. Waterhouse, P. Coggon and A. T. McPhail, *J. C. S. (A)* (1970) p. 2094.
680. G. G. Mather and A. Pidcock, *J. C. S. Dalton* (1973) p. 560.
681. N. N. Greenwood and N. F. Travers, *Chem. Comm.* (1967) p. 216.
682. N. N. Greenwood and N. F. Travers, *J. chem. Soc. (A)* (1971) p. 3257.
683. N. N. Greenwood and D. N. Sharrocks, *J. chem. Soc. (A)* (1969) p. 2334.
684. K. A. Hooton, *Prep. inorg. Reacts.*, 4 (1968) p. 85.
685. F. Glockling, *The Chemistry of Germanium* (Academic Press, New York, 1968), p. 171.
686. A. F. Clemmit and F. Glockling, *J. C. S. (A)* (1971) p. 1164.
687. K. A. Hooton, *J. C. S. (A)* (1971) p. 1251.
688. D. Seyferth, Annual Survey of Organometallic Chemistry. Mercury. *J. organometallic Chem.*, 41 (1972) p. 248; 62 (1973) p. 33; 75 (1974) p. 99.
689. U. Blaukat and W. P. Neumann, *J. organometallic Chem.*, 63 (1973) p. 27.
690. O. A. Kruglaya, G. S. Kalinina, B. I. Petrov and N. S. Vyazankin, *J. organometallic Chem.*, 46 (1972) p. 51.
691. G. Deganello, G. Carturan and P. Uguagliati, *J. organometallic Chem.*, 18 (1964), p. 216.
692. V. V. Udovenko and V. I. Zotov, *Ukr. Khim. Zh.*, 32 (1966) p. 8; *Chem. Abs.*, 64 (1966) p. 19394e.
693. V. V. Udovenko and V. I. Zotov, *Ukr. Khim. Zh.*, 38 (1972) p. 1214, *Chem. Abs.*, 78 (1972) p. 63039u.
694. H. M. Haendler and G. McP. Smith, *J. Am. chem. Soc.*, 63 (1941) p. 1164.
695. D. Venkappayya and G. Aravamudan, *Curr. Sci.*, 37 (1968) p. 12, *Chem. Abs.*, 68 (1968) p. 73752x.
696. I. S. Ahuja and P. Rastogi, *Inorg. nucl. chem. Letts.*, 5 (1969) p. 255.
697. E. Asmus and K. Ohls, *Z. Anal. Chem.*, 177 (1960) p. 100.
698. R. A. Majeste and L. M. Trefonas, *Inorg. Chem.*, 11 (1972) p. 1834.
699. J. S. Shukla and P. Bhatia, *J. inorg. nucl. Chem.*, 36 (1974) p. 1422.

700. H. Gehlen and H. Waeschke, *J. prakt. Chem.*, 312 (1970) p. 408.
701. M. K. Koul and K. P. Dubey, *J. inorg. nucl. Chem.*, 35 (1973) p. 2567.
702. O. Prochazkova, J. Podlakova and J. Podlaha, *Coll. Czech. Chem. Comm.*, 38 (1973) p. 1128.
703. D. R. Dakternieks and D. P. Graddon, *Aust. J. Chem.*, 27 (1974) p. 1351.
704. R. S. McEwan and G. A. Sim, *J. C. S. (A)* (1967) p. 1552.
705. L. Sindellari and G. Deganello, *Ric. Sci.*, 35 (1965) p. 744. *Chem. Abs.*, 64 (1965) p. 19669c.
706. S. S. Sandhu and S. S. Parmar, *Z. anorg. allg. Chem.*, 373 (1970) p. 64.
707. L. F. Lindoy, S. E. Livingstone and T. N. Lockyer, *Aust. J. Chem.*, 19 (1966) p. 1391.
708. V. K. Akimov, A. I. Buser, B. E. Zaitsev and S. I. Bragina, *Zh. Obsch. Khim.*, 40 (1970) p. 1331.
709. V. I. Suprunovich, Z. B. Kulikovskaya and Y. I. Usatenko, *Ukr. Khim. Zh.*, 38 (1972) p. 122; *Chem. Abs.*, 76 (1972) p. 135307v.
710. Y. I. Usatenko, V. I. Suprunovich and Z. B. Kulinkovskaya, *Zh. neorg. Khim.*, 16 (1971) p. 3200.
711. A. Ozola, J. Ozols and A. Levins, *Latv. PSR. Zinat. Akad. Vestis. Kim. Ser.*, 1 (1973) p. 8; *Chem. Abs.*, 78 (1973) p. 165579u.
712. D. Negoui and N. Mureson, *An. Univ. Bucuresti. Chim.*, 20 (1971) p. 133; *Chem. Abs.*, 78 (1973) p. 131536a.
713. G. Sutton, *Aust. J. Chem.*, 16 (1963) p. 1137.
714. F. Lions and K. V. Martin, *J. Am. chem. Soc.*, 80 (1958) p. 1591.
715. G. Bahr and E. Scholz, *Z. anorg. allg. Chem.*, 299 (1959) p. 281.
716. D. Negoui and V. Mureson, *An Univ. Bucuresti. Chim.*, 20 (1971) p. 121; *Chem. Abs.*, 78 (1973) p. 66308r.
717. D. Negoui, V. Mureson and C. Fulea, *An. Univ. Bucuresti. Chim.*, 22 (1973) p. 39; *Chem. Abs.*, 81 (1974) p. 98799n.
718. P. S. K. Chia, S. E. Livingstone and T. N. Lockyer, *Aust. J. Chem.*, 19 (1966) p. 1835.
719. B. P. Kennedy and A. B. P. Lever, *Can. J. Chem.*, 50 (1972) p. 3488.
720. D. S. Makadevappa and A. S. A. Murthy, *Aust. J. Chem.*, 25 (1972) p. 1565.
721. M. Nardelli and I. Chierici, *Ric. Sci.*, 30 (1960) p. 276.
722. G. R. Evans, *Inorg. Chem.*, 7 (1968) p. 277.
723. H. Irving, G. Andrew and E. J. Risdon, *J. chem. Soc.* (1949) p. 541.
724. M. M. Harding, *J. chem. Soc.* (1958) p. 4136.
725. W. J. Kozarek and Q. Fernando, *Inorg. Chem.*, 12 (1973) p. 2129.
726. S. N. Poddar and P. Ray, *J. Ind. chem. Soc.*, 29 (1952) p. 279.
727. I. P. Khullar and U. Agarwala, *Ind. J. Chem.*, 12 (1974) p. 1096.
728. B. S. Lakshmi and U. Agarwala, *Inorg. Chem.*, 8 (1969) p. 2341.
729. D. de Filippo, A. Lai, E. F. Trogu and G. Valenti, *J. inorg. nucl. Chem.*, 36 (1974) p. 73.
730. D. de Filippo, F. Devillanova, C. Prei and G. Verani, *J. chem. Soc. (A)* (1971) p. 1465.
731. G. Columbini and C. Prei, *J. inorg. nucl. Chem.*, 37 (1952) p. 1159.
732. W. V. Dahlhoff, T. R. Dick, G. H. Ford, W. S. J. Kelly and S. M. Nelson, *J. chem. Soc. (A)* (1971) p. 3495.
733. L. K. Peterson and E. W. Ainscough, *Inorg. Chem.*, 9 (1970) p. 2699.
734. L. W. Houk and G. R. Dobson, *J. chem. Soc. (A)* (1968) p. 1846.
735. C. A. McAuliffe and S. G. Murray, *Inorg. Chim. Acta Rev.*, 6 (1972) p. 103.
736. K. Brenzinger, *Z. physiol. Chem.*, 16 (1892) p. 552.
737. J. C. Andrews and P. D. Wyman, *J. biol. Chem.*, 87 (1930) p. 427.

738. K. Shinohara, *J. biol. Chem.*, 111 (1935) p. 435.
739. P. W. Preisler and D. W. Preisler, *J. biol. Chem.* , 95 (1932) p. 181.
740. H. Shindo and T. L. Brown, *J. Am. chem. Soc.*, 87 (1965) p. 1904.
741. G. A. Neville and T. Drakenberg, *Can. J. Chem.*, 52 (1974) p. 616.
742. G. R. Lenz and A. E. Martell, *Biochem.*, 3 (1964) p. 745.
743. R. W. Hay and L. J. Porter, *Aust. J. Chem.*, 20 (1967) p. 675.
744. L. J. Porter, D. D. Perrin and R. W. Hay, *J. chem. Soc. (A)* (1969) p. 118.
745. G. A. Neville and M. Berlin, *Can. J. Chem.* , 51 (1973) p. 3970.
746. R. C. Kapoor, G. Doughty and G. Gorin, *Biochem. Biophys. Acta*, 100 (1965) p. 376.
747. W. Stricks and I. M. Kolthoff, *J. Am. chem. Soc.*, 75 (1953) p. 5673.
748. C. A. McAuliffe, J. V. Quagliano and L. M. Vallarino, *Inorg. Chem.*, 5 (1966) p. 1996.
749. S. E. Livingstone and J. D. Nolan, *Inorg. Chem.*, 7 (1968) p. 1447.
750. B. Bergersson, T. Drakenberg and G. A. Neville, *Acta Chem. Scand.*, 27 (1973) p. 3953.
751. D. F. S. Natusch and L. J. Porter, *Chem. Comm.* (1970) p. 596.
752. N. N. Ghosh and A. Bhattacharyya, *J. Ind. chem. Soc.*, 48 (1971) p. 889.
753. N. N. Ghosh and A. Bhattacharyya, *J. Ind. chem. Soc.*, 50 (1973) p. 796.
754. N. N. Ghosh and A. Bhattacharyya, *J. Ind. chem. Soc.*, 49 (1972) p. 1053.
755. M. J. Adams, D. C. Hodgkin and U. A. Raeburn, *J. chem. Soc. (A)* (1970) p. 2632.
756. M. C. Baird, *Prog. inorg. Chem.*, 9 (1968) p. 1.
757. A. T. T. Hsieh and M. J. Mays, *M. T. P. Rev. inorg. Chem.*, Ser. 1, Vol. 6 (1972) p. 43.
758. W. Hieber and W. Schropp, *Chem. Ber.*, 93 (1960) p. 455.
759. R. S. Nyholm and K. Vrieze, *J. chem. Soc.*, (1965) p. 5337.
760. A. J. Layton, R. S. Nyholm, G. A. Pneumaticakis and M. L. Tobe, *Chem. Ind.* (1967) p. 465.
761. D. J. Cook and R. D. W. Kemmit, *Chem. Ind.* (1966) p. 946; D. J. Cook, J. L. Davies and R. D. W. Kemmit, *J. chem. Soc. (A)* (1967) p. 1547.
762. I. W. Nowell and D. R. Russel, *J. C. S. Dalton* (1972) p. 2393.
763. I. W. Nowell and D. R. Russel, *J. C. S. Dalton* (1972) p. 2396.
764. M. Casey and A. R. Manning, *J. chem. Soc. (A)* (1970) p. 2258.
765. A. N. Nesmeyanov, L. G. Makarova and V. N. Vinogradova, *Izv. Akad. Nauk. S.S.S.R. ser Khim.* (1969) p. 1398.
766. W. Hieber and W. Klingshirn, *Z. anorg. allg. Chem.*, 323 (1963) p. 292.
767. R. B. King, *Inorg. Chem.*, 9 (1963) p. 936.
768. E. O. Fischer and R. Böttcher, *Z. Naturforsch.*, 10B (1955) p. 600.
769. J. M. Burtlitch, *J. organometallic Chem.*, 9 (1967) p. P9.
770. W. Hieber, E. O. Fischer and E. Böckley, *Z. anorg. allg. Chem.*, 269 (1952) p. 308.
771. T. J. Meyer, *Prog. inorg. Chem.*, 19 (1975) p. 1.
772. K. A. Keblys and M. Dubeck, *Inorg. Chem.*, 3 (1964) p. 1646.
773. E. O. Fischer, W. Hafner and H. O. Stahl, *Z. anorg. allg. Chem.*, 282 (1955) p. 47.
774. A. R. Manning and D. J. Thornhill, *J. chem. Soc. (A)* (1971) p. 637.
775. M. J. Mays and J. D. Robb, *J. chem. Soc. (A)* (1968) p. 329.
776. K. Edgar, B. F. G. Johnson, J. Lewis and S. B. Wild, *J. chem. Soc. (A)* (1968) p. 2851.
777. A. T. T. Hsieh and M. J. Mays, *J. chem. Soc. (A)* (1971) p. 2648.
778. A. T. T. Hsieh and M. J. Mays, *J. chem. Soc. (A)* (1971) p. 729.

779. P. N. Bier, A. A. Chalmers, J. Lewis and S. B. Wild, *J. chem. Soc. (A)* (1967) p. 1889.
780. W. Hieber, G. Faulhaber, and F. Theubert, *Z. anorg. allg. Chem.*, 314 (1962) p. 125.
781. W. Hieber, M. Höfler and J. Muschi, *Chem. Ber.*, 98 (1965) p. 311.
782. D. J. Parker, *J. chem. Soc. (A)* (1969) p. 246.
783. H. Hock and H. Stuhlmann, *Chem. Ber.*, 61 (1928) p. 2097.
784. J. Lewis and S. B. Wild, *J. chem. Soc. (A)* (1966) p. 69.
785. H. W. Baird and L. F. Dahl, *J. organometallic Chem.*, 7 (1967) p. 503.
786. S. V. Dighe and M. Orchin, *J. Am. chem. Soc.*, 86 (1964) p. 3895.
787. W. Hieber and H. Beutner, *Z. anorg. allg. Chem.*, 320 (1963) p. 101.
788. R. B. King, *Inorg. Chem.*, 2 (1963) p. 1275.
789. J. P. Collman and W. R. Roper, *Chem. Comm.* (1966) p. 244.
790. W. H. Morrison and D. N. Hendrickson, *Inorg. Chem.*, 11 (1972) p. 2912.
791. W. Hieber and U. Teller, *Z. anorg. allg. Chem.*, 249 (1942) p. 43.
792. W. Hieber and R. Breu, *Chem. Ber.*, 90 (1957) p. 1259.
793. A. R. Manning, *J. chem. Soc. (A)* (1968) p. 1018.
794. J. Newman and A. R. Manning, *J. C. S. Dalton* (1972) p. 241.
795. A. R. Manning and J. R. Millar, *J. chem. Soc. (A)* (1970) p. 3352.
796. A. J. Cleland, S. A. Fieldhouse, B. H. Freeland, C. D. M. Mann and R. J. O'Brien, *Inorg. Chim. Acta*, 4 (1970) p. 479.
797. J. M. Burtlich, R. B. Petersen, H. L. Conder, and W. R. Robinson, *J. Am. chem. Soc.*, 92 (1970) p. 1783.
798. M. A. Bennett and D. J. Patmore, *Inorg. Chem.*, 10 (1971) p. 2387.
799. G. M. Intille and M. J. Braithwaite, *J. C. S. Dalton* (1972) p. 645.
800. J. P. Collman, F. D. Vastine and W. R. Roper, *J. Am. chem. Soc.*, 88 (1966) p. 5035.
801. P. D. Brotherton, J. M. Epstein, A. H. White and S. B. Wild, *Aust. J. Chem.*, 27 (1974) p. 2667.
802. W. Clegg and P. J. Wheatley, *J. chem. Soc. (A)* (1971) p. 3572.
803. M. L. Katcher and G. L. Simon, *Inorg. Chem.*, 11 (1972) p. 1651.
804. R. F. Bryan and A. R. Manning, *Chem. Comm.*, (1968) p. 1316.
805. G. M. Sheldrick and R. N. F. Simpson, *J. chem. Soc. (A)* (1968) p. 1005.
806. F. S. Stephens, *J. C. S. Dalton* (1972) p. 2257.
807. R. F. Bryan and H. P. Weber, *Acta Cryst.*, A21 (1966) p. 138.
808. H. W. Baird and L. F. Dahl, *J. organometallic Chem.*, 7 (1967) p. 503.
809. R. W. Baker and P. Pauling, *Chem. Comm.*, (1970) p. 573.
810. G. A. Bentley, K. R. Laing, W. R. Roper and J. M. Walters, *Chem. Comm.* (1970) p. 998.
811. A. A. Chalmers, J. Lewis and S. B. Wild, *J. chem. Soc. (A)* (1968) p. 1013.
812. T. H. Wirth and N. Davidson, *J. Am. chem. Soc.*, 86 (1964) p. 4314, 4318, 4325.
813. Y. Farhangi and D. P. Graddon, *Aust. J. Chem.*, 26 (1973) p. 983.
814. Y. Farhangi and D. P. Graddon, *Aust. J. Chem.*, 27 (1974) p. 2103.
815. D. R. Dakternieks and D. P. Graddon, *Aust. J. Chem.*, 24 (1971) p. 2077.

PART 3

The Organic Chemistry of Mercury

A. J. Bloodworth

Department of Chemistry, University College London

18 Introduction

There is general agreement as to what constitutes an organic compound, but what is encompassed by the term 'organometallic' is more controversial. We have taken the view that organomercury compounds are species in which there is bonding between mercury and organic carbon. Thus compounds such as $Hg(OPr)_2$, $Hg(NEt_2)_2$ or $Hg(SiMe_3)_2$ in which mercury is attached to some other atom of an organic moiety are regarded as inorganic mercurials and are not considered here. Further, in the belief that it is the metal–carbon interaction that is of central interest in organometallic chemistry, our treatment of organomercury reactions is concentrated on those involving the mercury–carbon bond. Thus reactions of the Hg—X bond in organomercurials of the type RHgX are considered only in as much as they provide a route to new compounds in the same series (RHgY) and to the fully organic mercurials (R_2Hg).

These two series of organomercury compounds are clearly related to the inorganic mercury(II) salts (HgX_2) from which they are usually prepared. Thus the existence of a series of binuclear mercury(I) salts (Hg_2X_2) naturally leads us to contemplate the possibility of preparing organomercury(I) compounds of type R_2Hg_2. However, the alkylation of mercury(I) salts with, for example, organotin[1] or organochromium[2] compounds affords only dialkylmercury(II) products. Both binuclear and mononuclear organomercury(I) species have been postulated as intermediates in some chemical[3] and electrical reductions of organomercury(II) compounds and in their exchange with mercury metal,[4] but none have been isolated. It appears that both types are unstable with respect to the formation of metallic mercury; the binuclear species gives organomercury(II) compounds (equation 18.1) and the mononuclear species either decomposes via the dimer or generates organic radicals (equation 18.2).

$$[RHgHgR] \rightarrow R_2Hg + Hg \qquad (18.1)$$

$$[RHg] \rightarrow R\cdot + Hg \qquad (18.2)$$

Another type of organomercury compound can be envisaged in which π-electrons are involved in binding carbon to mercury, that is, a π-complex. Again such species have not been isolated. However, a few have been characterised spectroscopically in super-acid media[5] and they are believed to play a vital role in the reaction of alkenes with mercury(II) salts in protic solvents (see equation 18.5 below).

Currently then there are just the two classes of σ-bonded organomercury(II) compounds, R_2Hg and RHgX, that can be obtained 'in a bottle'. Despite this an enormous number of organomercury compounds are known, thanks primarily to the availability of simple preparative routes that permit great diversity in the nature of the organic radical R. The most important of these routes involves the reaction of organic substrates with mercury(II) salts in electrophilic substitution (equations 18.3 and 18.4; M is a metal, which may have other groups attached to it) and addition (equation 18.5) processes.

$$R–M + HgX_2 \rightarrow RHgX + MX \tag{18.3}$$

$$R–H + HgX_2 \rightarrow RHgX + HX \tag{18.4}$$

$$C{=}C + HgX_2 + R_nYH \rightarrow R_nY–C–C–HgX + HX \tag{18.5}$$

The substitution processes provide both aliphatic and aromatic mercurials. The addition is obviously limited to the preparation of aliphatic compounds but the substituent R_nY is simultaneously introduced into the molecule. In fact a wide range of functional substituents in the organic substrates can be tolerated in all three reactions.

Transmetallation (equation 18.3) is the most convenient and sometimes the only route to simple aliphatic mercurials such as PrHgCl or MeCH: CHHgOAc. Mercuration (equation 18.4) is of limited application in the aliphatic series but is the main method of synthesising aromatic mercurials, despite the frequent formation of mixtures of isomers and of polymercurated species. Virtually all aromatic systems participate in the reaction and the recent examples shown in figure 18.1 are representative of the various types that can be made in this way. The kinds of β-substituted alkylmercury(II) salts that can be made by addition are determined by the nature of the nucleophile that captures the cationic intermediate formed from alkene and mercury(II) salt. The most common nucleophiles are protic reagents (R_nYH in equation 18.5) such as alcohols or amines, which often serve

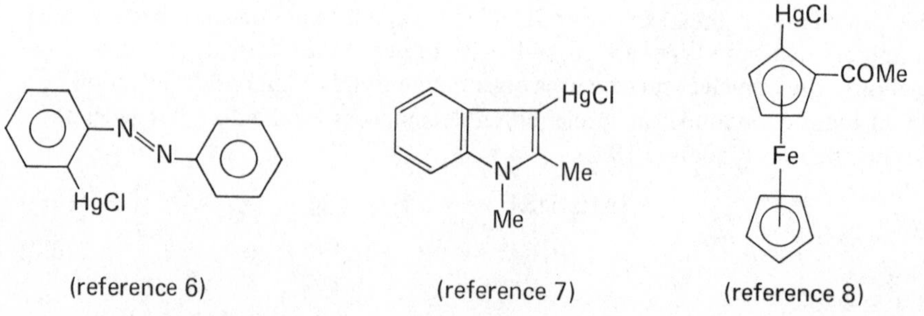

(reference 6)   (reference 7)   (reference 8)

Figure 18.1

also as the solvent. However, if the protic species is a suitably positioned substituent in the alkene, intramolecular addition can occur to give heterocyclic
products. The examples in figure 18.2 are taken from recent literature to illustrate
compounds formed from external and internal nucleophiles.

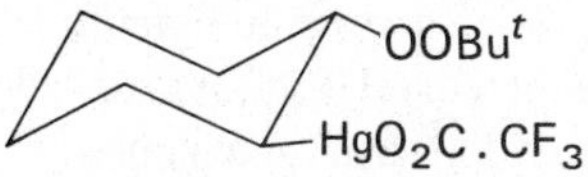

(reference 9) 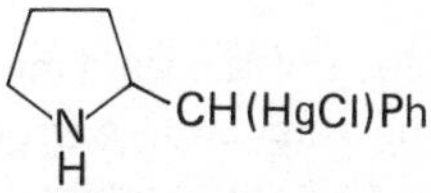   (reference 10)

Figure 18.2

It might be anticipated that reactions analogous to equations 18.3–18.5, in
which $RHgX$ replaces HgX_2, would similarly provide routes to diorganomercurials.
However, organomercury(II) salts in general are much less electrophilic than
mercury(II) salts and the reactions proceed less readily. Yields from transmetallation are often lower, the scope for mercuration is considerably reduced, and
comparable addition processes are virtually non-existent. For this reason it is
preparatively important that organomercury(II) salts can be transformed into
fully organic mercurials by a process known as symmetrisation (equation 18.6).

$$RHgX + RHgX \; \rightleftharpoons \; R_2Hg + HgX_2 \qquad\qquad (18.6)$$

The position of the equilibrium is normally well to the left and the reaction is
driven to the right by removing the mercury(II) salt, usually by reduction to
metallic mercury or by formation of insoluble complexes of the type $HgX_2 . L_2$.

It should be mentioned that organomercury compounds can themselves form
complexes, such as $R_2Hg . L$ and $RHgX . L_2$, by coordination of suitable anionic
or neutral ligands (L). The coordination chemistry of mercury has been covered
systematically in chapter 2 and complexation of organic mercury will be mentioned
here only in the context of how it may affect the reactivity of the parent
organomercurials.

If the availability of simple and versatile preparative routes is one factor that
accounts for the abundance of organomercury compounds, their remarkable
inertness towards atmospheric conditions is undoubtedly another. This property
distinguishes organomercury compounds from the organic derivatives of other
group II metals and almost certainly contributes to their being popular reactants
or products in mechanistic studies; the mechanistic background is one of the most
extensive in organometallic chemistry.

Our current understanding of the mechanism of electrophilic substitution at a
saturated carbon atom derives much from studies involving organomercury compounds. Mercury(II) salts have been the electrophiles in many investigations; hence
quite a lot is known about the mechanism of formation of organomercurials by
transmetallation (equation 18.3). Organomercury compounds (RHgX; X can be R)
have also been among the principal substrates used, and consequently have helped

to provide many details of the mechanism of acidolysis (equation 18.7) and halogenolysis (equation 18.8) of carbon–metal bonds.

$$RHgX + HA \quad \rightarrow \quad RH + AHgX \tag{18.7}$$

$$RHgX + Y_2 \quad \rightarrow \quad RY + YHgX \tag{18.8}$$

Much of the fundamental work involved reactions in which an organomercury
substrate underwent electrophilic substitution with a mercury(II) salt, the so-called
one-alkyl (equation 18.9) and two-alkyl (equation 18.10) mercury exchanges.

$$RHgX + Hg^*X_2 \quad \rightleftharpoons \quad HgX_2 + RHg^*X \tag{18.9}$$

$$R_2Hg + HgX_2 \quad \rightleftharpoons \quad RHgX + RHgX \tag{18.10}$$

The backward reaction in equilibrium (18.10) shows that organomercury(II) salts
can function as electrophiles as well as substrates. Thus a second symmetrical process exists, known as the three-alkyl exchange (equation 18.11).

$$R_2Hg + RHg^*X \quad \rightleftharpoons \quad RHgX + R_2Hg^* \tag{18.11}$$

Organomercury compounds do not react exclusively by heterolytic
mechanisms. Halogenolysis in non-polar solvents and metal hydride reduction,
for example, involve intermediate alkyl radicals.

The metal hydride reduction has assumed importance with a growing interest
in using organomercury compounds in organic synthesis. One of the two procedures
that dominate this area involves combining reductive demercuration with oxy- or
amino-mercuration (equation 18.12). This is used to introduce the oxygen or
nitrogen substituent, usually regiospecifically, or to synthesise heterocyclic
compounds.

$$R^1R^2C = CR^3R^4 + R_nYH \xrightarrow[\text{OH}^-]{\text{(i) HgX}_2 \quad \text{(ii) NaBH}_4} R^1R^2CYR_n \cdot CHR^3R^4 \tag{18.12}$$

The second procedure involves using halogenoalkylmercury compounds to make
three-membered ring compounds (equation 18.13) and for inserting divalent
carbon into single bonds (equation 18.14).

$$RHgCXYZ + A{=}B \longrightarrow \underset{X \quad Y}{\overset{A{-\!-}B}{\diagdown \diagup}} + RHgZ \tag{18.13}$$

$$RHgCXYZ + A{-}B \quad \rightarrow \quad A{-}CXY{-}B + RHgZ \tag{18.14}$$

In the past the synthetic role of organomercurials has been largely in the preparation of other organometallic compounds through reaction with free metals
(equation 18.15) or metal salts (equation 18.16).

$$nR_2Hg + 2M \quad \rightarrow \quad 2R_nM + nHg \tag{18.15}$$

$$nR_2Hg + 2MX_n \rightarrow 2R_nM + nHgX_2 \tag{18.16}$$

Related to this traditional field, a new synthetic method that couples mercury-to-metal transfer with subsequent functionalising demetallation is growing up. Thus, for example, a new route to phenols combines aromatic mercuration with oxidative deboronation (equation 18.17).[11]

$$ArH \xrightarrow{HgX_2} ArHgX \xrightarrow{BH_3} Ar-B< \xrightarrow{H_2O_2/OH^-} ArOH \qquad (18.17)$$

The conversion of easily formed but rather inert organomercury compounds into new organometallics capable of undergoing synthetically useful demetallation is obviously of great potential value in organic synthesis.

The embryonic organomercury chemist is well provided for as far as background reading material is concerned. Early synthetic endeavours are summarised in Whitmore's *Organic Compounds of Mercury*, published in 1921.[12] A book of the same name by Makarova and Nesmeyanov (1967)[13] exhaustively covers the literature up until the end of 1963 and contains some later references. The appearance in 1974 of the volume of Houben-Weyl's *Methoden der Organischen Chemie* devoted to organomercury compounds[14] brings the coverage of synthetic aspects almost up to date. Most of the mechanistic aspects have been reviewed in books published in 1968[15,16] and 1973.[17] Hence together these six books provide a balanced background to the subject.

19 Formation of the Mercury–Carbon Bond

19.1 Mercury-for-Metal Substitution (Transmetallation)

Transmetallation is the most general and universally applied method in organometallic synthesis. It can be used to prepare a wide variety of organomercury salts (equation 19.1), and since these products can themselves be alkylated, the method is also applicable to the synthesis of dialkylmercurials (equations 19.2 and 19.3).

$$RM + HgX_2 \rightarrow RHgX + MX \qquad (19.1)$$

$$R'M + RHgX \rightarrow RHgR' + MX \qquad (19.2)$$

$$2RM + HgX_2 \rightarrow R_2Hg + 2MX \qquad (19.3)$$

Organic derivatives of many main-group and transition metals will participate in these reactions and new examples are being discovered each year, as illustrated by recently reported alkyl transfers from zirconium,[18] silver[19] and platinum[20] (equations 19.4–19.6 respectively).

$$(\pi\text{-fluorenyl})_2 ZrCl_2 + HgCl_2 \rightarrow \text{fluorenyl HgCl} \qquad (19.4)$$

$$[4\text{-MeC}_6 H_4 N : C(OEt)Ag]_3 + (Ph_3 P)_2 HgCl_2 \rightarrow [4\text{-MeC}_6 H_4 N : C(OEt)]_2 Hg \qquad (19.5)$$

$$cis\text{-Me}_2 Pt(PPh_3)_2 + CF_3 HgCl \rightarrow CF_3 HgMe \qquad (19.6)$$

Most of the known reactions are not preparatively important simply because the organometallic reagents have themselves been obtained by transmetallation from more readily available compounds, and these can be used directly to transfer organic groups to mercury. However, some reactions do have specific synthetic value such as that yielding fluorenylmercury(II) chloride (equation 19.4) for which the more common metal reagents fail.

The organometallic reagents most widely used in organomercury synthesis are derived from magnesium or lithium for the usual reasons of availability, reactivity, and ease of handling, but the use of organoboranes, which have become readily accessible through hydroboration, is a developing area. As mentioned in the introduction, transmetallations involving mercury(II) salts have been the subject of many mechanistic investigations. The most detailed studies have been concerned with organotin compounds, while recent work has concentrated largely on transition-metal alkyls. There has been particular interest in organocobalt compounds stimulated no doubt by possible relevance to the origin of methylmercury pollution of inland and coastal waters of industrial countries.

Thus a discussion of this method of forming the mercury–carbon bond divides naturally into the synthetic aspect, which is concerned mainly with organic derivatives of magnesium, lithium and boron, and the mechanistic aspect where evidence has been derived largely from reactions involving tin and transition-metal compounds.

19.1.1 Synthesis

Details of the conversion of Grignard reagents into organomercury compounds have been summarised in some of the principal reference works[13,14] and only the salient points will be mentioned here. It is preferable to use the mercury(II) halide corresponding to the organic halide from which the Grignard reagent was prepared, since this avoids the complication of anion exchange, which can cause problems in separation. Diethyl ether and tetrahydrofuran are the usual solvents but some reactions have been carried out satisfactorily in xylene or heptane. Free magnesium must be removed from the Grignard reagent before it is used, and it is often convenient to introduce the mercury(II) salt via a Soxhlett extractor.

The Grignard route is suitable for the synthesis of primary, secondary or tertiary alkylmercury(II) halides, or cycloalkylmercury(II) halides. It can be used to prepare arylmercury(II) halides, alkenylmercury(II) halides, and alkylmercury(II) halides containing substituents compatible with the Grignard synthesis, but better alternatives are often available here. The yields of methyl, ethyl and propyl compounds (equation 19.7) are about 90 per cent but those of the C_4-C_7 homologues are only 50 per cent.[21]

$$RMgBr + HgBr_2 \xrightarrow{Et_2O} RHgBr + MgBr_2 \qquad (19.7)$$

In general, yields of secondary and tertiary alkyl compounds are lower than those of primary alkyl compounds. This trend is found again in the reactions (equation 19.8) leading to dialkylmercurials where, for example, the yield of diethylmercury is typically 60 per cent[22] whereas that for di-t-butylmercury is only 9 per cent.[23]

$$2RMgBr + HgBr_2 \xrightarrow{Et_2O} R_2Hg + 2MgBr_2 \qquad (19.8)$$

The pre-Grignard practice of using organozinc compounds (equation 19.9) has been recommended as preferable for the synthesis of organomercury chlorides or bromides where R is a group such as $CH_2:CH.CH_2$, $PhC\overset{..}{:}C.CH_2$ or $PhCH:C: CH.CH_2$.[24]

$$R_2Zn + 2HgX_2 \xrightarrow{THF} 2RHgX + ZnX_2 \qquad (19.9)$$

However, the main alternative to the Grignard route is to use organolithiums. This change increases the range of transferable organic groups since the conversion of simple alkyl metals into functionally substituted organometallic reagents via metallation (equation 19.10) has far more scope with lithium than with magnesium.

$$RH \xrightarrow[-BuH]{BuLi} RLi \xrightarrow{HgCl_2} R_2Hg \qquad (19.10)$$

Thus, for example, the alkyllithium can be derived by lithiation of a CH centre activated by adjacent unsaturated groups such as $C=C$,[25] SO_2 [26] or $P=O$.[27]

In recent years the organolithium route has found particular application in the synthesis of fluoroalkyl-, ferrocenyl-, and carboranylmercurials (for example equations 19.11,[28] 19.12[29] and 19.13[30]).

$$3\text{-}HC_6F_4Li + EtHgCl \rightarrow 3\text{-}HC_6F_4HgEt \qquad (19.11)$$

$$2\ 2\text{-}Me_2N(CH_2)_2 \cdot FerrocenylLi + HgCl_2 \rightarrow [2\text{-}Me_2N(CH_2)_2 \cdot Ferrocenyl]_2Hg \qquad (19.12)$$

$$2\ 1\text{-}Ph\text{-}1,6\text{-}B_8C_2H_8\text{-}6\text{-}Li + HgBr_2 \rightarrow (1\text{-}Ph\text{-}1,6\text{-}B_8C_2H_8\text{-}6\text{-})_2Hg$$

Since the Grignard reagents and simple organolithiums are usually prepared from alkyl halides and the appropriate metal, it is natural to enquire whether organomercury salts can be obtained directly from alkyl halides by an analogous route (equation 19.15)

$$RX + Hg \rightarrow RHgX \qquad (19.15)$$

In fact the first organomercury compounds were prepared by this reaction (for example $RX = MeI$[31]), but few halides are sufficiently reactive to render the procedure synthetically viable. Alkyl chlorides do not react, bromides react only with difficulty and even iodides do not react readily. Thus for the conversion of simple alkyl halides into alkylmercurials, the indirect route via transmetallation is always preferred to reaction with mercury. Reasonable yields have been obtained by the direct method from a range of α-bromoesters (for example equation 19.16[32]).

$$4\text{-}Pr^iC_6H_4CHBr \cdot CO_2Et + Hg \rightarrow 4\text{-}Pr^iC_6H_4CH(HgBr)CO_2Et \quad 89 \text{ per cent} \qquad (19.16)$$

Although mercury itself is poorly reactive towards alkyl halides, sodium amalgam readily affords dialkylmercurials in good yield (equation 19.17; $X = Br, I$).[13]

$$2RX + \underset{xs}{Na_2Hg} \xrightarrow[EtOAc]{0-5\,^\circ C} R_2Hg + 2NaX \qquad (19.17)$$

The range of the method is similar to that of the organomagnesium route but yields are sometimes higher. More vigorous conditions are needed to prepare diaryl-mercurials.

The transfer of groups from organoboronic acids or esters to mercury (equation 19.18) is a long-established reaction.[33]

$$RB(OH)_2 + HgX_2 + H_2O \rightarrow RHgX + B(OH)_3 + HX \qquad (19.18)$$

It has found application mainly in the preparation of aromatic mercury compounds (R = Ar), but has been used in the aliphatic series particularly to afford *gem*-poly-mercurated species; a recent example is shown in equation 19.19.[34]

$$\underset{Me}{\overset{Me}{>}}C=C\left[B\overset{O}{\underset{O}{<}}\right]_2 + HgCl_2 \longrightarrow \underset{Me}{\overset{Me}{>}}C=C\overset{HgCl}{\underset{HgCl}{<}} \qquad (19.19)$$

$$90 \text{ per cent}$$

It should be remembered that organoboronic acids are usually prepared from Grignard reagents so that these transmetallations do not employ 'primary' organo-metallic reagents. The most noteworthy development in mercurideboronation is the recent discovery that trialkylboranes, which can be regarded as 'primary' by virtue of their ready formation via hydroboration of alkenes and alkynes,[35] are suitable transmetallating reagents (equation 19.20).

$$R_3B + 3HgX_2 \rightarrow 3RHgX + BX_3 \qquad (19.20)$$

The reactions are generally carried out in ether solvents at or below room tempera-ture and provide the alkylmercury salts in yields of greater than 60 per cent. Where mercury(II) acetate is used (X = OAc) the boron reagents derived from alkenes are only suitable for transferring primary alkyl groups[36,37] (equation 19.21), but if mercury(II) methoxide[38] or better mercury(I)t-butoxide[39] is employed, secondary alkyl transfer can be achieved in high yield (equation 19.22).

$$3RCH:CH_2 \xrightarrow[THF]{BH_3} (RCH_2.CH_2)_3B \xrightarrow[THF]{3Hg(OAc)_2} 3RCH_2.CH_2HgOAc \qquad (19.21)$$

$$3RCH:CHR \xrightarrow[THF]{BH_3} (RCH_2.CHR)_3B \xrightarrow[Bu^tOH]{2Hg_2(OBu^t)_2} 2RCH_2.CHR.HgOBu^t \qquad (19.22)$$

Terminal alkynes undergo dihydroboration and therefore can be used to prepare *gem*-dimercurated hydrocarbons (equation 19.23).[40]

$$RC:CH \xrightarrow[THF]{2BH_3} RCH_2.CH(BH_2)_2 \xrightarrow{MeOH} RCH_2.CH[B(OMe)_2]_2$$

$$\xrightarrow{HgCl_2} RCH_2.CH(HgCl)_2 \qquad (19.23)$$

Alkenyl products are obtained from both terminal and medial alkynes by using more selective hydroborating reagents such as dicyclohexylborane[41] (equation 19.24) or catecholborane.[42]

$$RC:CR^1 \xrightarrow{(C_6H_{11})_2BH} RCH:CR^1.B(C_6H_{11})_2 \xrightarrow{Hg(OAc)_2} RCH:CR^1.HgOAc$$

$$(19.24)$$

Another advantage of these reagents is that they can tolerate a wide range of functional groups, so that functionally substituted organomercurials can be prepared.

Dialkylmercurials can be prepared by hydroboration-transmetallation but excess of the trialkylborane is usually required to obtain high yields. This can be avoided by symmetrising the initially formed alkylmercury(II) acetates with zinc, making use of the boron triacetate by-product as the Lewis-acid catalyst that is needed[43] (equation 19.25).

$$R_3B + 3Hg(OAc)_2 \rightarrow 3RHgOAc + B(OAc)_3 \xrightarrow{3Zn} \tfrac{3}{2}R_2Hg \qquad (19.25)$$

The transmetallations occur with retention of configuration (see later) so that the *cis*-stereospecificity of hydroboration can be exploited to prepare organo-mercury compounds of particular stereochemistry.

19.1.2 Mechanism

A detailed assessment of the experimental evidence concerning the mechanism of transfer of alkyl groups from various metals to mercury(II) salts can be found in an excellent survey by Abraham.[17] Many studies establish that for alkyls of a wide range of metals, for example tin,[44] gold[45] (equation 19.26), chromium[46] (equation 19.27), manganese,[47] iron[48] and cobalt,[49] the transmetallations follow second- (or higher-[50]) order kinetics, being first order in mercury(II) salt and first order in organometallic reagent; no exceptions are known.

$$RAuPPh_3 + HgX_2 \rightarrow RHgX + XAuPPh_3 \qquad (19.26)$$

$$\overset{+}{H}NC_5H_4 . CH_2Cr(H_2O)_5{}^{2+} + HgX_2 \xrightarrow{H_2O} \overset{+}{H}NC_5H_4 . CH_2HgX + \overset{-}{X} + Cr(H_2O)_6{}^{3+}$$

$$(19.27)$$

Other investigations show that the normal stereochemical course of these reactions is retention of configuration. This has been established for example for transfers from lithium,[51] magnesium,[52] boron[53] (equation 19.28), tin,[54] iron (equation 19.29; R = Bu^t[55] or Ph[56]) and palladium.[57]

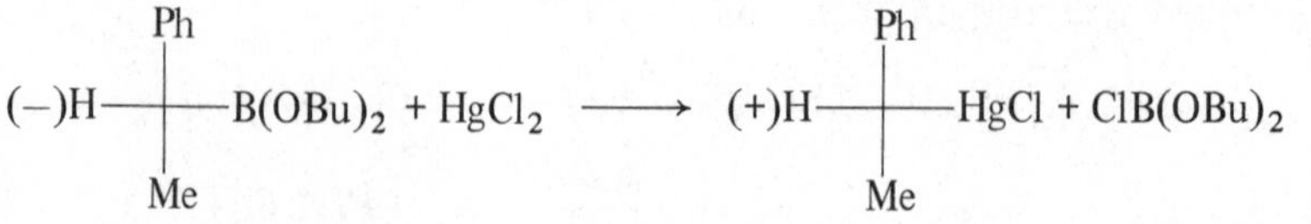

$$(-)H\underset{Me}{\overset{Ph}{|}}B(OBu)_2 + HgCl_2 \longrightarrow (+)H\underset{Me}{\overset{Ph}{|}}HgCl + ClB(OBu)_2 \qquad (19.28)$$

$$threo\text{-}RCHD . CHD . Fe(CO)_2Cp + HgCl_2 \rightarrow threo\text{-}RCHD . CHD . HgCl$$
$$+ ClFe(CO)_2Cp \qquad (19.29)$$

Exceptionally, reactions with alkylcobaloximes (equation 19.30; R = *trans*-4-t-butylcyclohexyl[58] or *erythro*-Bu^{t}CHD . CHD[59]) involve inversion.

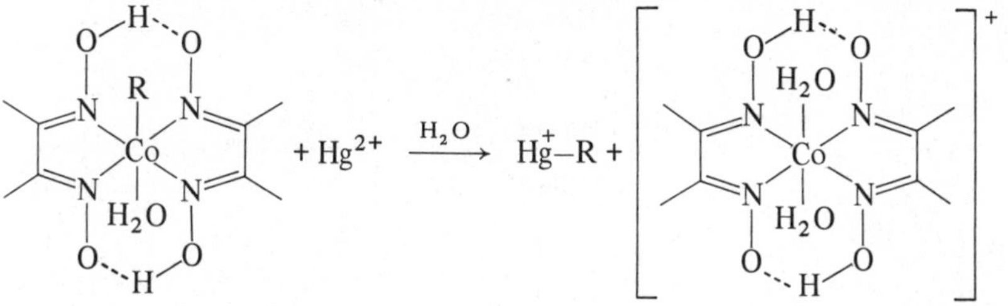

$$(19.30)$$

All the data are consistent with a mechanism for transmetallation involving bimolecular electrophilic substitution (S_E2). Normally the mercury(II) salt attacks the alkyl–metal bond from the front but presumably it is prevented from doing so in the alkylcobaloxime system by steric hindrance of the macrocyclic ligand. In all the alkyl transfers, other than those from cobalt, the only point of controversy is whether or not the transition state involves significant interactions between the metal and one of the ligands on mercury; that is, whether the transition state is open or cyclic (figure 19.1).

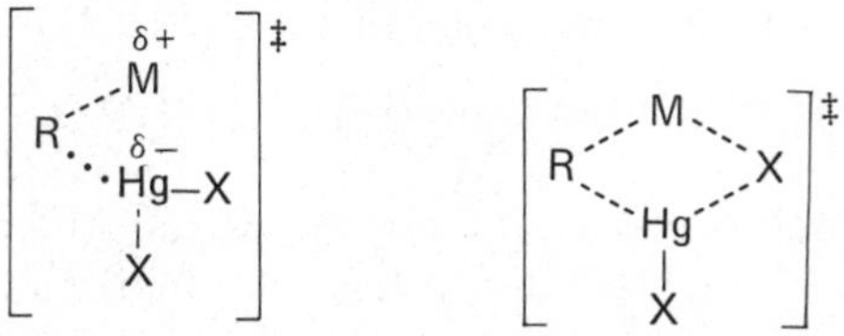

S_E2 (open) S_E2 (cyclic)

Figure 19.1

This question has been thoroughly examined by Abraham *et al.*[17] for transmetallations with tetraalkyltins in methanol–water mixtures (equation 19.31; X = Cl, I).

$$R_4Sn + HgX_2 \rightarrow RHgX + R_3SnX \qquad (19.31)$$

On the basis of solvent effects and kinetic salt effects they conclude that the S_E2 (open) mechanism obtains. The interpretation of the experimental data to arrive at this conclusion is not simple, for it was shown that the observed rate enhancements, which are expected for the S_E2 (open) mechanism on the basis of the Hughes–Ingold postulate, actually arise because of destabilisation of the reactants.

Transmetallations involving mercury(II) acetate are potentially different in that a six-centred rather than a four-centred cyclic transition state can be envisaged (figure 19.2 1; R' = Me). In a series of reactions between tetraethyl- or tetrabutyltin and mercury(II) carboxylates (equation 19.32),[60] the second-order rate coefficients increase with the introduction of electron-attracting substituents into R', thus again supporting an S_E2 (open) mechanism.

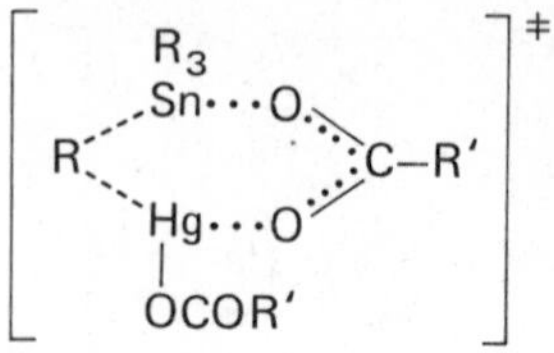

1

Figure 19.2

$$R_4Sn + Hg(OCOR')_2 \xrightarrow{MeOH} RHgOCOR' + R_3SnOCOR' \qquad (19.32)$$

An open transition state was also suggested for the transfer of pyridiomethyl groups from chromium to mercury, partly because the inorganic product contained no X-ligand.[46]

From these results it appears likely that the S_E2 (open) mechanism is the normal pathway by which alkyl groups are transferred from metals to mercury. Allyl groups are transferred much more readily than comparable alkyls.[17] The reactions with germanium[61] and tin[62] compounds follow second-order kinetics and afford unrearranged products (equation 19.33; M = Ge, Sn).

$$Et_3MCH_2 . CH:CHR + HgX_2 \rightarrow RCH:CH.CH_2HgX + Et_3MX \quad (19.33)$$

However an S_E2' mechanism, that is, attack by the mercury(II) salt on the double bond, cannot be rigorously excluded since it is possible for the rearranged allylmercury(II) salt to be formed initially and then itself to undergo rearrangement.

Less is known about aryl transfer, where the possibility of an addition–elimination mechanism, that is, normal aromatic electrophilic substitution, must also be considered. However, an S_E2 (open) transition state was suggested as a possibility in the cleavage of aryl groups from tin by mercury(II) acetate,[63] whereas an S_E2 (cyclic) transition state (cf. 1) with charge transfer to the ring (that is, aromatic-substitution character) has been suggested for the mercuridesilylation of *p*-tolyltrimethylsilane.[64]

19.2 Mercury-for-Hydrogen Substitution (Mercuration)

The replacement of a C–H group by a C–Hg group through reaction of a mercury(II) salt with an organic substrate (equation 19.34) is the most versatile method of preparing organomercury salts.

$$RH + HgX_2 \rightarrow RHgX + HX \qquad (19.34)$$

Further reaction to afford dialkylmercurials (equation 19.35) is known but is very much more restricted.

$$RH + RHgX \rightarrow R_2Hg + HX \qquad (19.35)$$

Mercury is exceptional among metals for the vast range of organic compounds that can be metallated by readily available salts.[12-14] The method is universal in the aromatic series, being applicable to free hydrocarbons, hydrocarbons complexed with transition metals, hydrocarbons carrying substituents both activating and deactivating towards electrophilic substitution, and heterocyclic compounds. Since the substitution usually proceeds by electrophilic attack on carbon (see below), it is much less generally applicable in the aliphatic series, being limited to structures containing substituents, such as carbonyls, that are capable of enhancing nucleophilicity in neighbouring carbons.

With such a range of organic reagents it is not surprising that the conditions for mercuration vary enormously. Using mercury(II) acetate for example, aniline can be mercurated in aqueous solution at room temperature, benzene requires several hours at 110 °C in glacial acetic acid, and benzoic acid reacts in the melt at 130–170 °C. Mercury(II) acetate is the most commonly used salt partly because the resulting organomercury(II) acetates are not readily decomposed by the liberated acetic acid even at high temperatures. Mercury(II) oxide is also popular and the salts of other oxyacids such as perchloric and nitric acid have been used, though less extensively. A mixture of mercury(II) nitrate, mercury(II) oxide and anhydrous calcium sulphate has found particular application in the preparation of arylmercurials and in the mercuration of aliphatic ketones (equation 19.36).[65]

$$4Me_2CO + 2Hg(NO_3)_2 + 2HgO + CaSO_4 \rightarrow 4MeCOCH_2HgNO_3 + CaSO_4 . 2H_2O$$

$$(19.36)$$

Recent interest has been shown in developing the use of mercury(II) trifluoroacetate[66] and the compound $Hg[N(SiMe_3)_2]_2$[67] in organomercury synthesis. The latter reagent cleanly provides almost quantitative yields in the mercuration of a variety of aliphatic substrates but it is comparatively difficult to prepare and handle.

Mercury(II) halides, often the reagents of choice in the transmetallation route, are among the least reactive mercurating agents. However, for this very reason they are used, admixed with sodium acetate, in the mercuration of highly reactive heterocycles such as furan and thiophene. Polymercuration of reactive substrates occurs very readily and the choice of mercurating agent often determines the nature of the product. Thus selenophene is converted into 2-chloromercurioselenophene by buffered mercury(II) chloride (equation 19.37),[68] but gives the dimercurated product with mercury(II) oxide in acetic acid (equation 19.38).[69]

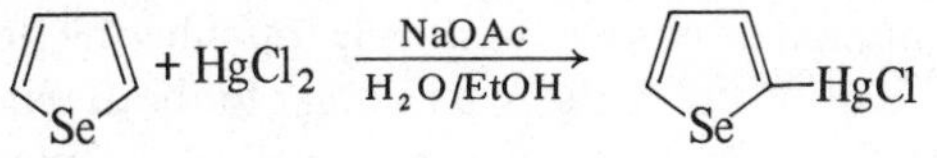

$$(19.37)$$

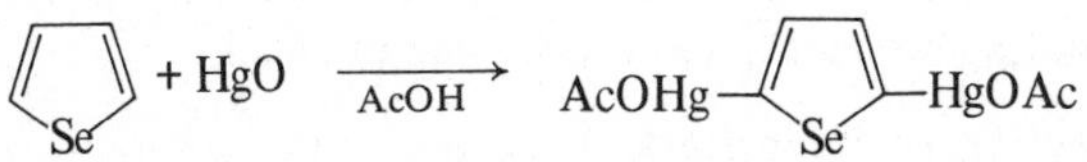

$$(19.38)$$

The synthesis of arylmercury salts is complicated not only by polymercuration but also by the formation of isomeric monomercurated compounds. The usual orientation rules for aromatic electrophilic substitution are obeyed provided that the mercuration is carried out under mild conditions, but low regioselectivity can be observed at high reaction temperatures where homolytic mechanisms or thermo- dynamic control may assume importance. The aromatic substrates in which most interest has been shown recently are ferrocene and related complexes[70,71] and polyfluoroarenes.[72,73]

The mercuration of simple aliphatic compounds was studied extensively around the turn of the century. It requires vigorous conditions and often leads to the formation of ill-defined products that are apparently polymercurated species. Thus ethanol, acetaldehyde, and acetone and higher homologues afford such products on prolonged boiling with alkaline mercury(II) oxide. The structural formulae originally assigned to many of these compounds (see reference 12) are incompatible with modern knowledge of bonding in organomercurials, and are probably no more than attempts by the early workers to accommodate their analytical data. Most of the products are insoluble amorphous powders and are probably mixtures of polymers.

As yet little light has been shed on this confusing area of organomercury chemistry. However, the product of composition $C_{12}H_{14}O_{12}Hg_5$, obtained from the mercuration of acetic acid, has been shown to have the structure $MeCO_2$-$(HgCH_2CO_2)_n H$ (mainly with $n = 5$)[74] rather than one with two tri-mercurated carbon atoms, $[AcOC(HgOAc)_2]_2 Hg$, as suggested earlier. It has been argued from this and theoretical considerations that many compounds previously thought to be di- and tri-mercurated are probably polymeric arrangements of monomercurated repeat units.[75] Our discussion of aliphatic mercuration is restricted to systems that give well-defined mono- or di-mercurated products. This has been one of the areas in mercuration of greatest activity during the past few years and particular effort has been devoted to determining the structure of mercurials obtained from 1,3-dicarbonyl compounds[76-78] and cyclopentadiene derivatives.[25]

19.2.1 Aromatic Mercuration

The mercuration of benzene and its derivatives under aqueous (heterogeneous) conditions or in acetic or trifluoroacetic acid (homogeneous conditions) is an electrophilic substitution process. The evidence available up until 1968 has been reviewed[75,79] and little more has been added. Apart from the observation that aqueous mercuration is *strongly* catalysed by anions, an effect that has been attributed to the formation of more powerful electrophiles by deaquation of solvated mercury(II) ions,[80] the findings are similar for both sets of conditions. For convenience therefore our discussion is restricted to the more recent and more detailed work, which is that carried out in non-aqueous media.

Three mercurating systems have been studied in detail

(i) mercury(II) acetate in acetic acid

(ii) mercury(II) acetate in acetic acid containing perchloric acid (or mercury(II) perchlorate in acetic acid, which can amount to the same thing)

(iii) mercury(II) trifluoroacetate in trifluoroacetic acid.

The first two systems are complicated by departure from second-order kinetics at more than low conversions, competitive mercuration of the solvent, and isomerisation of the products, but none of these problems are encountered with mercury(II) trifluoroacetate in trifluoroacetic acid.[81] It has been established that the equilibria (19.39) and (19.40) are important in system ii and that both the ion pair 2 and the ion triplet 3 are active electrophiles.[82]

$$Hg(OAc)_2 + HClO_4 \rightleftharpoons Hg(OAc)^+ClO_4^- + AcOH \qquad (19.39)$$
$$2$$

$$Hg(OAc)^+ClO_4^- + HClO_4 \rightleftharpoons Hg^{2+}(ClO_4^-)_2 + AcOH \qquad (19.40)$$
$$3$$

Unionised mercury(II) acetate is the main electrophile in system i, but the species $Hg(OAc)^+$ is also active in the absence of added acetate ions.[64] The dramatic rate enhancement produced by added perchloric acid is accounted for by the reactivity of ion pair 2 relative to $Hg(OAc)_2$, which is estimated to be about four orders of magnitude towards benzene at 25 °C.

In all three systems, partial rate factors for mercuration of substituted benzenes such as monoalkylbenzenes, polymethylbenzenes, anisole, halogenobenzenes, and nitrobenzene correlate with electrophilic substituent constants (σ^+).[79] In each case the attacking mercury(II) species discriminates rather poorly between substrates and between positions in a substrate, but surprisingly selectivity *increases* with increasing reactivity. Thus the rates of mercuration go up along the series: system i < system ii < system iii and the corresponding ρ-factors are −4.0, −5.1 and −5.7.

The familiar mechanistic picture for electrophilic aromatic substitution can therefore be applied to mercuration and equation 19.41 shows the gross features for reaction of benzene with the acetoxymercury(II) cation.

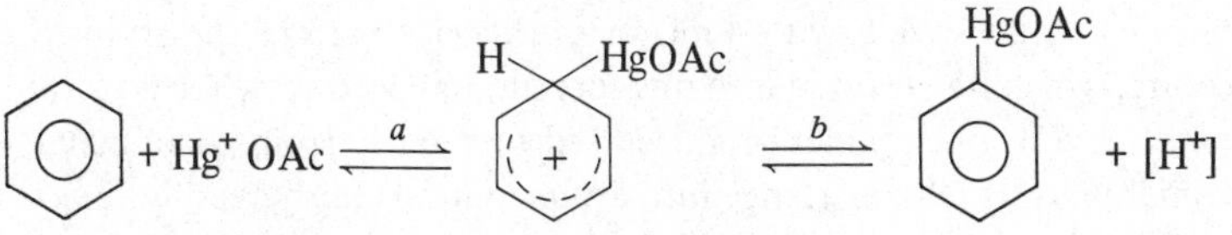

$$4 \qquad (19.41)$$

In the mercuration of pentamethylbenzene by mercury(II) trifluoroacetate in trifluoroacetic acid, the intermediate σ-complex analogous to 4 has been detected in the reaction mixture by n.m.r. spectroscopy.[83]

In most electrophilic substitutions of benzene the formation of the σ-complex (here equation 19.41a) is rate determining. However, under conditions where ion pair 2 is essentially the sole electrophile [$Hg(OAc)_2$ (0.051 mol dm^{-3}) in acetic acid containing $HClO_4$ (0.32 mol dm^{-3}) and H_2O (0.23 mol dm^{-3})], the rate of mercuration of hexadeuteriobenzene at 25 °C is about six times slower than that of benzene,[84] indicating that proton loss (equation 19.41b) is rate controlling. This being so, the nature of the base that accepts the proton assumes importance, and intramolecularly assisted proton removal by the acetate group of the electrophile has been suggested for mercuration both by 2 and by unionised mercury(II) acetate.[64]

If fast reversible formation of a σ-complex followed by slow proton loss was the universal mechanism of aromatic mercuration, the differing reaction rates for different mercurating electrophiles could be accommodated by changes in the equilibrium constant for the first step. That an identical rate-determining step for all mercurating species (crudely represented by equation 19.41b) is not the case is indicated by two pieces of information. Firstly the previously mentioned fact that different electrophiles have different reaction constants, and secondly the fact that the kinetic isotope effect for uncatalysed mercuration of benzene is only about half of that for catalysed mercuration.

If catalysed and uncatalysed reactions proceed by rate-determining proton loss, these data can be compatible with the Hammond postulate only if the transition state for the uncatalysed reaction involves more proton–base than ring–proton bonding. An alternative explanation is that the formation of the σ-complex becomes partially rate controlling in the uncatalysed mercuration. The ρ-factor would then result from an appropriate combination of the selectivities of the transition states for each stage and these are not necessarily related. If we assume lower selectivity for σ-complex formation than for proton loss then the kinetic isotope effect and selectivity data are again consistent.

Thus, although it appears beyond doubt that these aromatic mercurations are electrophilic substitutions, the finer mechanistic details are by no means fully established. Furthermore, under certain conditions other pathways may be operative with some benzene derivatives. Thus in the mercuration of phenols and amines, which are activated towards electrophilic substitution, there are indications that the initial products are Hg—O- and Hg—N-bonded species, which are readily transformed into mercury aryls, particularly at lower pH. Thus, for example, aqueous mercury(II) acetate reacts with 4-nitronaphth-1-ol to afford the aryloxymercury(II) acetate, which is converted into ring-mercurated product when dissolved in acetic acid; mercury(II) phenoxides produced independently from alkali metal derivatives and mercury(II) salts rearrange into arylmercurials fairly readily. The mercuration of aniline and related arylamines is reported to proceed via rearrangement of N-mercurated products (equation 19.42).

$$PhNH_2 + Hg(OAc)_2 \rightarrow Ph\overset{+}{N}H_2HgOAc . \overset{-}{O}Ac \rightarrow PhNHHgOAc$$

$$\rightarrow H_2NC_6H_4HgOAc \qquad (19.42)$$

In some cases the intermediates have been isolated,[85] and again they are favoured by a deficiency of acid in the reaction medium. Thiophenols afford Hg—S-bonded products but these do not undergo rearrangement, presumably because of the greater strength of the Hg–S bond. Most of the evidence concerning these systems was collected before 1930[13] and the area has yet to receive the benefit of detailed investigation using modern techniques.

However, related reactions with phenylmercury(II) hydroxide have been studied in recent years. This reagent is a much weaker electrophile than a mercury(II) salt and does not effect *C*-mercuration of aniline even in refluxing benzene,[86] but some sterically hindered phenols containing ortho-triorganosilyl groups do afford ring-mercurated products.[87] The suggested mechanism is shown in equation 19.43 and the intermediate aryloxyl radicals have been detected by e.s.r. spectroscopy.

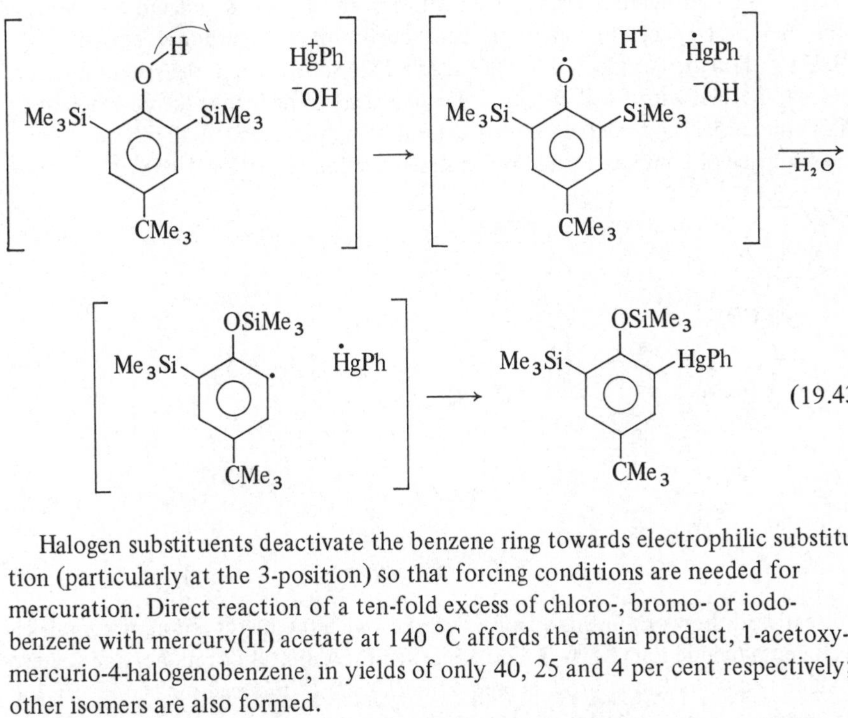

$$(19.43)$$

Halogen substituents deactivate the benzene ring towards electrophilic substitution (particularly at the 3-position) so that forcing conditions are needed for mercuration. Direct reaction of a ten-fold excess of chloro-, bromo- or iodo-benzene with mercury(II) acetate at 140 °C affords the main product, 1-acetoxy-mercurio-4-halogenobenzene, in yields of only 40, 25 and 4 per cent respectively; other isomers are also formed.

Drastic conditions are similarly required for the mercuration of more highly halogenated benzenes, but with penta-substituted compounds the problems of isomerism and polymercuration cannot arise. The best results are obtained using mercury(II) trifluoroacetate (equation 19.44) and although the yield is poor with pentachlorobenzene (X = Y = Cl),[88] it varies from 85 to 47 per cent for poly-fluoroarenes (X = F: Y = F,[89] 4-MeO,[72] 2-NO$_2$[72]).

$$C_6X_4YH + Hg(O_2C.CF_3)_2 \xrightarrow{\approx 150\,°C} C_6X_4YHgO_2C.CF_3 + HO_2C.CF_3$$

$$(19.44)$$

The presence of many halogens strongly deactivates arenes towards electrophilic attack, but it also greatly enhances their acidity and, as a result, pentafluorophenyl mercurials are conveniently prepared under basic conditions (equations 19.45 and 19.46; Ar = Ph, 4-MeC$_6$H$_4$).[73]

$$2C_6F_5H + HgBr_4{}^{2-} + 2^-OH \xrightarrow[\text{H}_2\text{O}]{\text{Bu}^t\text{OH}} (C_6F_5)_2Hg + 4Br^- + 2H_2O \quad (19.45)$$

$$C_6F_5H + ArHgCl + {}^-OH \xrightarrow[\text{H}_2\text{O}]{\text{Bu}^t\text{OH}} C_6F_5HgAr + Cl^- + H_2O \quad (19.46)$$

The reactions presumably proceed by formation of pentafluorophenyl carbanion, which then carries out nucleophilic displacement at mercury(II).

Aromatic hydrocarbons complexed with transition metals undergo mercuration. Thus mercury(II) acetate mercurates benzenechromium tricarbonyl (equation 19.47)[90] though, as expected for an electrophilic substitution, the reaction occurs less readily than with free benzene. It is interesting to note that the corresponding diaryl mercurial (and related derivatives) can be prepared essentially by reversing the sequence of complex formation and mercuration (equation 19.48).[91]

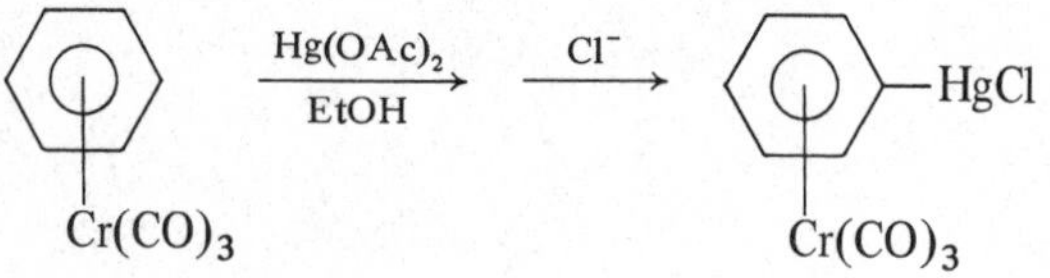

$$(19.47)$$

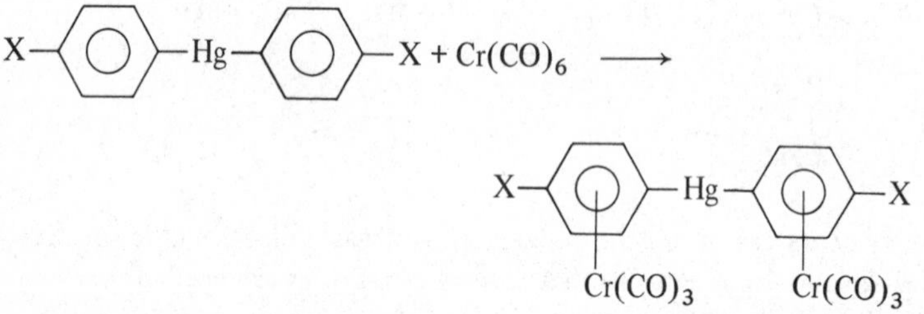

$$(19.48)$$

Nearly all the work in this area is concerned with the mercuration of complexed cyclopentadienide ion (C$_5$H$_5{}^-$). Ferrocene affords a mixture of mono- and (hetero-annularly) di-substituted products (equation 19.49);[70] the reaction can be carried out in acetic acid, but better yields are obtained using ether–alcohol or benzene–alcohol solvents.

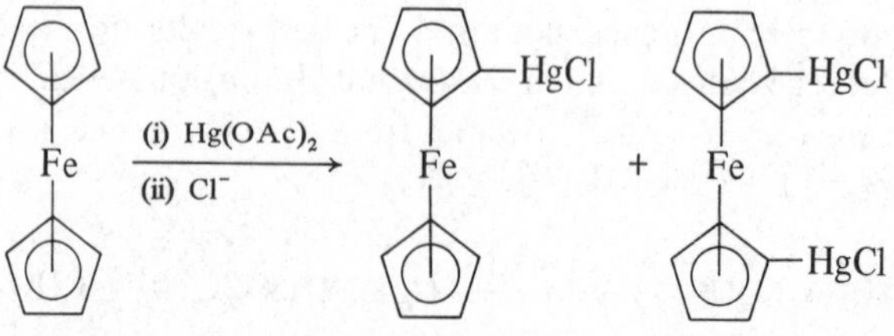

$$(19.49)$$

A 1 : 1 mixture of reagents favours dimercuration (64 per cent of **6** and 19 per cent of **5**), but 50 per cent of **5** (with 11 per cent of **6**) can be obtained by using a four-fold excess of ferrocene. Mechanistically electron transfer from iron to bind the mercury(II) electrophile can be envisaged as preceding σ-complex formation,[75] but no experimental evidence either for or against this is available.

Monomercuration products have been isolated from the reaction of mercury(II) acetate with a number of substituted ferrocenes. Formyl-, acetyl-, and methoxy-carbonyl-ferrocene undergo both 2- and 1'-mercuration,[8] whereas halogenoferrocenes are substituted in the 3- and 1'-positions;[92] 1-chloro-2-formylvinylferrocene is mercurated only in the unsubstituted ring.[93] The following cyclopentadienide complexes similarly react to afford mixtures of mono- and di-mercurated products: $(C_5H_5)_2Ru$,[94] $(C_5H_5)Mn(CO)_3$,[71] $(C_5H_4Me)Mn(CO)_3$,[71] $(C_5H_5)Re(CO)_3$[95] and $(C_5H_5)Co(C_4Ph_4)$.[96] Unlike cyclopentadienide, cyclobutadiene is not a Hückel aromatic system but its complex with iron tricarbonyl is extremely reactive towards electrophilic substitution and with mercury(II) acetate affords all five possible mercuration products.[97]

The pattern of reactivity of heterocyclic aromatic compounds towards mercuration is the same as in other electrophilic substitutions. The five-membered ring systems such as pyrrole, furan, and thiophene in which the heteroatom contributes two electrons to the aromatic sextet are mercurated extremely readily, particularly at the 2- and 5- positions. On the other hand the six-membered ring systems analogous to benzene derivatives are markedly less reactive than the corresponding hydrocarbons, but they can be mercurated, preferentially at the 3-position, under sufficiently vigorous conditions. Thus whereas mercury(II) chloride effects both mono- and di-substitution of furan at room temperature,[98] pyridine has to be heated at 155 °C with the more powerful mercury(II) acetate to obtain 3-acetoxymercuriopyridine.[99]

Not surprisingly therefore the greatest interest has been shown in the five-membered ring heterocycles for which a host of mercuration products have been characterised.[13,14] For example, the range of compounds derived from mono-substituted furans (equation 19.50; X = O, Y = alkyl, CH_2OH, CHO, I) and thiophenes (equation 19.50; X = S, Y = Me, Ph, CH(OH)Me, NHAc, Cl, Br, I) are just a few of the known mercurials based on these heterocyclic systems.

$$Y \diagdown X + HgCl_2 \xrightarrow[H_2O/EtOH]{NaOAc} Y \diagdown X \diagdown HgCl \qquad (19.50)$$

Polymercuration occurs with many heterocycles, particularly if mercury(II) acetate is used. Thus two acetoxymercurio groups are introduced into indole in aqueous solution; isotopic labelling experiments[100] indicate that the pattern of substitution is 1,3 (equation 19.51a) and not 2,3 as originally suggested. Mono-mercuration can be accomplished by carrying out the reaction in dimethyl-formamide, which affords a mixture of isomers (equation 19.51b) or in acetic acid (equation 19.51c).[101]

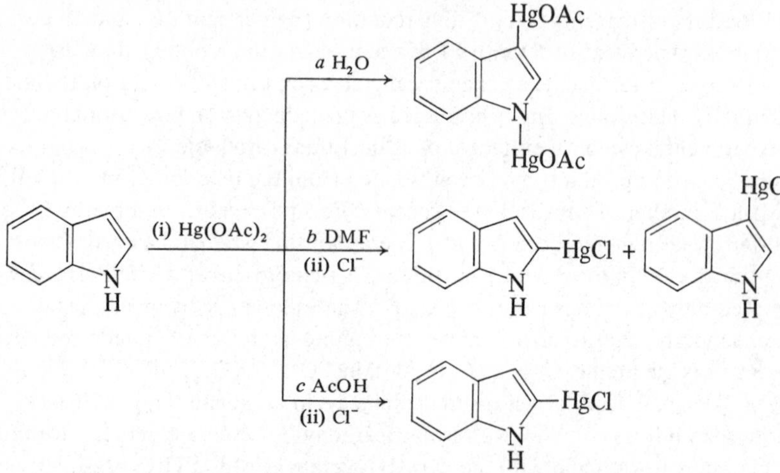

$$(19.51)$$

Of the systems containing more than one heteroatom, derivatives of 1,2- and 1,3-oxazole, 1,3-thiazole and pyrazole have been mercurated with mercury(II) acetate. Imidazolium[102] and benzimidazolium[103] salts can be mercurated by mercury(II) chloride under basic conditions (for example equation 19.52; X = H, Me, Cl, NO$_2$).

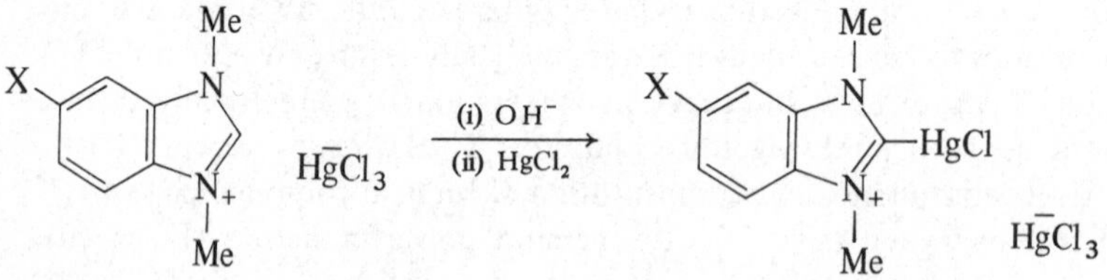

$$(19.52)$$

These reactions presumably involve the formation of carbenes, which effect nucleophilic displacement at mercury(II) and as such are related to the basic mercuration of pentafluorobenzene (p. 156).

19.2.2 Aliphatic Mercuration

Aliphatic compounds with suitably high CH acidities can be monomercurated under mild conditions. 1,3-dicarbonyl compounds are particularly interesting substrates because in principle the mercury derivatives can adopt C-bonded (**7**), O-bonded (**8**), or chelated (**9**) structures (figure 19.3). The structural implications

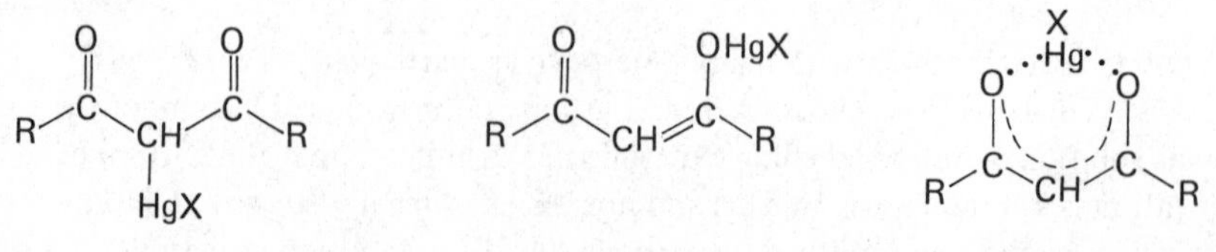

Figure 19.3

of formulae used in describing the formation of these derivatives are justified (and modified) in the ensuing discussion.

The preparative aspects of this subject go back a long way; for example the mercuration of dimethyl malonate was described in 1908 (equation 19.53).[104]

$$2 \ CH_2(CO_2Me)_2 + HgO \xrightarrow[H_2O]{} Hg[CH(CO_2Me)_2]_2 \qquad (19.53)$$
$$91 \ per \ cent$$

In recent years interest has centred on 1,3-diketones with a view to answering the structural question. These compounds react with mercury(II) acetate in aqueous ethanol to afford the organomercury acetates in yields of about 70 per cent (equation 19.54; R = Me,[105] Bu^t[105,106]).

$$CH_2(COR)_2 + Hg(OAc)_2 \rightarrow (RCO)_2CH.HgOAc \qquad (19.54)$$

Similar reactions with mercury(II) chloride afford lower yields (about 30 per cent) of the corresponding organomercury chlorides[105] but dimercuration (equation 19.55) also occurs.[105,107]

$$CH_2(COR)_2 + 2HgCl_2 \rightarrow (RCO)_2C(HgCl)_2 \qquad (19.55)$$

Unlike the aliphatic diketones, but similar to diethyl malonate and ethyl aceto-acetate,[108] dibenzoylmethane is dimercurated by mercury(II) acetate.[76]

The reaction of diketones with mercury(II) acetate also gives diorganomer-curials if an excess of diketone is used[105] or longer reaction times are employed[106] (equation 19.56; R = R' = Me or Bu^t), or if a perfluoropropyl group is present (equation 19.56; R = C$_3$F$_7$, R' = Bu^t).[77]

$$2 \ RCOCH_2COR' + Hg(OAc)_2 \rightarrow [RCOCH(COR')]_2Hg \qquad (19.56)$$

Better yields are obtained and the possibility of contamination with organo-mercury salts avoided by carrying out the mercuration using bis(hexamethylsilyl-amino)mercury (equation 19.57; R = Me, Bu^t).[105]

$$2 \ CH_2(COR)_2 + Hg[N(SiMe_3)_2]_2 \xrightarrow[Et_2O]{} [(RCO)_2CH]_2Hg \qquad (19.57)$$

X-ray diffraction studies show that in the solid state both dipivaloylmethyl-mercury acetate[78] and bis(dipivaloylmethyl)mercury[105] are C–Hg bonded (7; R = Bu^t, X = OAc or CH(COBu^t)$_2$); molecular parameters are given in chapter 21. These results confirm earlier suggestions based on the observation of a sharp i.r. absorption at 1678 cm^{-1} ($\nu_{C=O}$). Chelated acetylacetonate complexes of other metals exhibit a broad band at 1580 cm^{-1} and the presence of a similar band in dimercurated dibenzoylmethane is taken to indicate a structure analogous to 9 in which the position α to both CO groups is also mercurated.[76]

The structure of diorganomercurials in solution has been examined using ^{1}H n.m.r. spectroscopy where geminal ^{199}Hg–^{1}H coupling constants are parti-cularly valuable in diagnosing the species present. The solid-state structure of bis(dipivaloylmethyl)mercury is largely retained in deuteriochloroform but small amounts ($<$2 per cent) of a tautomer, probably the dienolate (Bu^tCOCH: $CBu^tO)_2Hg$, are also present.[105] However, the forms of the mercurial derived from $C_3F_7COCH_2COBu^t$ present in deuterioacetone have been clearly established (equation 19.58) and at $-60\,°C$ the equilibrium constant is 1.0.[77]

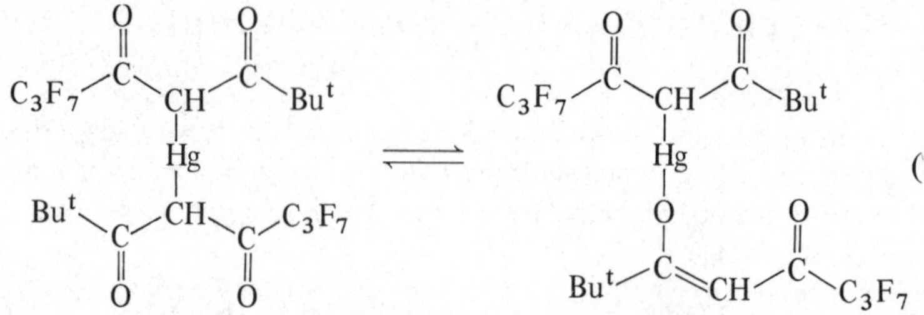

$$ \text{(19.58)} $$

This tautomerisation was studied between -60 and $+40\,°C$ and the activation parameters found to be $\Delta H^{\ddagger} = 9.35 \pm 0.4$ kcal mol^{-1} and $\Delta S^{\ddagger} = -3.45 \pm 1.6$ e.u.

Monocarbonyl compounds undergo α-mercuration under mild conditions provided that the α-position also carries a diazo substituent (equation 19.59; $Y = R$ or OR, $X_2 = O$[109] or $X = N(SiMe_3)_2$ [67,110])

$$ 2\,HC(N_2)COY + HgX_2 \rightarrow Hg[C(N_2)COY]_2 \qquad (19.59) $$

The corresponding reaction with organomercury reagents provides unsymmetrical diorganomercurials (equation 19.60; $X = N(SiMe_3)_2$,[67] OH,[111] or OEt[111])

$$ HC(N_2)COY + RHgX \rightarrow RHgC(N_2)COY \qquad (19.60) $$

Diazoalkanes, $HC(N_2)Z$, with other electron-withdrawing substituents (Z) such as CF_3,[109] CN,[109] NO_2,[112] $P(O)R_2$,[113] and $P(O)(OR)_2$ [114] react similarly to give the diorganomercurials, $Hg[C(N_2)Z]_2$, in high yield. In the absence of the diazo group the dipolar multiply bonded substituents can activate the α-position sufficiently for mercuration to occur under mild conditions provided, as with carbonyl groups, that at least two such substituents are present. Examples of the mercuration of compounds containing NO_2,[115] PO,[116] and SO_2 [117] groups are shown in equations (19.61) to (19.63) respectively.

$$ 2\,HCF(NO_2)_2 + HgO \rightarrow Hg[CF(NO_2)_2]_2 \qquad (19.61) $$

$$ CH_2[P(O)(OEt)_2]_2 + 2Hg(OAc)_2 \rightarrow [(EtO)_2P(O)]_2C(HgOAc)_2 \qquad (19.62) $$

$$ CH_2(SO_2Ar)_2 + Hg(OAc)_2 \rightarrow (ArSO_2)_2CH\cdot HgOAc \qquad (19.63) $$

Carbon–carbon double bonds can be responsible for facilitating mercuration. Kinetics and product analysis indicate that allylic oxidation of alkenes by mercury(II) acetate proceeds via rate-determining formation of allylmercury acetate followed by product-forming solvolysis (equation 19.64).[118]

$$ RCH_2\cdot CH{:}CH_2 \xrightarrow{Hg(OAc)_2} RCH{:}CH\cdot CH_2HgOAc \xrightarrow{HOAc} RCH{:}CH\cdot CH_2OAc $$
$$ + RCH(OAc)CH{:}CH_2 \qquad (19.64) $$

As implied by the equation it is envisaged that the mercury attacks the double bond rather than the allylic position so that an S_E' rather than an S_E process is involved.

The mercuration of cyclopentadiene can also be envisaged as S_E' process, although direct replacement of an allylic hydrogen should now be more accessible in view of the higher acidity. Dicyclopentadienylmercury is obtained if the mercurating agent is bis(hexamethylsilylamino)mercury,[119] or mercury(II) oxide in the presence of a primary amine (equation 19.65).[120]

$$2\,C_5H_6 + HgO \xrightarrow{PrNH_2} (C_5H_5)_2Hg \qquad (19.65)$$
$$\text{85–90 per cent}$$

With mercury(II) acetate or chloride, mixtures of non-reproducible stoichiometry are produced, but exhaustive mercuration takes place if a large excess of chloride is used in a buffered medium (equation 19.66).[121]

$$\underset{xs}{C_5H_6 + HgCl_2} \xrightarrow{NaOAc} C_5(HgCl)_6 \qquad (19.66)$$

Pentachlorocyclopentadiene should be even more acidic than the hydrocarbon and certainly it is readily mercurated (equations 19.67 and 19.68).[122]

$$2\,C_5Cl_5H + Hg(OMe)_2 \xrightarrow{0\,^{\circ}C} (C_5Cl_5)_2Hg \qquad (19.67)$$

$$C_5Cl_5H + PhHgOH \xrightarrow[Me_2C(OMe)_2]{0\,^{\circ}C} PhHgC_5Cl_5 \qquad (19.68)$$

Cyclopentadienylmercurials, which can also be prepared by the transmetallation route, pose an interesting bonding problem that has been investigated extensively. At various times structures involving monohapto (**10**), trihapto (**11**), and penta-hapto (**12**) rings have been proposed (figure 19.4). No crystal structures have been reported, probably because single crystals of these compounds are difficult to grow and they decompose rapidly when exposed to light. However, it has been firmly established by i.r.[123–125] and ^{1}H n.m.r.[126] spectroscopic means that in solutions of these derivatives the rings are σ-bonded to mercury (**10**). The n.m.r. data further show that the compounds h^1-C_5H_5HgX are 'fluxional', that is, the point of attachment of the mercury to the ring is changing rapidly on the n.m.r. time scale,

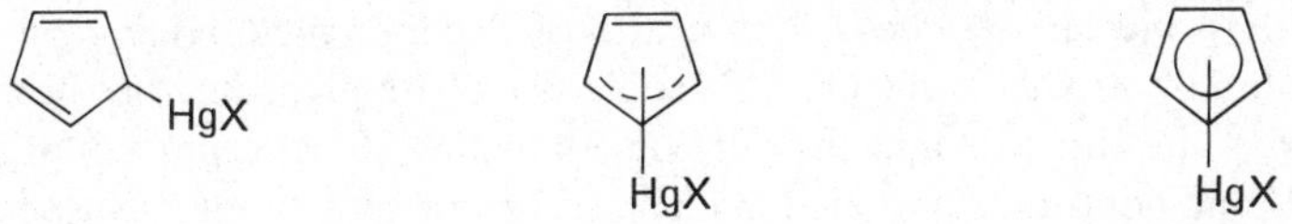

10   **11**   **12**

Figure 19.4

even at temperatures as low as $-100\,^{\circ}$C. A spectrum corresponding to the instantaneous structure of cyclopentadienylmercury chloride (**10**; X = Cl) was obtained at $-113\,^{\circ}$C, but the ease of 'ring whizzing' increases as the electronegativity of X decreases. Wide-line n.m.r. spectroscopic evidence indicates that these molecules also reorient in the solid state.[127]

It is clear that the introduction of one substituent into the cyclopentadienyl ring does not greatly affect the fluxional behaviour of organomercurials. Thus rapid intramolecular migration of mercury occurs in the compounds $(h^1\text{-MeC}_5\text{-}H_4)_2\text{Hg}$[128] and $h^1\text{-Bu}^t\text{C}_5\text{H}_4\text{HgCl}$[25] (equation 19.69), and not until the temperature is lowered to $-96\,^{\circ}$C is the latter compound 'frozen' in one of its lowest free energy configurations (**13** or **14**).

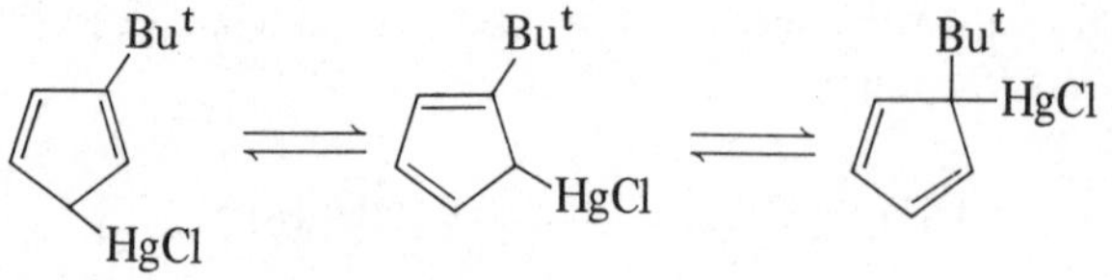

$$(19.69)$$

However the replacement of *all* hydrogens by methyl groups leads to a considerable reduction in the mobility of the mercury and the ^{1}H n.m.r. spectrum of the compound $(h^1\text{-C}_5\text{Me}_5)_2\text{Hg}$ shows it to be 'static' at room temperature.[129] It has not yet been determined how the introduction of five chlorines[122] affects fluxional behaviour.

Halogen substituents increase the susceptibility of substrates towards mercuration under basic conditions and reactions with pentafluorobenzene (equation 19.46) and pentachlorocyclopentadiene (equation 19.68) have already been mentioned. The mercuration of haloforms and related compounds is the principal route to the 'Seyferth reagents' (equation 19.70)[130], which are important in synthesis for their ability to transfer divalent carbon groups to a variety of acceptors (see chapter 23).

$$\text{RHgCl} + \text{HCXYZ} + \text{Bu}^t\text{OK} \xrightarrow[-60\,^{\circ}\text{C}]{\text{THF/Et}_2\text{O}} \text{RHgCXYZ} \qquad (19.70)$$

Phenylmercury compounds (R = Ph) make up the most extensive series but several examples in which R is cyclohexyl are also known.[131] Two halogens (X and Y) are always present in the substrate but while the third substituent (Z) is often also a halogen, it can be a group such as CF_3,[132] CO_2R,[133] $CONR_2$,[134,135] or SO_2Ph,[135] and can even be H.[136] Occasionally sodium methoxide is preferred to potassium t-butoxide as the base,[137] and a two-phase procedure has been developed[138] in which an aqueous solution containing 20 per cent each of sodium hydroxide and potassium fluoride is added at room temperature to a suspension of phenylmercury(II) chloride in the haloform and a small amount of triethylbenzylammonium chloride.

These reactions are generally thought of as proceeding by nucleophilic attack of the carbanion $:\overset{\ominus}{C}XYZ$ on mercury but it is known that organomercury alkoxides are formed from the corresponding chlorides and alkali-metal alkoxides[139] and that they mercurate chloroform under mild conditions (equation 19.71; R = Me[140] or Bu^t[139]).

$$PhHgOR + HCCl_3 \rightarrow PhHgCCl_3 + ROH \qquad (19.71)$$

The scope for using phenylmercury(II) alkoxides in these mercurations has not been investigated even though simple alternative methods of preparing them have been described (equations 19.72 and 19.73).[140]

$$(PhHg)_2O + (MeO)_2CO \rightarrow 2\, PhHgOMe + CO_2 \qquad (19.72)$$

$$PhHgOMe + ROH \rightarrow PhHgOR + MeOH \qquad (19.73)$$

Terminal alkynes are another class of comparatively acidic hydrocarbons and their mercuration under basic conditions (equations 19.74[141] and 19.75[142]) has long been established.

$$2RC\vdots CH + HgI_4{}^{2-} + 2^-OH \rightarrow (RC\vdots C)_2Hg + 4I^- + 2H_2O \qquad (19.74)$$

$$HC\vdots CH + 2RHgCl + 2^-OH \rightarrow RHgC\vdots CHgR + 2Cl^- + 2H_2O \qquad (19.75)$$

Aryl-, alkyl-, and various functionally substituted alkyl-acetylenes are known to participate in reaction 19.74.[13] Alternative mercurating agents are mercury(II) oxide in benzene[143] and mercury(II) acetate[144] or chloride[143] in the presence of an organic base.

It has been suggested[145] that the first step in these mercurations is the formation of a π-complex involving the triple bond. Evidently this complex is converted to a σ-bonded species by proton loss under basic conditions but by addition of a nucleophile to form a β-substituted alkenylmercurial in acidic media (see the following section). Addition is the usual outcome of the reaction between alkenes and mercury(II) salts but a few examples of vinylic mercuration are known, particularly where steric factors make addition unfavourable, for example in 1,7,7-trimethyl-2-phenylbicyclo[2.2.1]hept-2-ene[146] and some $\alpha\beta$-unsaturated steroidal ketones.[147]

19.3 Addition (Oxymercuration and Related Reactions)

Aliphatic organomercury compounds containing selected oxygen, nitrogen, and, to a lesser extent, carbon substitutents β to the mercury are readily prepared by the reaction of alkenes with mercury(II) salts in the presence of appropriate nucleophiles (equation 19.76).

$$R^1R^2C\vdots CR^3R^4 + HgX_2 + R_nYH \rightleftharpoons R^1R^2C(YR_n)C(HgX)R^3R^4 + HX$$

$$(19.76)$$

Similar addition processes are possible with alkynes and cyclopropanes (section 19.3.4) but the scope is very much more limited. Because of the ease and versatility of the reaction, the unusual chemical behaviour of the adducts compared with that of simpler organomercurials, and the potential value of the system in organic synthesis (chapter 23), a very extensive literature exists, which spans over seventy years yet continues to flourish today. Sizeable parts of the major reference works[13,14] are devoted to the subject and several specialised reviews exist. The most valuable reviews are the two by Kitching,[148] which concentrate on mechanistic aspects and correlate the reaction with similar systems involving salts of other metals, and an earlier one by Chatt.[149]

19.3.1 Scope

Considerable variation is possible in each of the three components of the reaction mixture. Of the mercury(II) salts the acetate is by far the most widely used. This is partly because while the reaction is generally reversible it proceeds essentially to completion when X = OAc. The more strongly electrophilic nitrate and perchlorate react more rapidly but with most alkenes the equilibrium position is unfavourable so that concurrent neutralisation of the generated acid is necessary. Because mercury(II) chloride is a relatively weak electrophile and halogenoacids are particularly effective in promoting deoxymercuration (the back reaction), the chloride is rarely used; the main exception is in aminomercuration where the nucleophile can also serve as the base to remove hydrogen chloride. This is unfortunate because the organomercury chlorides are usually more amenable to purification than the corresponding acetates, for example they crystallise and chromatograph more readily. However the acetates undergo rapid anion exchange when treated with aqueous sodium or potassium chloride.

An alternative route to organomercury chlorides is to use mixed salts (equation 19.77;[150] for example Y = $NHCOCH_2NHCONH_2$) but this has rarely been followed.

$$YCH_2 . CH : CH_2 + AcOHgCl + MeOH \rightarrow YCH_2 . CH(OMe)CH_2HgCl + HOAc$$

$$(19.77)$$

Mercury salts of other carboxylic acids can be useful. In particular the trifluoroacetate has been employed recently to circumvent the problem of acetate or acetic acid competing with the desired nucleophile.[9] Salts of chiral acids, especially mercury(II) tartrate, have been used to promote asymmetric synthesis of secondary alcohols;[151,152] optical purities of up to 32 per cent have been obtained (equation 19.78).

$$R^1CH : CH_2 \xrightarrow[H_2O/THF]{Hg(O_2CR^*)_2} \xrightarrow[OH^-]{NaBH_4} R^1{}^*CH(OH)Me \qquad (19.78)$$

An extremely wide range of alkenes participate in these addition reactions. The relative rates of hydroxymercuration ($R_n YH = H_2O$) of many simple alkenes reveal the following reactivity trend[153]

$$R_2C : CH_2 > RCH : CH_2 > cis\text{-}RCH : CHR > trans\text{-}RCH : CHR$$

$$> R_2C : CHR > R_2C : CR_2$$

Reactivity is reduced by branching in R or by conjugation of the double bond with phenyl groups. A similar study[154] for methoxymercuration ($R_n YH = MeOH$) mainly of cyclic and bicyclic alkenes shows that, contrary to earlier feelings, there is no correlation between reactivity and angle strain energy in the alkene.

Dienes with cumulative, conjugated, or isolated double bonds often behave like simple alkenes, but cyclisation reactions can occur if the isolated double bonds are suitably positioned (see below). Many functional substituents can be tolerated. Thus CO_2H, CO_2R, COR, or CN can be attached directly to the double bond and merely reduce its reactivity. Addition to alkenyl halides, ethers, and acetates also occurs but can be complicated by subsequent elimination, particularly in aqueous media (equation 19.79[155] for example).

$$MeCCl : CH_2 + Hg(NO_3)_2 + H_2O \rightarrow [MeCCl(OH)CH_2HgNO_3]$$

$$\rightarrow Me_2CO + ClHgNO_3 \qquad (19.79)$$

All these substituents and others such as OH or NHR can be located anywhere in an alkyl group attached to the double bond (equation 19.77 for example) and they will not interfere with the reaction unless they are positioned suitably to function as internal nucleophiles.

A large variety of nucleophiles are active. The oldest and by far the most abundant examples are of reactions involving oxygen-centred reagents and hence called 'oxymercurations'. Reactions with water and alcohols (especially methanol) have had the lion's share of attention within this class, but those with carboxylic acids (mainly acetic acid), hydrogen peroxide, and alkyl hydroperoxides are also known. The nitrogen-centred nucleophiles include the non-protic species NO_2^- (equation 19.80[156]), N_3^- (equation 19.81[157]), and RCN[158,159] (equation 19.82) in addition to primary and secondary amines.

$$EtCMe : CH_2 + HgCl_2 + NO_2^- \rightarrow EtCMe(NO_2)CH_2HgCl \qquad (19.80)$$

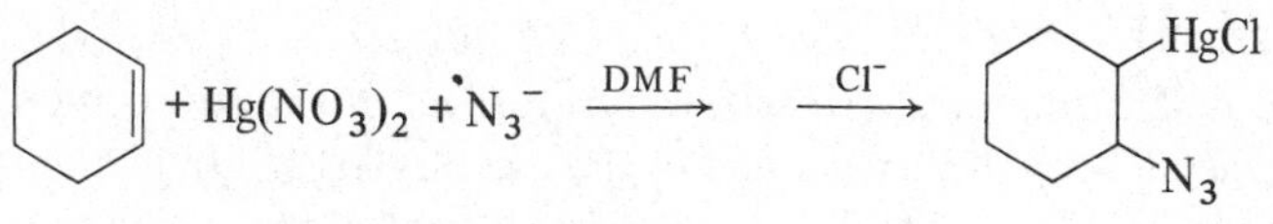

$$(19.81)$$

$$BuCH : CH_2 + Hg(NO_3)_2 + MeCN \rightarrow [BuCH(NCMe)CH_2HgNO_3]NO_3$$

$$\xrightarrow[H_2O]{Cl^-} BuCH(NH.COMe)CH_2HgCl \qquad (19.82)$$

Where the nucleophile is water or a liquid alcohol, carboxylic acid, or amine it often serves also as the solvent and the term 'solvomercuration' has been coined for such systems. Alternatively the reaction can be carried out in an inert solvent and tetrahydrofuran or dichloromethane are particularly suitable; mercury(II) trifluoroacetate is considerably more soluble than the acetate in these solvents. Hydroxymercuration is conveniently carried out with mercury(II) acetate in a water–tetrahydrofuran mixture if the subsequent replacement of mercury by hydrogen is intended (for example equation 19.78).

Reactions involving carbon nucleophiles are of considerable synthetic importance since they result in chain extension. Suitable aromatic derivatives and β-dicarbonyl compounds are reactive and the original examples described by Ichikawa are shown in equations 19.83[160] and 19.84[161] respectively; reactions with aromatic species have been extended to alkenes such as propene, styrene, but-2-ene, and cyclohexene.[162]

$$CH_2 : CH_2 + Hg(OAc)_2 + C_6H_5OMe \xrightarrow{AcOH}$$

$$4\text{-MeOC}_6H_4 . CH_2CH_2HgOAc \rightarrow demercuration \quad (19.83)$$

$$CH_2 : CH_2 + Hg(OAc)_2 + MeCOCH_2CO_2Et \xrightarrow[BF_3]{AcOH}$$

$$MeCOCH(CO_2Et)CH_2CH_2HgOAc \rightarrow demercuration \quad (19.84)$$

Although some of these adducts have been isolated they demercurate readily under the reaction conditions to give products of the type RCH_2CH_2OAc and $ArCH_2\text{-}CH_2Ar$. Since the reactions are carried out in acetic acid, these 'carbomercurations' are probably best regarded as β-substitutent exchanges. Related exchanges are known (for example equations 19.85[163] and 19.86[164]) and the carbon adducts can be prepared from the corresponding acetoxymercurials if a catalytic amount of perchloric acid is present[165,166] (for example equation 19.87; R = H, Me, Ph, R^1 and R^2 = Me or Ph).

$$C_6H_{13}CH(OMe)CH_2HgClO_4 + EtOH \rightarrow C_6H_{13}CH(OEt)CH_2HgClO_4$$
$$(19.85)$$

$$MeCH(OH)CH(HgX)Me + NO_2^- \rightarrow MeCH(NO_2)CH(HgX)Me \quad (19.86)$$

$$RCH(OAc)CH_2HgOAc + R^1COCH_2COR^2 \xrightarrow{Cl^-}$$

$$R^1COCH(COR^2)CHR . CH_2HgCl \quad (19.87)$$

β-substituted alkylmercury(II) salts can be formed from a *two*-component mixture if the nucleophile is the anion of the mercury(II) salt (equation 19.88, R = Me[167] or CF$_3$[168]) or alternatively is a substituent in the alkene so positioned that a five- or six-membered ring can be formed.

$$R^1R^2C:CR^3R^4 + Hg(O_2CR)_2 \underset{solvent}{\overset{inert}{\rightleftharpoons}} R^1R^2C(O_2CR)CR^3R^4 . HgO_2CR$$
$$(19.88)$$

Many examples of the second type are known where the internal nucleophilic group is OH, CO_2H, or NHR (section 19.3.3), and intramolecular iminomercuration was recently postulated to account for mercury(II) catalysis of a sigmatropic rearrangement (equation 19.89[169]).

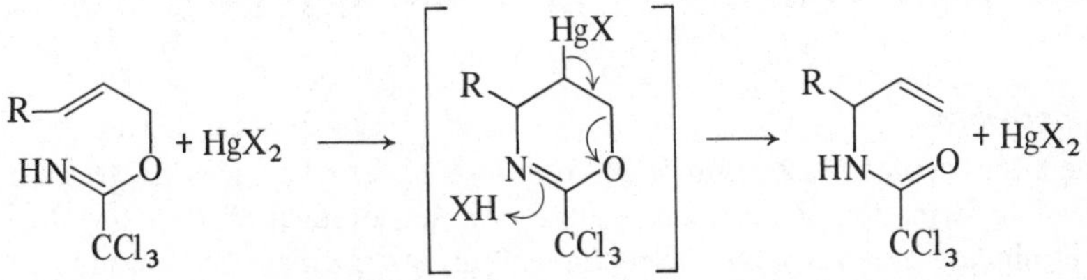

$$(19.89)$$

A comparatively new development is the discovery of reactions in which the carbon–carbon double bond of a diene[170,171] or triene[172] (equation 19.90) is the internal nucleophile. These differ in that except for one example (equation 19.91[173]) an external nucleophile again enters the molecule, now to saturate the carbon atom adjacent to that involved in ring closure.

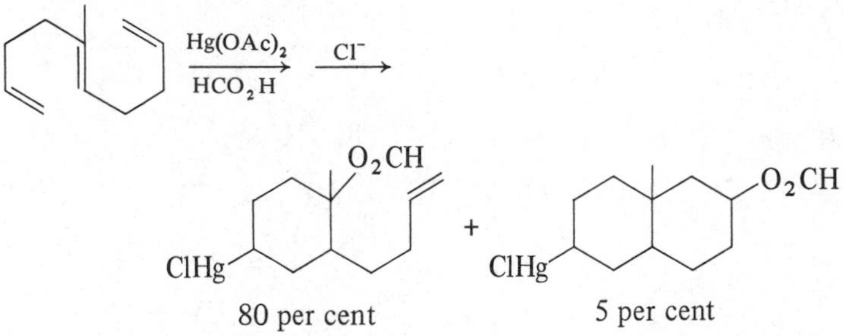

80 per cent   5 per cent

$$(19.90)$$

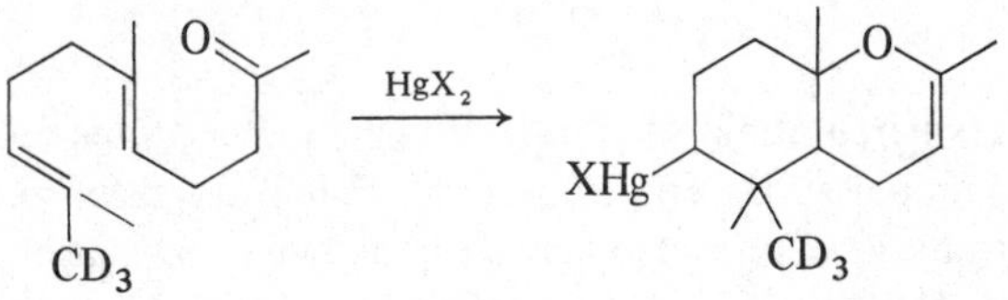

$$(19.91)$$

A re-investigation[174] of the oldest example of this reaction revealed that the kinetically controlled product from norbornadiene is the usual 1,2-adduct; the nortricyclic compound is formed in a subsequent rearrangement, which is catalysed by mercury(II) salts. It is possible that a similar mechanism might apply to some of the more recent examples, and it should be noted that there is no evidence for oligomerisation of simple alkenes under oxymercuration conditions.

Dienes can also give rise to $\beta\beta'$-dimercurated cyclic ethers if hydroxymercuration of one double bond affords an unsaturated alcohol capable of undergoing intramolecular oxymercuration. Analogous reactions with hydrogen peroxide produce cyclic peroxides.[175]

The breadth of the scope for preparing mercurials by addition has been emphasised and more details of the main types, that is oxy- and amino-mercuration, appear later. Other important features of the reaction are that it is highly regio- and stereo-selective and is very rarely accompanied by rearrangement. These observations have helped to build up our current picture of the mechanism of addition.

19.3.2 Mechanism

The generally accepted mechanism of addition to unhindered alkenes involves the fast reversible formation of a cationic bridged π-complex (equation 19.93) with rate determining *trans*-attack by nucleophile on this intermediate 'mercurinium ion' (equation 19.94).

$$HgX_2 \overset{fast}{\rightleftharpoons} \overset{+}{Hg}X + \bar{X} \tag{19.92}$$

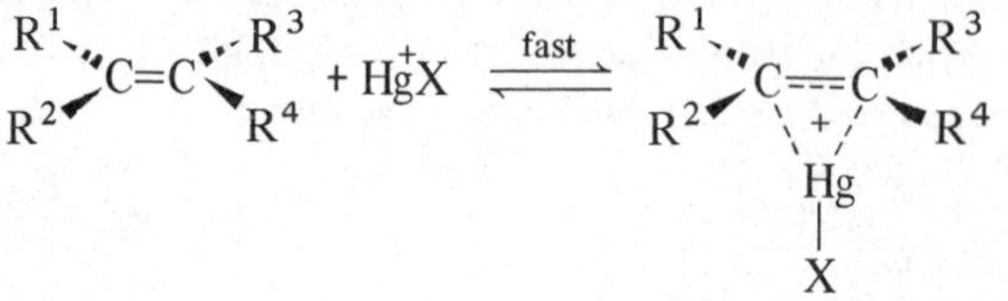

$$\begin{array}{c} R^1 \\ \diagdown \\ R^2 \end{array} C = C \begin{array}{c} R^3 \\ \diagup \\ R^4 \end{array} + \overset{+}{Hg}X \overset{fast}{\rightleftharpoons} \begin{array}{c} R^1 \\ \diagdown \\ R^2 \end{array} C \underset{Hg}{\overset{+}{=}} C \begin{array}{c} R^3 \\ \diagup \\ R^4 \end{array} \tag{19.93}$$

$$\begin{array}{c} R^1 \\ \diagdown \\ R^2 \end{array} C \underset{\underset{X}{\overset{|}{Hg}}}{\overset{+}{=}} C \begin{array}{c} R^3 \\ \diagup \\ R^4 \end{array} + R_n YH \overset{slow}{\rightleftharpoons} \begin{array}{c} \overset{+}{R_n YH} \quad R^3 \\ | \qquad | \\ R^{1} C - C - R^4 \\ | \qquad | \\ R^2 \quad HgX \end{array} \tag{19.94}$$

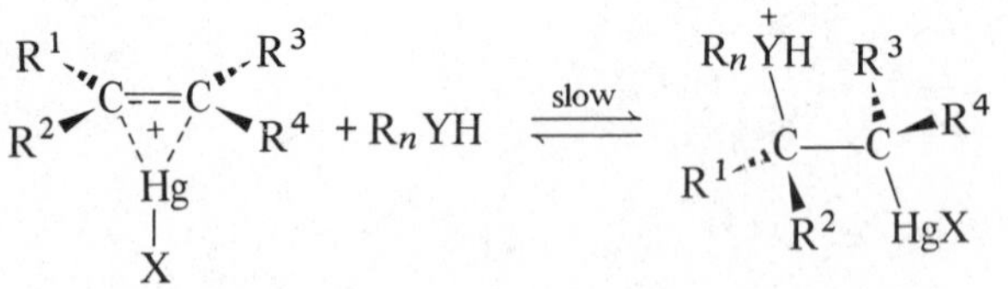

$$R^1R^2C(\overset{+}{Y}HR_n)C(HgX)R^3R^4 + Base \overset{fast}{\rightleftharpoons} R^1R^2C(YR_n)C(HgX)R^3R^4 + Base\,\overset{+}{H} \tag{19.95}$$

As with aromatic mercuration (section 19.2.1) a number of mercury(II) electrophiles including unionised HgX_2 may be active. The catalysis of mercury(II) acetate reactions by perchloric acid may again be due to formation of highly reactive ion pairs Hg^+OAc,ClO_4^-, but an alternative idea[176] is that species such as $HgOAc(HOAc)^+$ are involved.

Kinetic studies have been limited to hydroxymercuration[177,178] and methoxy-mercuration[176,179,180] under solvomercuration conditions where determination of the order with respect to the nucleophile is precluded. The observed rate law for terminal, medial, cyclic, and functionally substituted alkenes is first order in alkene and first order in mercury(II). This result is consistent not only with the picture we have presented but also with a mechanism in which formation of the π-complex (equation 19.93) is rate determining. However, the latter interpretation seems unlikely[177] in view of the very high rates of formation of other Hg(II) complexes and the high rates of formation of alkene complexes with Ag(I) ions for which the bonding mode is similar.[154] Furthermore a different order of alkene reactivities is observed in bromination where formation of the cyclic bromonium ion is considered to be rate determining. Thus tetramethylethylene is 10^7 times more reactive than ethylene in bromination but reacts very slowly in hydroxy-mercuration.[153]

In reactions with symmetrical alkenes, the observed reactivity trend of $CH_2=CH_2 > RCH=CHR > R_2C=CR_2$ is consistent with normal steric effects in an S_N2 process; that is, the primary carbon atom in the ethylene mercurinium ion is attacked by the nucleophile (equation 19.94) more readily than a corresponding secondary carbon atom, which in turn is attacked more readily than a tertiary carbon atom. Evidently, however, there is considerable S_N1 character in the transition state for reactions with unsymmetrical alkenes. Thus the extremely high regioselectivity for Markovnikov-type addition shows that electronic effects must here outweigh the steric factor. The influence of substituents on rates supports this view; a Taft plot of $\log k$ against $\sigma_R{}^*$ for R varying from CH_2CN to CH_2CH_3 in reaction 19.96 gives $\rho^* = -3.3$, which implies a high degree of positive-charge localisation, in the transition state, on the carbon bearing the substituent.[177]

$$RCH:CH_2 + HgX_2 + H_2O \xrightarrow{k} RCH(OH)CH_2HgX \qquad (19.96)$$

This raises doubts as to whether it is necessary to postulate mercurinium ions in these reactions and a brief review of the evidence supporting their intermediacy seems called for. The only kinetic evidence comes from a study[176] of the methoxy-mercuration of methyl 4-substituted cinnamates (equation 19.97).

$$4\text{-}XC_6H_4CH:CHCO_2Me + Hg(OAc)_2 + MeOH \rightarrow$$

$$4\text{-}XC_6H_4CH(OMe)CH(HgOAc)CO_2Me \qquad (19.97)$$

The rate increases by only a factor of about twenty on going from $X = NO_2$ to $X = MeO$ which, in a $\rho\sigma^+$ correlation, is equivalent to a ρ value of about -0.9. This implies that, in contrast to reaction 19.96, *little* positive charge is generated on the carbon undergoing nucleophilic attack and this is consistent with a transition state resembling a bridged ion rather than a carbonium ion.

That skeletal rearrangements rarely occur during oxymercuration (for example equation 19.98[181]) seems to speak against a carbonium ion intermediate.

$$Bu^tCH:CHBu^t + Hg(OAc)_2 + MeOH \rightarrow Bu^tCH(OMe)CH(HgOAc)Bu^t$$

$$(19.98)$$

The first case of rearrangement not involving a polycyclic system has recently been reported[182] (equation 19.99; $X = NO_3, ClO_4$)

$$ArC(OH)Me.CMe:CH_2 \xrightarrow[\text{(2) KI}]{\text{(1) } HgX_2/MeOH} MeCOC(Me)Ar.CH_2HgI \qquad (19.99)$$

Even with this alkene, which possesses a group of high migratory aptitude and a hydroxyl that can stabilise the rearranged cation by proton loss, rearrangement is not found when mercury(II) acetate is used.

The strongest evidence favouring the mercurinium ion is undoubtedly the observed high stereoselectivity for *trans*-addition. Thus methyl *cis*- and *trans*-4-methoxycinnamate yield specifically the products of *trans*-addition in reaction 19.97[176] and *cis*-di-*t*-butylethylene affords only the *threo*-adduct in reaction 19.98.[181] If a free carbonium ion were involved, the formation of some product corresponding to *cis*-addition would be expected either by *cis*-attack or by *trans*-attack after rotation about the $\overset{+}{C}$–CHg bond; such rotation would be particularly favourable with the highly strained *cis*-di-*t*-butylethylene.

Because of the importance of the stereochemical argument to the status of the mercurinium ion as an intermediate it is worth summarising the nature and extent of the evidence. It is important to remember that two quite separate pieces of information need to be established: (i) the selectivity of the reaction and (ii) the configuration of the (major) product. This has not always been recognised; thus the elegant experiments[183] that convincingly establish the configurations of aminomercurials formed from *cis*- and *trans*-but-3-ene cannot be taken to show that aminomercuration occurs *stereospecifically trans*, since the products investigated represented yields of only 27 and 4 per cent respectively.

To determine the selectivity the isomer distribution in the whole (crude) reaction mixture must be analysed. Provided that both stereoisomeric products are available, this is easily achieved with sensitive modern techniques such as spectroscopy and chromatography. For example ^{1}H n.m.r. spectroscopy has been used to show that t-butyl peroxymercuration of but-2-ene and stilbene (equation 19.100; R = Me, Ph[9]) and of methyl cinnamate[184] has a stereoselectivity of greater than 95 per cent.

$$RCH:CHR + Hg(O_2C.CF_3)_2 + Bu^tOOH \xrightarrow{CH_2Cl_2} RCH(OOBu^t)CH(HgO_2C.CF_3)R$$

$$(19.100)$$

The stereospecificity of both oxymercuration *and* deoxymercuration was neatly demonstrated some years ago by the fact that the appreciably different equilibrium constants for *cis*- and *trans*-but-2-ene (equation 19.101) do not drift towards a common value.[185]

$$MeCH:CHMe + HgCl_2 + H_2O \rightleftharpoons MeCH(OH)CH(HgCl)Me + H^+ + Cl^-$$

$$(19.101)$$

To determine the configuration of addition products is considerably more difficult. X-ray diffraction studies have revealed a *trans*-diequatorial disposition of chloromercuri- and methoxy- substituents in the cyclohexene product,[186] and the same configuration for the hydroxy-,[187,188] trifluoroacetoxy-,[189] nitro-[156] and t-butyl peroxy-[190] mercurials (see p. 141) was established by ^{1}H n.m.r. spectroscopy on the basis of vicinal proton–proton coupling constants. The latter technique has also been used to diagnose the configurations of oxymercurials from acyclic alkenes, namely the methoxymercurials from methyl 4-methoxycinnamate (equation 19.97[176]), *cis*-di-*t*-butylethylene (equation 19.98[181]) and ethylene-1,2-d_2[191] and the *t*-butyl peroxymercurials from but-2-ene and stilbene (equation 19.100[190]). The method uses the firmly established Karplus relationship, which predicts larger coupling constants for conformations with *anti*-disposed protons than for those with *gauche*-disposed protons, and its reliability depends upon the reasonableness of assumed conformational preferences. It has been pointed out that both repulsive and attractive vicinal interactions must be considered[190] and that preferences are solvent dependent.[192] The assignments least open to question are those of the ethylene-1,2-d_2 methoxymercurials but nevertheless the general conclusion that *trans*-addition occurs in these systems seems well founded.

Because of the importance of the stereochemical evidence to our mechanistic picture of addition, known exceptions to the rule of *trans*-addition require special attention. It is not surprising that *trans*-cyclooctene and *trans*-cyclononene exhibit *cis*-addition (equation 19.102; R = H, Me[193]) since rearside attack on the mercurinium ion is prevented sterically.

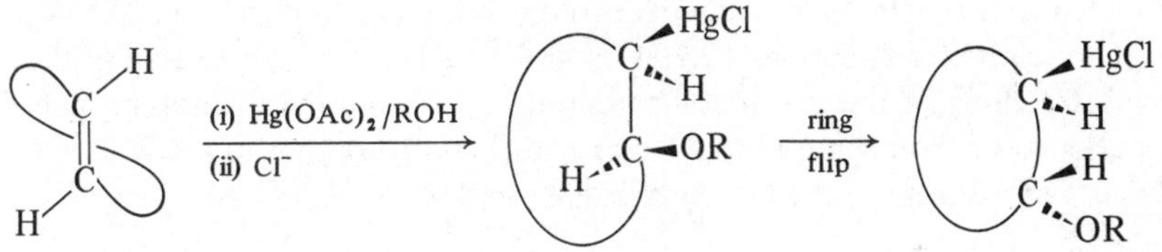

$$(19.102)$$

Stereospecific *cis*-addition is also found with norbornene (equation 19.103)[194,187] and a number of structurally related compounds such as norbornadiene, benzonorbornadiene, dicyclopentadiene[148] and bicyclo[3.2.1]octa-2,7-diene;[195] *trans*-di-*t*-butylethylene is the sole acyclic example.

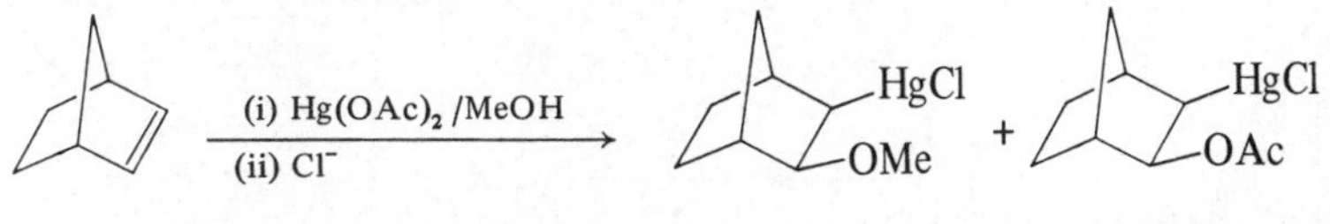

$$(19.103)$$

Mixtures of *trans* and *cis*-adducts are formed only with bicyclo[2.2.2]octene[194] and related bicyclooct-2-enes[154] and with *trans*-cyclodecene.[196]

The mercurinium ion is retained in the current view of these reactions and *cis*-addition is envisaged as proceeding by frontside collapse of the solvated intermediate; the effect of solvent on the balance between *cis*- and *trans*-addition to bicyclo[2.2.2]octene is nicely explained by this picture.[197] The feeling is that *trans*-addition is the preferred mode but that for the bicyclic alkenes, twist-strain makes the appropriate transition state energetically unfavourable and *cis*-addition prevails. It remains to be explained why the mercury(II) electrophile approaches only from the *exo*-side, and steric effects do not seem to be the answer since *exo*-attack still occurs on norbornene derivatives bearing a *syn*-7-substituent.[198]

Mercurinium ions derived from acyclic alkenes, cyclohexene and norbornene have been prepared in superacid media of low nucleophilicity and characterised by n.m.r. spectroscopy.[5] While this is not directly relevant as far as the question of whether such species are involved in oxymercuration is concerned, it greatly enhances their credibility as intermediates.

19.3.3 Oxy- and Aminomercuration

The aim of this section is to illustrate further the kinds of organomercury compounds that can be prepared when water, hydrogen peroxide, alcohols, alkyl hydroperoxides, carboxylic acids, or amines assume the role of external nucleophile or, where appropriate, of internal nucleophile in addition processes. Where possible examples will be taken from the more recent literature.

Reactions in aqueous solution normally produce β-mercurated alcohols. Current interest is mainly in obtaining the corresponding mercury-free alcohols by subsequent reduction (chapter 23) and very often this objective has been achieved without characterising the intermediate organomercury compounds. Two recent exceptions are given in equations 19.104[147] and 19.105.[199] A comparison of the hydroxyl stretching frequencies in the parent mercurials and the alcohols produced on reduction has proved a useful tool for assigning configurations to the hydroxymercurials derived from cyclic and bicyclic alkenes.[193]

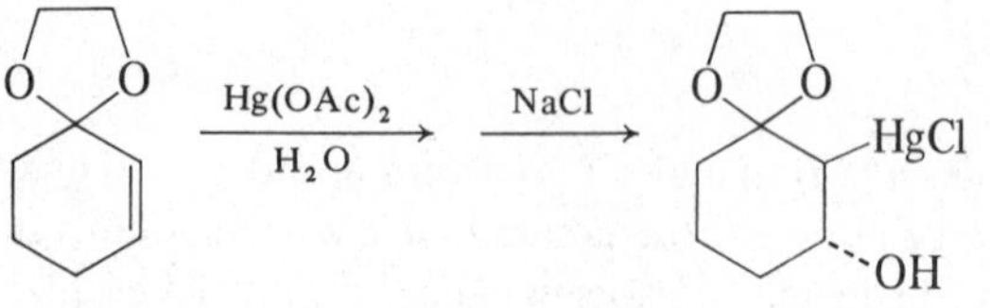

$$\text{(19.104)}$$

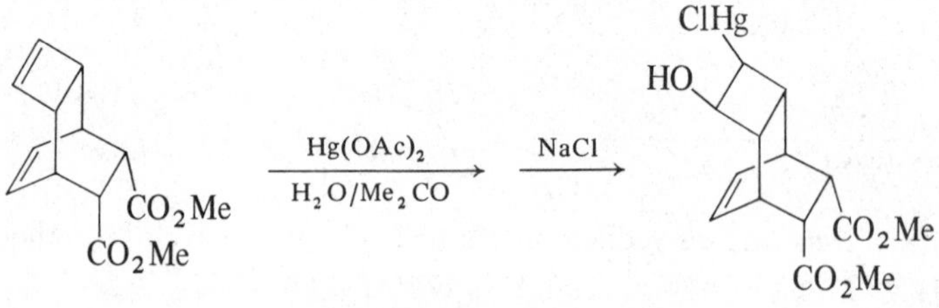

$$\text{(19.105)}$$

In the original work on hydroxymercuration, Hofman and Sand described the formation from ethylene of di-β-mercurioethyl ether in addition to β-mercurioethyl alcohol (equation 19.106; see references 12, 13, and 149).

$$CH_2 : CH_2 + HgSO_4 + H_2O \xrightarrow{(KOH)} \xrightarrow{Cl^-} HOCH_2CH_2HgCl + O(CH_2CH_2HgCl)_2$$

$$\text{(19.106)}$$

Production of the ether can be thought of as a β-substituent exchange or as an alkoxymercuration of ethylene with β-mercurioethyl alcohol functioning as the nucleophile; either way attack of β-mercurioethyl alcohol on an ethylene mercurinium ion (equation 19.107) is probably involved.

$$\begin{array}{c} CH_2 \cdots CH_2 \\ \diagdown \overset{+}{} \diagup \\ Hg \\ | \\ X \end{array} + HOCH_2CH_2HgX \longrightarrow$$

$$(XHgCH_2CH_2)_2\overset{+}{O}H \xrightarrow{-H^+} (XHgCH_2CH_2)_2O \quad (19.107)$$

Surprisingly the scope for making ethers from alkenes and water by two-stage oxymercuration appears to be unexplored other than that for making cyclic ethers from dienes (see below).

Not until seventy years after the discovery of hydroxymercuration was the analogous reaction with hydrogen peroxide reported[200,201] (equation 19.108).

$$PhCH:CH_2 + Hg(OAc)_2 + H_2O_2 \rightarrow PhCH(OOH)CH_2HgOAc \quad (19.108)$$

A large excess of concentrated hydrogen peroxide was employed but this potentially hazardous procedure can be avoided if mercury(II) trifluoroacetate is used and the reaction carried out in dichloromethane.[175] Under these conditions mixtures of hydroperoxides and di-β-mercurioalkyl peroxides are obtained (equation 19.108). This caused problems in isolating the hydroperoxides derived from ethylene, propene and styrene, but those from isobutylene and α-methylstyrene (equation 19.109) could be obtained in 50 per cent yield.

$$RCH:CH_2 + Hg(O_2C.CF_3)_2 + H_2O_2 \xrightarrow{CH_2Cl_2} \xrightarrow{Cl^-}$$
$$RCH(OOH)CH_2HgCl + [RCH(CH_2HgCl)O]_2$$
$$(19.109)$$

$$RCMe:CH_2 + Hg(O_2C.CF_3) + H_2O_2 \xrightarrow{CH_2Cl_2}$$
$$RCMe(OOH)CH_2HgO_2C.CF_3 \quad (19.110)$$

A mixed di-β-mercurioalkyl peroxide has been prepared by carrying out the reaction in a stepwise fashion (equations 19.109 and 19.111; $X = O_2C.CF_3$)

$$PhCH:CH_2 + HgX_2 + Me_2C(OOH)CH_2HgX \xrightarrow{CH_2Cl_2}$$
$$XHgCH_2CMe_2OOCHPhCH_2HgX \quad (19.111)$$

The only functionally substituted alkene yet to be used is dihydropyran, which gives a normal 1,2-adduct with the hydroperoxide group α to the ring oxygen.[201]

The first example of alkoxymercuration was the reaction of ethylene with mercury(II) acetate in methanol, which was reported in 1913.[202] The methoxymercuration of a wide range of unsaturated systems has been studied subsequently[13,14] and recent examples include simple alkenes (equation 19.112[203]), a bicyclic diene,[204] and tetraene[205] (equation 19.113), where the addition is accompanied by rearrangement, and cyclohexene carrying one or two cyanide substituents[206] (equation 19.114; R = H or CN). Conformational analysis in the last system led to the conclusion that, in contrast to an earlier suggestion,[207] there is no preliminary orienting coordination of the mercury(II) electrophile by the substituent.

$$Bu^tCH:CH_2 + Hg(OAc)_2 + MeOH \rightarrow Bu^tCH(OMe)CH_2HgOAc$$

$$(19.112)$$

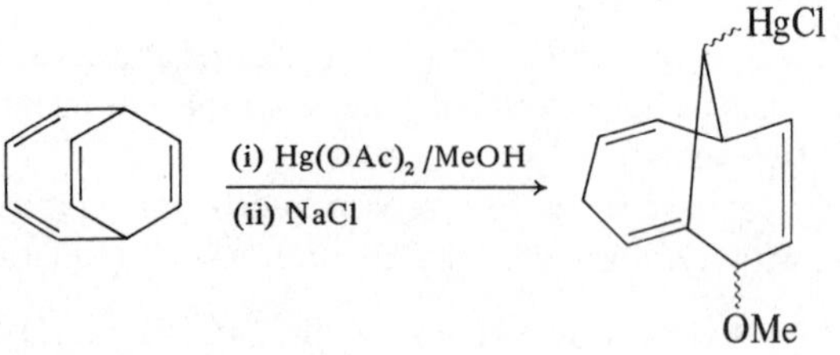

$$(19.113)$$

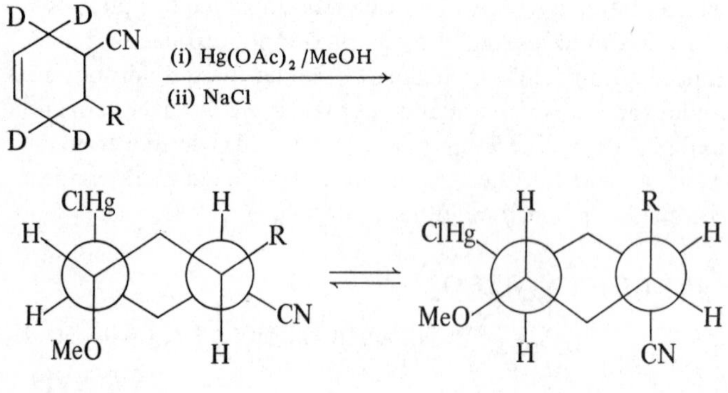

$$(19.114)$$

The methoxymercuration of 1,2-dienes and $\alpha\beta$-unsaturated carbonyl compounds has also attracted attention recently, mainly with the aim of identifying its regio- and stereo-selectivity. Whereas allene gives a terminally mercurated double addition product[208] (equation 19.115), penta-2,3-diene and 2-methylpenta-2,3-diene (equation 19.116) afford alkenylmercurials.[208–210]

$$CH_2:C:CH_2 \xrightarrow[MeOH]{Hg(OAc)_2} AcOHgCH_2C(OMe)_2CH_2HgOAc \xrightarrow{H_2O} (AcOHgCH_2)_2CO$$

$$(19.115)$$

$$MeCH:C:CMe_2 \xrightarrow[MeOH]{Hg(OAc)_2} \textit{trans-}MeCH:C(HgOAc)CMe_2OMe$$
$$72 \text{ per cent}$$

$$+ MeCH(OMe)C(HgOAc):CMe_2 \quad (19.116)$$
$$18 \text{ per cent}$$

The methoxymercuration of $\alpha\beta$-unsaturated ketones,[211] esters,[211] and aldehydes[212] is also regiospecific with the position of mercuration depending on the
pattern of alkylation at the carbon–carbon double bond. The mercury becomes
attached to the carbon α to the carbonyl group (equation 19.117) with all compounds *except* those alkylated *solely* at the α-carbon atom for which β-mercuration
takes place (equation 19.118).

$$R^2R^3C : CR^1.COY + Hg(OAc)_2 + MeOH \xrightarrow[HClO_4]{cat.} R^2R^3C(OMe)CR^1(HgOAc)COY$$

$$(19.117)$$

$$CH_2 : CR^1.COY + Hg(OAc)_2 + MeOH \xrightarrow[HClO_4]{cat.} AcOHgCH_2.CR^1(OMe)COY$$

$$(19.118)$$

The reactions with crotonaldehyde and methacrolein are complicated by the additional formation of products arising from conversion of the aldehyde group into
a dimethoxymethyl substituent.[212] That there is stereospecifically *trans*-addition to
$\alpha\beta$-unsaturated esters has already been mentioned (section 19.3.2). However, the
same mixture of diastereoisomeric methoxymercurials is obtained from either *cis*-
or *trans*-$\alpha\beta$-unsaturated ketones (equation 19.119, R = Me, Ph)[176] because the
diastereoisomers equilibrate under the reaction conditions. The isomerisation
probably takes place by enolisation of the acetoxymercury group followed by
re-ketonisation.

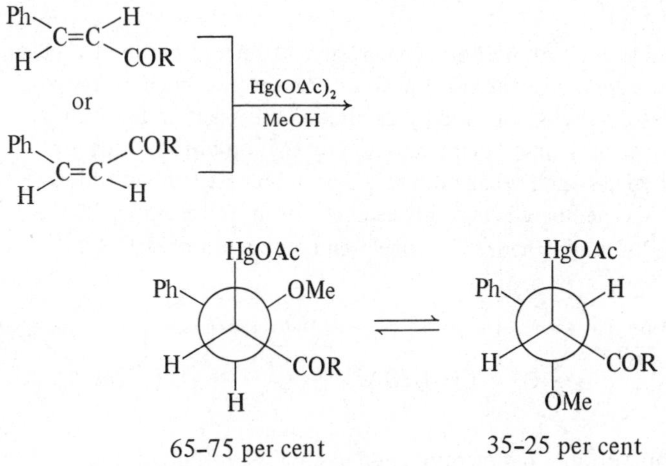

$$(19.119)$$

65–75 per cent   35–25 per cent

Reports of additions with alcohols other than methanol are much less widespread
but the range and behaviour of ethoxymercuration appear similar; for example
isobutylene,[213] norbornadiene,[214] and cyclic 1,2-dienes[209,215] give products
analogous to those from methoxymercuration. With the less-nucleophilic *t*-butyl
alcohol, competing acetoxymercuration assumes a major role but the alkoxy-

mercuration product can be obtained if mercury(II) trifluoroacetate is used instead of acetate.[216]

Intramolecular oxymercuration can occur with suitably unsaturated alcohols. The formation of five-membered rings is most common as exemplified for acyclic (equation 19.120),[207,217] cyclic (equation 19.121),[218] and bicyclic (equation 19.122)[207,217] systems.

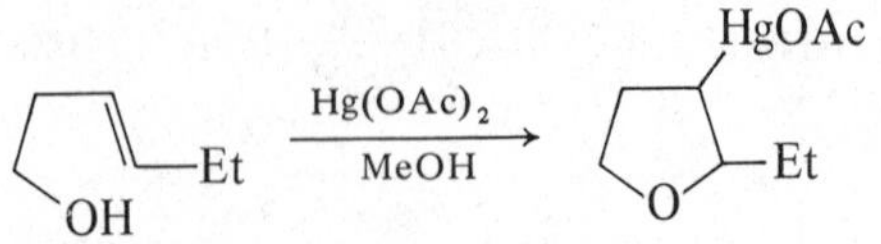

$$\text{(19.120)}$$

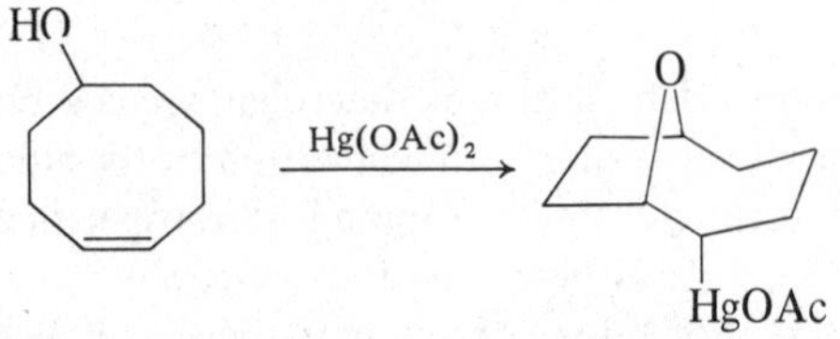

$$\text{(19.121)}$$

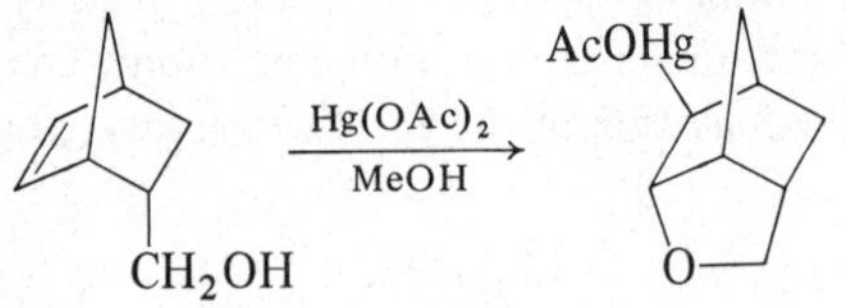

$$\text{(19.122)}$$

For cylisation to occur with norbornene derivatives, the hydroxymethyl group must be in either the *endo*-5 or the *syn*-7 position. An early example of these ring closures involved 2-allylphenol and by combining this reaction using chiral mercury(II) salts with subsequent demercuration by sodium borohydride, optically active 2,3-dihydro-2-methylbenzofuran has been prepared, albeit of low optical purity.[219] Six-membered rings can also be formed, for example from 7-*syn*-(2-hydroxyethyl)norbornene[220] or hex-5-en-l-ol (equation 19.123[221]).

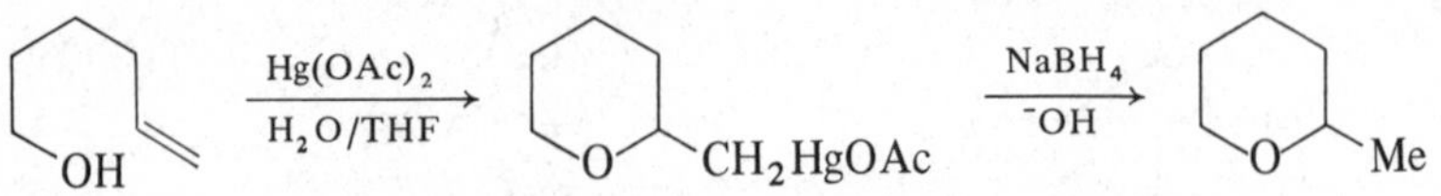

$$\text{(19.123)}$$

The cyclic product from 2,6-dimethylhept-5-en-2-ol is a tetrahydropyran derivative[222] because intramolecular oxymercurations also exhibit Markovnikov-type orientation.

The unsaturated alcohol can itself be generated by hydroxymercuration of an appropriate diene (equation 19.124; Y = CH$_2$,[221] NH,[223] O,[223] S,[224] or Y absent[221,223]). Analogous reactions with hydrogen peroxide afford mercurated cyclic peroxides (equation 19.125).[175]

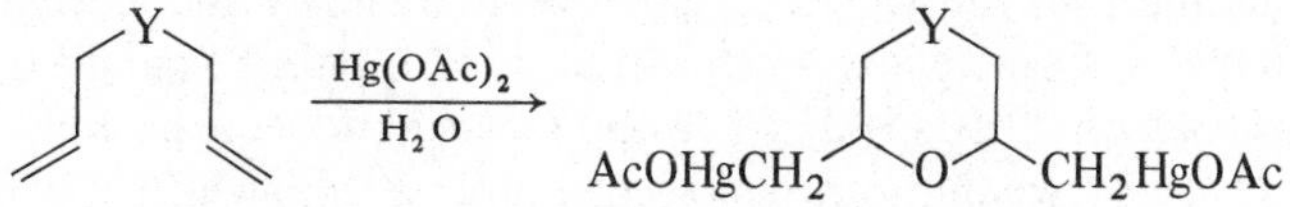

$$\text{[alkene with Y substituent]} \xrightarrow[\text{H}_2\text{O}]{\text{Hg(OAc)}_2} \text{AcOHgCH}_2\text{-O-ring-}\text{CH}_2\text{HgOAc} \qquad (19.124)$$

$$\text{[diene]} \xrightarrow[\text{H}_2\text{O}_2/\text{CH}_2\text{Cl}_2]{\text{Hg(NO}_3)_2} \xrightarrow{\text{Cl}^-} \text{ClHgCH}_2\text{-[ring with O-O]-CH}_2\text{HgCl} \qquad (19.125)$$

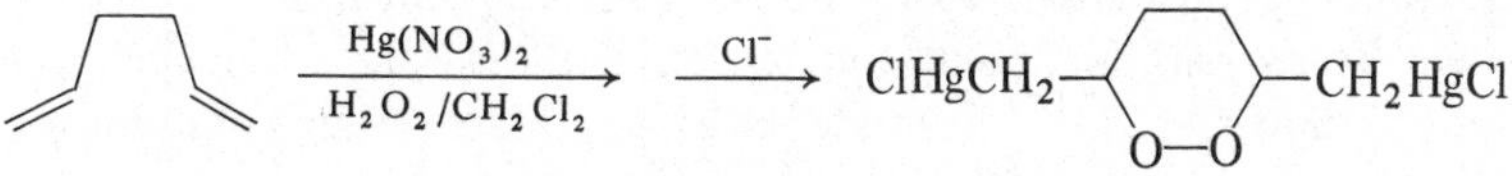

The use of alkyl hydroperoxides in oxymercuration dates only from 1969[225] and except for single examples with the propyl, isopropyl, and 2-phenylprop-2-yl compounds,[200] *t*-butyl hydroperoxide alone has been employed. The reaction is usually carried out in dichloromethane using a one-fold excess of hydroperoxide. If mercury(II) acetate is used under these conditions, acetoxymercuration of simple alkenes competes with the desired reaction but can be eliminated by using the trifluoroacetate; adducts with terminal alkenes (equation 19.126), medial alkenes (equation 19.100), cyclohexane, and norbornene have been characterised.[9]

$$R^1R^2C:CH_2 + Hg(O_2C.CF_3)_2 + Bu^tOOH \xrightarrow{CH_2Cl_2}$$
$$R^1R^2C(OOBu^t)CH_2HgO_2C.CF_3 \qquad (19.126)$$

The reaction has been extended to functionally substituted alkenes. With $\alpha\beta$-unsaturated ketones,[211] esters,[211] and aldehydes[212] the addition is regiospecific and the same orientation rules apply as for methoxymercuration (see p. 175). The stereochemical behaviour of the reaction also parallels that of methoxy-mercuration; whereas methyl *cis*- and *trans*-cinnamate give stereoisomerically pure products, *cis*- and *trans*-chalcone (equation 19.127) each afford the same 4:1 mixture of diastereoisomers.[184]

$$PhCH:CH.COPh + Hg(OAc)_2 + Bu^tOOH \xrightarrow{CH_2Cl_2} PhCH(OOBu^t)CH(HgOAC)COPh$$
$$(19.127)$$

The peroxymercurials from vinyl ether and acetate (equation 19.128; R = Et, Ac[226]) and from $\alpha\beta$-unsaturated peroxyacetals (equations 19.129 and 19.130[212]) have also been prepared.

$$CH_2:CH.OR \xrightarrow[Bu^tOOH]{Hg(OAc)_2} AcOHgCH_2.CH(OOBu^t)OR \qquad (19.128)$$

$$PhCH:CH.CH(OOBu^t)_2 \xrightarrow[Bu^tOOH]{Hg(OAc)_2} PhCH(OOBu^t)CH(HgOAc)CH(OOBu^t)_2 \qquad (19.129)$$

$$CH_2:CMe.CH(OOBu^t)_2 \xrightarrow[Bu^tOOH]{Hg(OAc)_2} \xrightarrow{Br^-} BrHgCH_2CMe(OOBu^t)CH(OOBu^t)_2 \qquad (19.130)$$

Additions in carboxylic acid media have received much less attention than that afforded to hydroxy- and alkoxy-mercuration.[13,14] A recent example is the analogue of reaction 19.105 with acetic acid replacing aqueous acetone.[199] More interest has been shown in the intramolecular process from which mercurated lactones are produced. Reactions with pent-4-en-1-oic acid (equation 19.131),[227] norborn-5-ene-2-*endo*-carboxylic acid and derivatives,[207,217] and a tricyclic compound (equation 19.132)[199] illustrate the scope. Under certain conditions lactonisation also occurs with related esters[217,228] and anhydrides.[199]

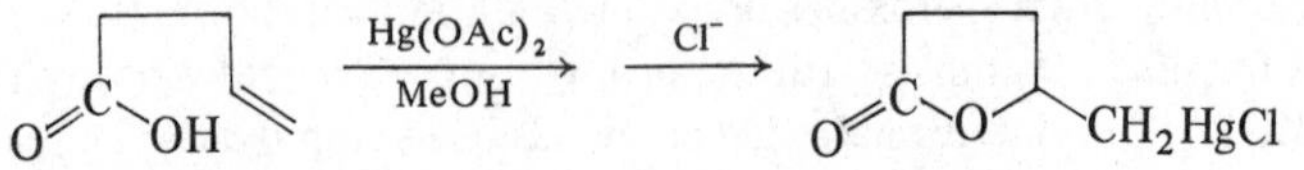

$$(19.131)$$

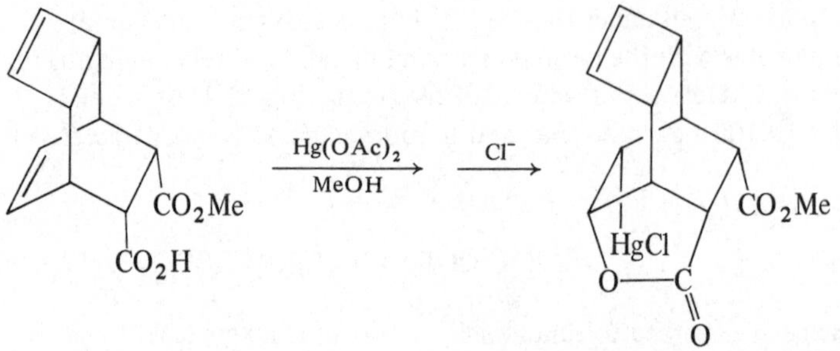

$$(19.132)$$

Aminomercuration was first described in 1945[229] in the form of the reaction of ethylene with mercury(II) chloride in piperidine (equation 19.133).

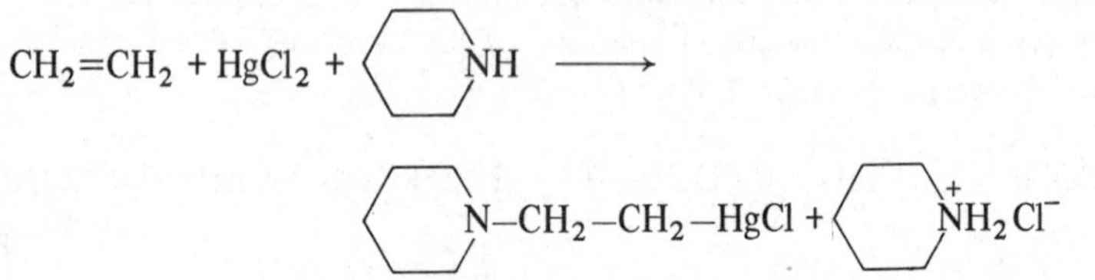

$$(19.133)$$

The reaction with ethylene has been extended to other secondary amines, namely diethylamine,[230] for which the structure of the adduct has been determined recently by X-ray crystallography,[231] and *N*-methylaniline, *N*-ethylaniline and *N*-methylcyclohexylamine.[232] Rather less is known about the reaction with primary amines. In aniline the adduct $PhNH . CH_2CH_2 . HgCl$ is obtained from mercury(II) chloride, but if the acetate is used the amine's bi-functionality manifests itself (equation 19.134[232]) in a way analogous to that of water (equation 19.106) and hydrogen peroxide (equation 19.110).

$$CH_2 : CH_2 + Hg(OAc)_2 + PhNH_2 \rightarrow$$

$$PhNH . CH_2 . CH_2HgOAc + PhN(CH_2CH_2 . HgOAc)_2 \quad (19.134)$$
$$\text{60 per cent} \qquad\qquad\qquad \text{40 per cent}$$

Aminomercuration of other simple alkenes and of functionally substituted alkenes is also known. Additions with piperidine and pyrrolidine have been extended to styrene and propene[232,233] where Markovnikov-type orientation is observed. The reaction of a wider range of secondary amines with allyl ureas[234] and allyl benzo-ate[235] (equation 19.135; X = RCONH and $PhCO_2$ respectively) has been reported and a number of alkenes containing carbonyl, hydroxyl, or alkoxyl substituents undergo aminomercuration in piperidine solution.[235] With pent-4-en-1-ol, amino-mercuration (equation 19.136) is preferred to intramolecular oxymercuration.

$$XCH_2 . CH : CH_2 + HgCl_2 + R_2NH \rightarrow XCH_2 . CH(NR_2)CH_2HgCl$$

$$(19.135)$$

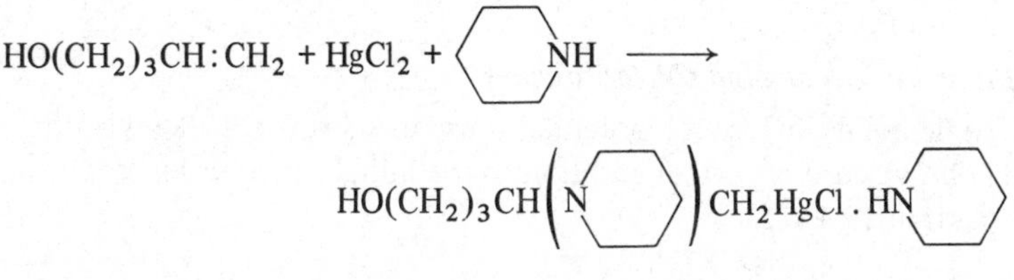

$$HO(CH_2)_3CH : CH_2 + HgCl_2 + \langle \quad NH \longrightarrow$$

$$HO(CH_2)_3CH\left(N\langle \quad \rangle\right)CH_2HgCl . HN\langle \quad \rangle \quad (19.136)$$

As with OH and CO_2H groups, the substituents NHR and NH_2 can participate in intramolecular addition to give five- or six-membered ring products;[10,236,237] the scope of the reaction is illustrated in equations 19.137–19.139.

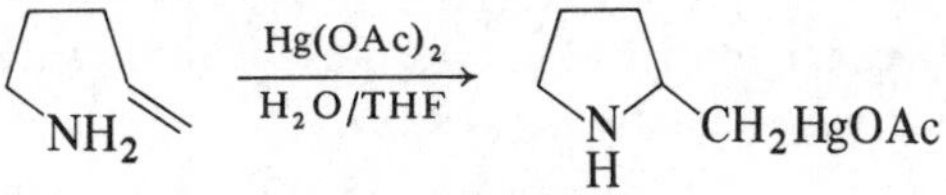

$$(19.137)$$

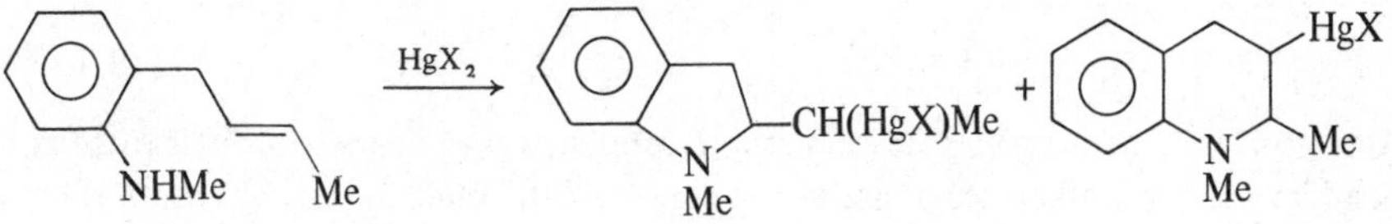

$$(19.138)$$

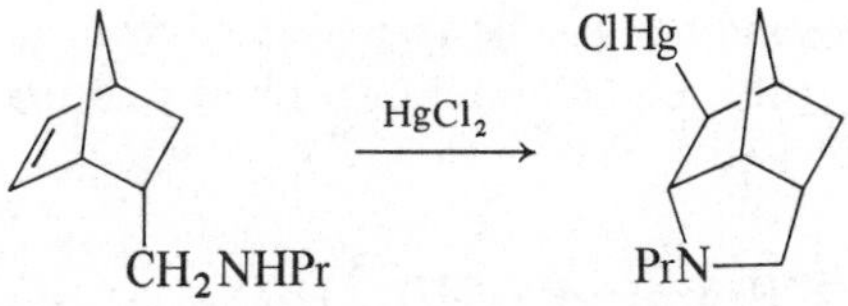

$$(19.139)$$

The ring size is determined primarily by the requirement of Markovnikov-type addition but where this is not selective, as in reaction 19.138, the choice of mercury(II) salt and of solvent is important in determining the balance. Thus mercury(II) acetate in pyridine gives a mixture containing 71 per cent of **15** and

29 per cent of **16** whereas with mercury(II) chloride in acetone the distribution is 14 per cent of **15** and 86 per cent of **16**.

In reactions analogous to those of water (equation 19.124) and hydrogen peroxide (equation 19.125), primary aromatic amines combine with hexa-1,5-diene (equation 19.140) and cycloocta-1,5-diene in the presence of mercury(II) acetate to afford nitrogen-containing heterocyclic compounds.[475]

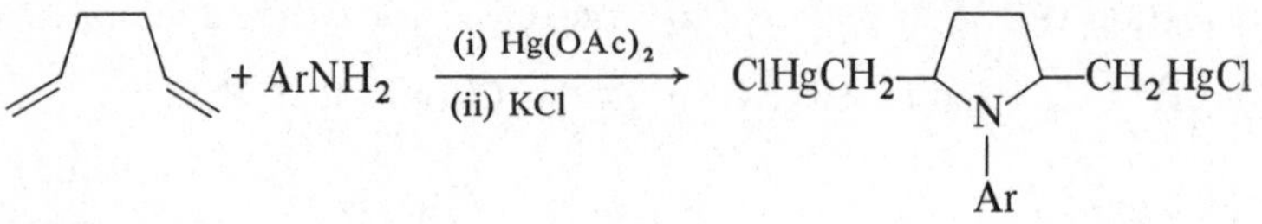

$$\text{\Large/\!\!\!\!\!\diagup} + ArNH_2 \xrightarrow[\text{(ii) KCl}]{\text{(i) Hg(OAc)}_2} ClHgCH_2\text{-}\underset{\underset{Ar}{|}}{\boxed{}}\text{-}CH_2HgCl \qquad (19.140)$$

19.3.4 Additions to Alkynes and Cyclopropanes

Additions to triple bonds represent a potential route to substituted alkenyl-mercurials (**17**) provided that further addition to the initial adduct is not too favourable (equation 19.141).

$$R^1C{\equiv}CR^1 \xrightarrow[R_nYH]{HgX_2} \underset{R_nY}{\overset{R^1}{>}}C{=}C\underset{R^1}{\overset{HgX}{<}} \xrightarrow[R_nYH]{HgX_2} R^1C(YR_n)_2C(HgX)_2R^1$$

$$\mathbf{17}$$

$$(19.141)$$

It should also be recognised that where the nucleophile is water the adduct can be expected to ketonise (equation 19.142).

$$R^1C{\equiv}CR^1 + HgX_2 + H_2O \rightarrow R^1C(OH){:}C(HgX)R^1 \rightleftharpoons R^1CO{.}CH(HgX)R^1$$
$$\qquad\qquad\qquad\qquad\qquad\qquad \mathbf{18a} \qquad\qquad\qquad\qquad \mathbf{18b}$$

$$(19.142)$$

Additions in aqueous solution were studied some twenty years before Hofman and Sand's work with alkenes and are at the centre of the once industrially important process in which mercury(II) sulphate catalyses the conversion of acetylene into acetaldehyde. Yet the area remains comparatively unexplored, probably for two main reasons. Firstly alkynes are generally less susceptible than alkenes to electrophilic addition; secondly much of the early work was carried out with terminal alkynes where addition can be accompanied by mercuration, and in aqueous solution where the product mixture is likely to be most complex (equation 19.143).

$$RC{\equiv}CH \xrightarrow{HgX_2} RC{\equiv}CHgX \xrightarrow[H_2O]{HgX_2} RC(OH){:}C(HgX)_2 \rightleftharpoons RCO{.}CH(HgX)_2$$
$$\qquad\qquad\quad \mathbf{19} \qquad\qquad\qquad\qquad \mathbf{20a} \qquad\qquad\qquad\qquad \mathbf{20b}$$
$$\qquad\qquad\qquad\qquad\qquad\qquad\qquad \Big\downarrow HgX_2 \qquad\qquad\qquad\qquad (19.143)$$
$$\qquad\qquad\qquad RC(OHgX){:}C(HgX)_2 \rightleftharpoons RCO{.}C(HgX)_3$$
$$\qquad\qquad\qquad\qquad \mathbf{21a} \qquad\qquad\qquad\qquad \mathbf{21b}$$

An organomercury compound of the simplest type (17) has been obtained from but-2-yne (equation 19.144[238]) although under more vigorous conditions the *cis*-isomer and the product $MeCH(HgCl)C(OAc):CH_2$ were isolated. Diphenylacetylene affords a similar adduct but its stereoisomeric composition is in question.[239,240]

$$MeC\equiv CMe \xrightarrow[24\,h/R.T.]{Hg(OAc)_2/HOAc} \xrightarrow{Cl^-} \textit{trans-}MeC(OAc):C(HgCl)Me \quad (19.144)$$
$$70 \text{ per cent}$$

The formation of an intermediate adduct of type 18 is implicated in the conversion of hex-3-yne into the ketone by hydroxymercuration–reduction (equation 19.145[241]), but the mercurial was not isolated.

$$EtC\equiv CEt \xrightarrow[H_2O/THF]{Hg(OAc)_2} \xrightarrow{NaBH_4} EtCOPr \quad (19.145)$$

A number of tri-mercurated species have been described (see reference 242 for example) but the arguments[13] about whether they possess structure 21a or 21b are based on chemical conversions that cannot be diagnostic. Adducts of type 17 or analogous to type 20a are presumably intermediates in the formation of alkenyl ethers (equation 19.146) and amines (equation 19.147), but again they were not characterised; with the weakly nucleophilic *t*-butyl alcohol only the mercuration product (19) was obtained.[243]

$$C_6H_{13}C\equiv CH \xrightarrow[MeOH]{Hg(OAc)_2} \xrightarrow{NaBH_4} C_6H_{13}C(OMe):CH_2 \quad (19.146)$$

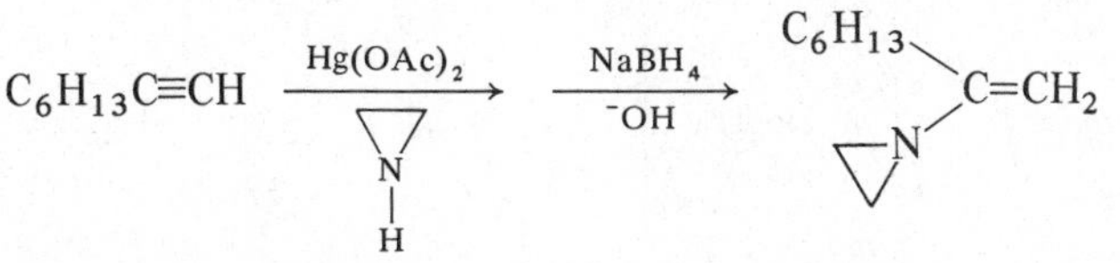

$$C_6H_{13}C\equiv CH \xrightarrow{Hg(OAc)_2} \xrightarrow[^-OH]{NaBH_4} \underset{\underset{H}{|}}{\overset{C_6H_{13}}{\diagdown}}C=CH_2 \quad (19.147)$$

The tri-mercurated adducts (21) react with hydrochloric acid to yield methyl ketones, thereby providing a rationalisation of how mercury(II) salts catalyse terminal alkyne hydration in acidic media. Under *strongly* acidic conditions chloromercuration (equation 19.148)[244] is observed, a process without parallel in alkene chemistry. Similar products formed by $HgCl_2$-addition to propynyl alcohols (equation 19.149) and propynoic acids and esters (equation 19.150; R = H, Me, Ph; R^1 = H, Me, Et) have also been characterised.[245]

$$HC\equiv CH + HgCl_2 \cdot \xrightarrow[(50\text{ per cent})]{HCl} \underset{Cl}{\overset{H}{\diagdown}}C=C\underset{H}{\overset{HgCl}{\diagup}} \quad (19.148)$$

$$R_2C(OH)C\equiv CH \xrightarrow[NaCl]{HgCl_2} R_2C(OH)CCl:CHHgCl \quad (19.149)$$

$$RC\equiv C.CO_2R^1 \xrightarrow[NaCl]{HgCl_2} RCCl:C(HgCl)CO_2R^1 \quad (19.150)$$

Intramolecular alkoxymercuration can occur with acetylenic glycols and if a secondary alcohol group is involved, the products aromatise by dehydration (equation 19.151).[246] The organomercury compound obtained from 1,8-*bis*-(phenylethynyl)naphthalene can be rationalised in terms of an intramolecular carbomercuration (equation 19.152).[247]

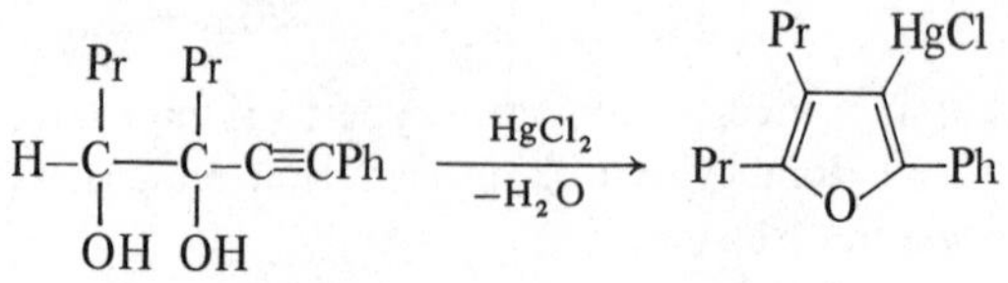

$$\text{(19.151)}$$

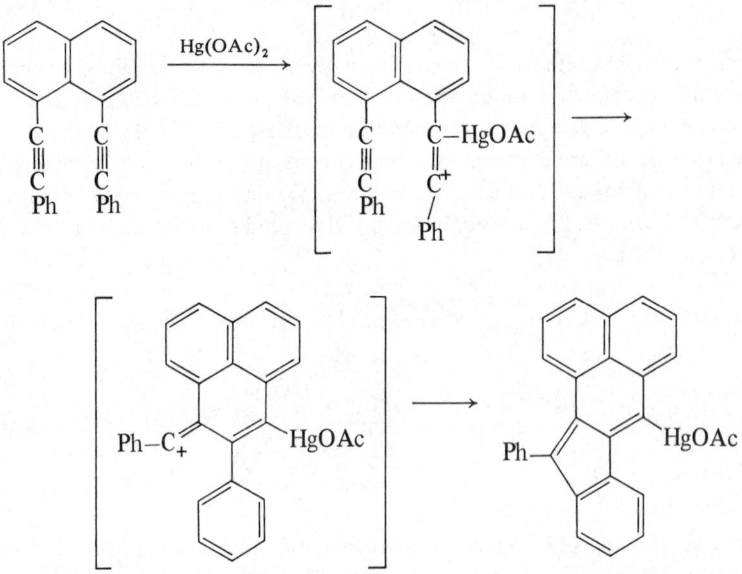

$$\text{(19.152)}$$

56 per cent

Cyclopropane derivatives also undergo addition reactions that are formally similar to those with alkenes but that now yield γ-functionally substituted organomercury compounds (equation 19.153).

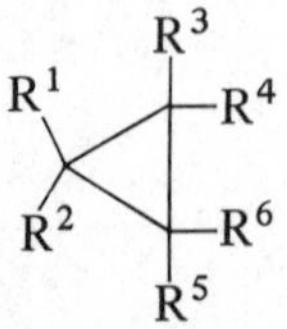

$$R^1R^2C(YR_n)CR^3R^4 \cdot C(HgX)R^5R^6 + HX \quad \text{(19.153)}$$

This reaction was discovered by Levina in 1950,[248] considerably later than the corresponding additions with alkenes and alkynes. However the system is much better behaved than that with alkynes since the complications of mercuration and tautomerisation are absent. Consequently modern investigations are comparatively abundant. Cyclopropanes are markedly less reactive than alkenes, as demonstrated by the outcome of an internal competition (equation 19.154).[249]

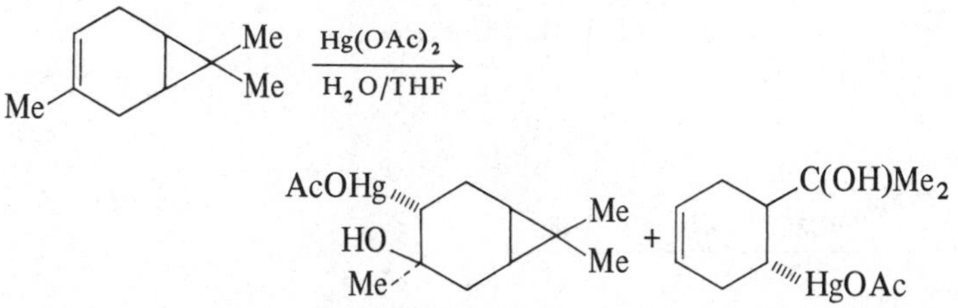

$$(19.154)$$

82 per cent   18 per cent

By the same token, however, the back reaction is less favourable and cyclopropane adducts have a considerably higher acid tolerance than the corresponding β-substituted organomercurials.

In principle ring opening could occur by fission of any of the three bonds and might lead to complicated mixtures of isomers. However, if one carbon carries no substituents the mercury always becomes attached to it ($R^5 = R^6 = H$) and the nucleophile adds according to Markovnikov's rule. Recent examples are shown in equations 19.155[250] and 19.156.[251]

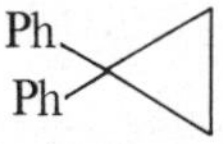

$$Ph_2C(OMe)CH_2CH_2HgOAc \quad (19.155)$$

30 per cent

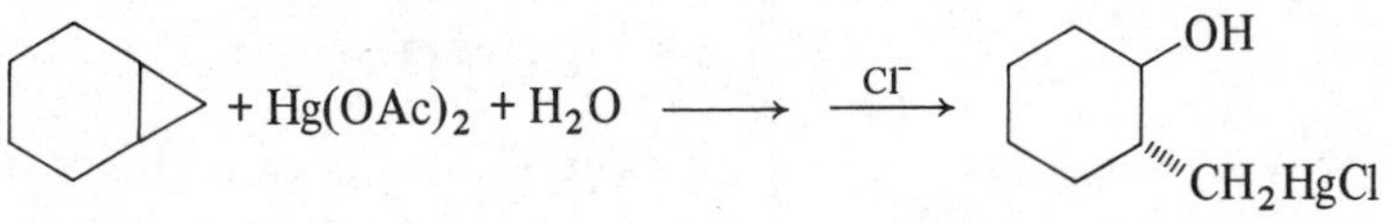

$$(19.156)$$

83 per cent

The reactivity of the ring increases along the following series: alkyl- < 1,2-dimethyl- < phenyl- < 1,1-dimethyl < 1,1,2-trimethyl < 1,1,2,2-tetramethyl-cyclopropane. Where *all three* ring carbons bear substituents the addition is still regiospecific.[252] Here the mercury(II) electrophile attacks the least-substituted ring bond or, for bonds of equal substitution, at the *cis*-substituted rather than the *trans*-substituted bond; the nucleophile again becomes attached to the site best able to stabilise a positive charge (equations 19.157 and 19.158).

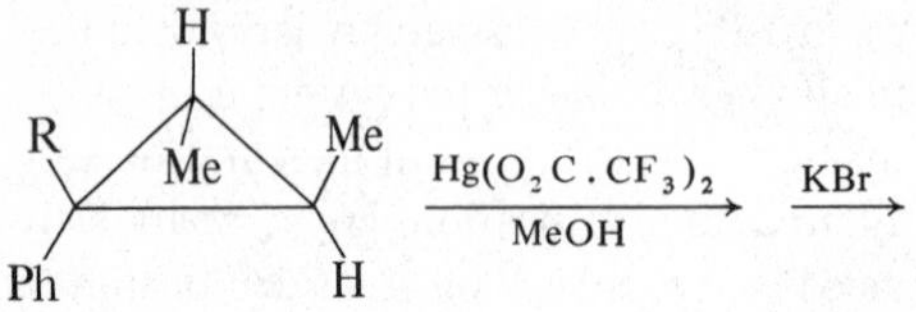

PhCR(OMe)CHMe . CH(HgBr)Me   (19.157)

R = H   88 per cent
R = Me   85 per cent

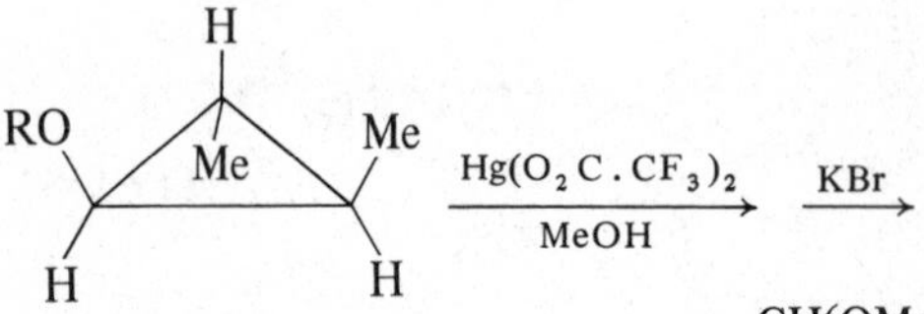

CH(OMe)$_2$CHMe . CH(HgBr)Me   (19.158)

R = H   88 per cent
R = Me   68 per cent

If methanol is replaced by acetic acid, hydroxy- and methoxy-cyclopropanes afford β-mercurio-aldehydes or -ketones.[253]

As with alkenes most is known about reactions in water or methanol but additions where the nucleophile is hydrogen peroxide (equation 19.159[201]) or a nitrile (equation 19.160[254]) have been reported, and tricyclo-[3.1.1.0^{6,7}]heptane undergoes acetoxymercuration with rearrangement (equation 19.161[255]).

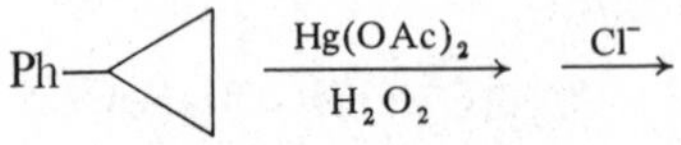

PhCH(OOH)CH$_2$CH$_2$HgCl   (19.159)

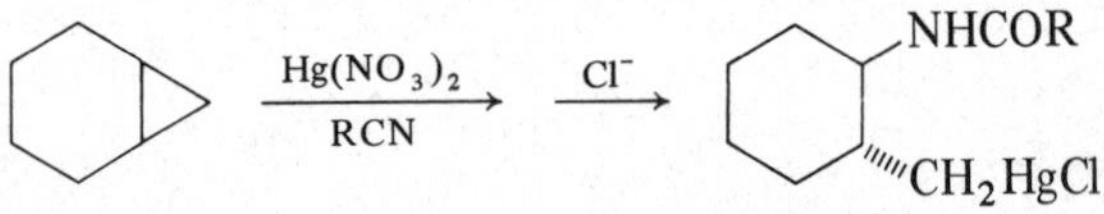

(19.160)

R = Me, Ph

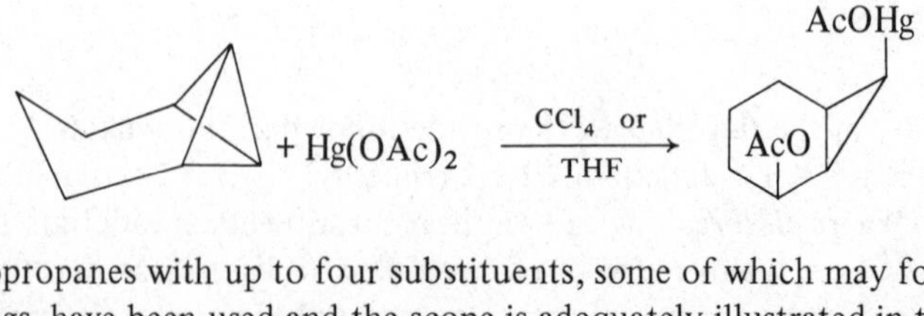

(19.161)

Cyclopropanes with up to four substituents, some of which may form part of other rings, have been used and the scope is adequately illustrated in the preceding discussion; a related reaction with benzocyclopropene (equation 19.162) has also been described.[256]

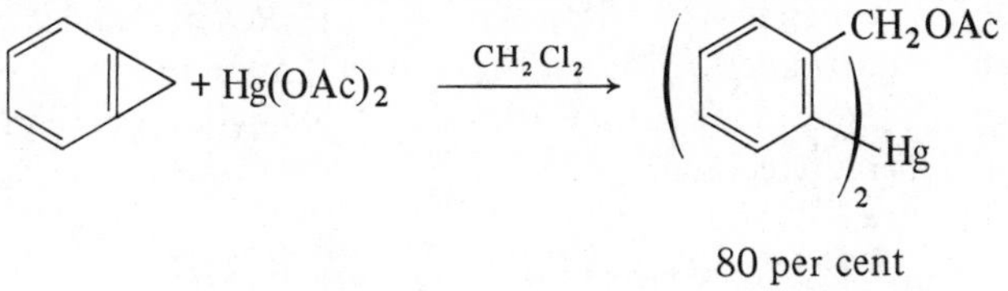

$$\text{(structure)} + \text{Hg(OAc)}_2 \xrightarrow{\text{CH}_2\text{Cl}_2} \left(\text{structure with } \text{CH}_2\text{OAc, Hg} \right)_2 \qquad (19.162)$$

80 per cent

The oxymercuration of cyclopropanes is highly stereoselective. For unsymmetrically substituted systems electrophilic attack can occur with predominant inversion[251–253] or retention,[252] but there is *always* substantially complete inversion of configuration at the site of nucleophilic attack. In the symmetrical compound *cis,cis*-1,2,3-trimethylcyclopropane, inversion predominates slightly at the mercuration site while methanol enters with 100 per cent inversion. On the basis of these results a mechanism involving the formation of a *corner*-mercurated intermediate (**22**) has been proposed (equation 19.163[252]) although the initial attack is presumably at an edge.

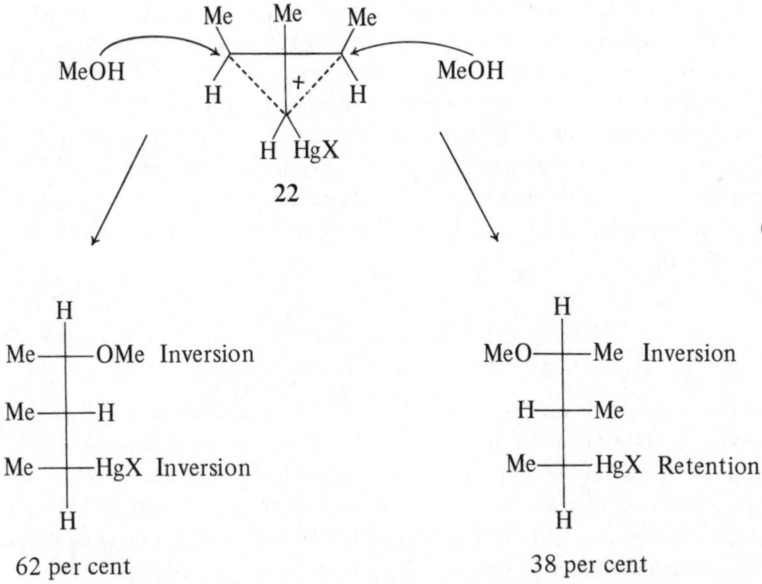

$$(19.163)$$

22

	H	
Me—	—OMe	Inversion
Me—	—H	
Me—	—HgX	Inversion
	H	

62 per cent

	H	
MeO—	—Me	Inversion
H—	—Me	
Me—	—HgX	Retention
	H	

38 per cent

19.4 Other Methods

In addition to the three main routes to organomercury compounds there are a number of other preparative methods that can prove useful in more restricted areas. These include reactions in which loss of carbon dioxide (equation 19.164), sulphur dioxide (equation 19.165), or sulphur trioxide (equation 19.166) occurs from an appropriate mercury(II) salt, reactions of mercury(II) halides with aryl-diazonium salts (equation 19.167), and the electrochemical reduction of aryl-diazonium salts,[257] ketones (equation 19.168)[258] or alkyl halides (equation 19.169)[259] at a mercury cathode.

$$Hg(O_2CR)_2 \xrightarrow{-CO_2} RHgO_2CR \xrightarrow{-CO_2} R_2Hg \qquad (19.164)$$

$$Hg(O_2SR)_2 \xrightarrow{-SO_2} RHgO_2SR \xrightarrow{-SO_2} R_2Hg \qquad (19.165)$$

$$Hg(O_3SR)_2 \xrightarrow{-SO_3} RHgO_3SR \xrightarrow{SO_3} R_2Hg \qquad (19.166)$$

$$ArN_2(HgCl_3) + 2Cu \rightarrow ArHgCl + N_2 + 2CuCl \qquad (19.167)$$

$$2\,PhCH_2COMe^- + Hg + 6H^+ + 6e \rightarrow (PhCH_2.CHMe)_2Hg + 2H_2O$$
$$(19.168)$$

$$2\,ArCH_2Br + Hg + 2e \rightarrow (ArCH_2)_2Hg + 2Br^- \qquad (19.169)$$

The electrochemical methods have not received the attention they probably deserve but this may soon be rectified if the present resurgence of interest in organic electrosynthesis is maintained. The 'diazo method' (equation 19.167) is a long-established and valuable alternative to transmetallation and mercuration for making arylmercury salts.[13,14] Little has been added to the area in the past few years and since it was reviewed recently[260] only the salient features will be mentioned here. The double salts of aryldiazonium chlorides and mercury(II) chloride are prepared by adding a solution of the latter to diazotised amines (equation 19.170).

$$ArNH_2 \xrightarrow[HCl]{NaNO_2} \xrightarrow[HCl]{HgCl_2} ArN_2(HgCl_3) \downarrow \qquad (19.170)$$

Copper-induced decomposition of the isolated double salt is then carried out in acetone, ethyl acetate, or water at −5 to −15 °C. Yields of the arylmercury(II) chlorides are typically 50–80 per cent, but are lower if the aromatic nucleus contains electronegative substituents. The advantage that the diazo method has over transmetallation (section 19.1) is that it permits the preparation of arylmercurials containing substituents such as CO_2R, SO_3H, or NO_2, which are sensitive to organo-magnesium and -lithium reagents. The advantage over mercuration (section 19.2) is that the product is obtained free of isomers with the mercury only occupying the position selected by the location of the NH_2 group in the starting amine.

Decarboxylation (equation 19.164) can be brought about under two quite different sets of conditions, which have complementary areas of application. Since the initial discovery in the mid-1950s, Ol'dekop and Maier have developed the *acyl-peroxide-initiated* decarboxylation into a viable synthetic route to alkylmercury(II) salts. The reactions are best carried out in boiling benzene and the products are isolated as the halides formed by treatment with potassium chloride, bromide, or iodide (equation 19.171).

$$Hg(O_2CR)_2 \xrightarrow[-CO_2]{(RCO_2)_2} RHgO_2CR \xrightarrow{KX} RHgX \qquad (19.171)$$

Yields of over 90 per cent are commonly obtained and recent additions to the scope include cycloalkyl-[261,262] and cycloalkylmethyl-[263] mercury(II) halides. The reaction is believed to proceed by a free radical chain mechanism (equations 19.172–19.174).

$$(RCO_2)_2 \rightarrow 2RCO.\dot{O} \qquad (19.172)$$

$$RCO.\dot{O} \rightarrow \dot{R} + CO_2 \qquad (19.173)$$

$$\dot{R} + Hg(O_2CR)_2 \rightarrow RHgO_2CR + RCO.\dot{O} \qquad (19.174)$$

Evidence for intermediate alkyl radicals is provided by the formation of approximately equimolar amounts of *cis*- and *trans*-4-methylcyclohexylmercury(II) salts from either *cis*- or *trans*-carboxylates,[264] and a chain mechanism is indicated since the decarboxylation of $Hg(O_2CR)_2$ in the presence of $(R^1CO_2)_2$ gives predominantly $RHgO_2CR.$[265] In some cases the alkyl radicals that initiate the reaction can be generated alternatively from the mercury(II) carboxylates themselves by u.v. irradiation.

Thermally induced decarboxylation (for a review see reference 266) is a much older reaction and one that has its major application in the synthesis of diarylmercurials. The reaction was discovered by Pesci in 1901 for mercury(II) phthalate and developed into a viable synthetic route to diarylmercurials by Kharasch in the early 1920s. Whereas electronegative substituents inhibit free-radical decarboxylation they promote the thermal reaction and much of the more recent work has involved using the method to prepare polyhalogenoorganomercury compounds.

Pyrolysis of the mercury(II) carboxylates is carried out either directly at temperatures usually in excess of 200 °C, or in refluxing polar solvents such as water, pyridine, dimethoxyethane, or hexamethylphosphorus triamide, which appear to aid the decarboxylation. Examples involving monocarboxylic acid derivatives are given in equations 19.175[267] and 19.176,[268] and the 'fluoro-Pesci' reaction is depicted in equation 19.177.[269]

$$Hg(O_2C.C_6F_5)_2 \xrightarrow{210\text{-}212\,°C} (C_6F_5)_2Hg + 2CO_2 \qquad (19.175)$$
$$56 \text{ per cent}$$

$$Hg(O_2C.C_6Cl_5)_2 \xrightarrow[\text{pyridine}]{\Delta} (C_6Cl_5)_2Hg + 2CO_2 \qquad (19.176)$$
$$80 \text{ per cent}$$

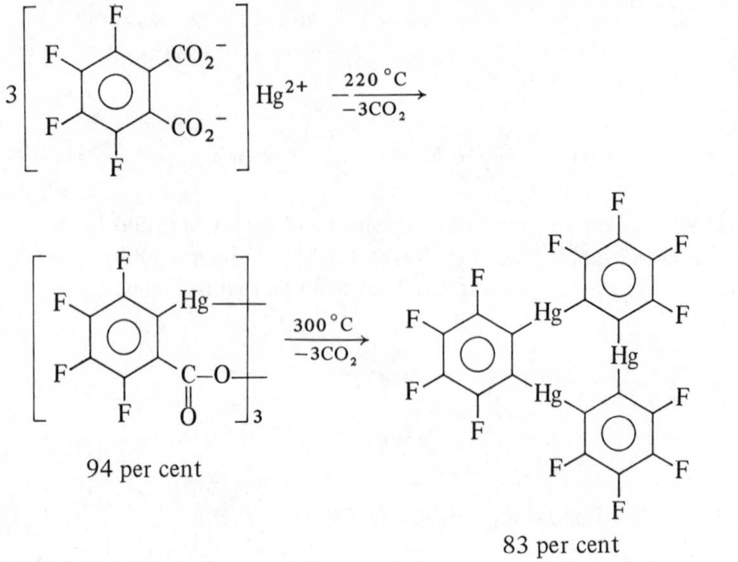

94 per cent

83 per cent

(19.177)

The aromatic group can also be heterocyclic and for the *N*-methylbenzimidazole derivative decarboxylation succeeded (equation 19.178) where transmetallation and mercuration failed.[270]

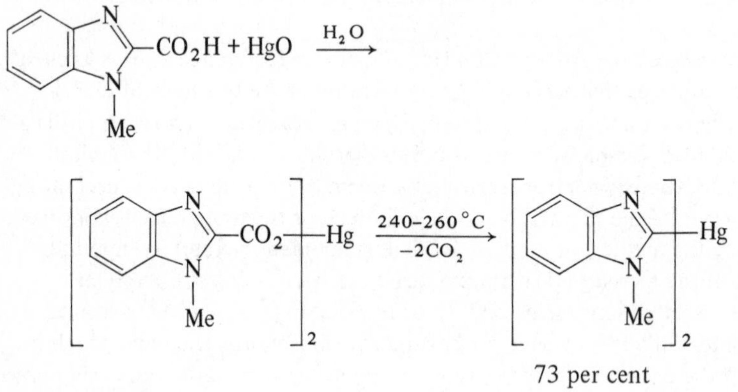

(19.178)

73 per cent

Application in the aliphatic series is essentially limited to reactions 19.179 and 19.180 (X = F, Cl, Br) where electronegative substituents again feature.

$$(MeCOCR_2 . CO_2)_2 Hg \xrightarrow{90\,^{\circ}C} (MeCOCR_2)_2 Hg + 2CO_2 \qquad (19.179)$$

$$(CX_3 . CO_2)_2 Hg \xrightarrow[-CO_2]{\Delta} CX_3 HgO_2C . CX_3 \xrightarrow[-CO_2]{\Delta} (CX_3)_2 Hg \qquad (19.180)$$

The trifluoroacetate loses only one unit of carbon dioxide even at 300 °C,[271] but complete decarboxylation can be accomplished at 200 °C if fused potassium carbonate is present[272] or a complex with a bidentate ligand is used.[273]

Thermal decarboxylation of suitable organomercury carboxylates can be used to prepare mixed diarylmercurials (for example equation 19.181[274]) or aryl(alkyl)-mercurials (for example equation 19.182[28]); electronegative substituents in the carboxylate group are again necessary.

$$4\text{-EtO}.C_6F_4.CO_2\text{HgPh} \xrightarrow[\text{pyridine}]{\Delta} 4\text{-EtO}.C_6F_4.\text{HgPh} + CO_2 \quad (19.181)$$
$$47 \text{ per cent}$$

$$1,3\text{-}(\text{MeHgO}_2\text{C})_2C_6F_4 \xrightarrow[\text{pyridine}]{\Delta} 1,3\text{-}(\text{MeHg})_2C_6F_4 + 2CO_2 \quad (19.182)$$
$$56 \text{ per cent}$$

Thermal desulphination (equation 19.165) closely resembles decarboxylation but generally occurs under milder conditions. The original Peters reaction (equation 19.183, M = H), which was discovered in 1905, was carried out in aqueous or alcoholic solution and was later improved by employing sodium salts (M = Na) rather than the free sulphinic acids.

$$MO_2SR + HgCl_2 \xrightarrow{\Delta} RHgCl + SO_2 + MCl \quad (19.183)$$

As with decarboxylation, electronegative substituents promote the elimination and this is illustrated dramatically by the fact that pentafluorophenylmercury(II) salts can be obtained at room temperature in a modified Peters reaction (equation 19.184).[275]

$$LiO_2S.C_6F_5 + HgX_2 \xrightarrow[\text{R.T.}]{H_2O} C_6F_5HgX + SO_2 + LiX \quad (19.184)$$

In the latest method, mercury(II) arenesulphinates (equation 19.185) are isolated and then pyrolysed in a separate step to afford diarylmercurials (equation 19.186).[276]

$$2 \text{ NaO}_2SAr + Hg(OAc)_2 \xrightarrow{H_2O} Hg(O_2SAr)_2\downarrow + 2NaOAc \quad (19.185)$$

$$Hg(O_2SAr)_2 \xrightarrow[\text{in vacuo}]{110\text{-}180\,°C} Ar_2Hg + 2SO_2 \quad (19.186)$$
$$30\text{-}90 \text{ per cent}$$

Desulphination has been used in the synthesis of 2-(arylazo)arylmercury(II) chlorides where the appropriate sodium arenesulphinates are conveniently obtained by base-induced rearrangement of 2-nitrobenzenesulphenanilide derivatives (equation 19.187).[277]

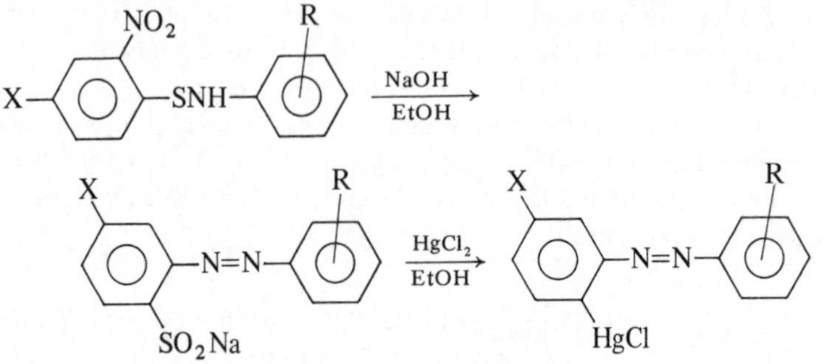

$$\text{(19.187)}$$

Several bispolyhalogenophenylmercurials have been obtained by thermal decomposition of the corresponding mercury(II) arenesulphonate dihydrates (equation 19.188).[278]

$$(4\text{-H}.C_6Cl_4.SO_3)_2Hg.2H_2O \xrightarrow{130\text{-}230\,^\circ C} (4\text{-H}.C_6Cl_4)_2Hg + 2SO_3 + H_2O$$

$$\text{(19.188)}$$

In general there is considerable competition from hydrolysis and hydrolytic desulphonation and better yields are obtained by using the corresponding pyridinates.[279]

The scope of thermal eliminations (CO_2, SO_2, and SO_3) as a route to organomercurials is limited by the vigorous conditions that are usually required, and in practice it is largely restricted to highly halogenated compounds. A problem that frequently arises is that mercuration competes with elimination and in extreme cases becomes the sole reaction. Thus in the pyrolysis of the compounds (2-H.$C_6X_4.SO_3)_2Hg(py)_2$ (equation 19.189; X = F, Cl) the polymeric species **23** are the only organomercury products;[279] at 300–400 °C the fluorine derivative affords a cyclic product identical with that from decarboxylation (equation 19.177).

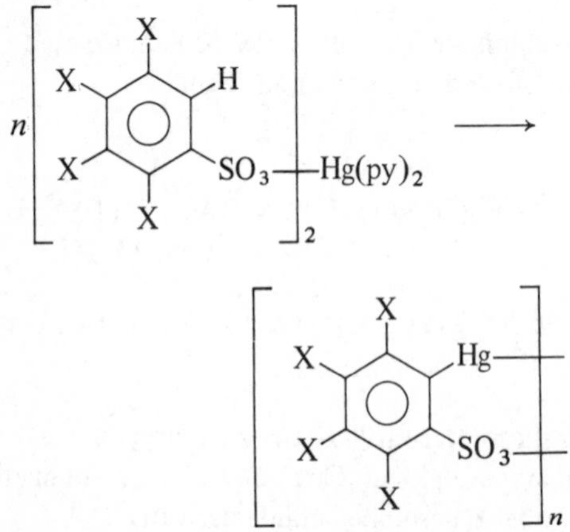

$$+ n\ 2\text{-H}.C_6X_4.SO_3H \qquad \text{(19.189)}$$

23

20 Formation of Organo-mercury Compounds from other Organomercury Compounds

Because the main preparative routes to organomercury compounds (sections 19.1 to 19.3) afford organomercury salts, RHgX, it is important to consider how these are transformed into the fully organic species, R_2Hg. The obvious routes to consider are the analogues of transmetallation, mercuration, and addition in which organomercury salts assume the role previously fulfilled by mercury(II) salts. However the organomercury salts are insufficiently electrophilic to promote additions and the scope for mercuration is considerably reduced; transmetallation is generally successful although yields are not always good. Transmetallation (equation 20.1; M is mainly Li or MgX) and mercuration with an organomercury base [equation 20.2; X = OH, OR or $N(SiMe_3)_2$] are the principal routes to unsymmetrical diorganomercurials, $RHgR^1$.

$$RHgX + MR^1 \rightarrow RHgR^1 + MX \tag{20.1}$$

$$RHgX + HR^1 \rightarrow RHgR^1 + HX \tag{20.2}$$

A special variation of transmetallation is the system in which the organometallic reagent is also an organomercury salt. Although of rather limited use in preparing mixed diorganomercury compounds, this is the major route to the symmetrical class (equation 20.3).

$$RHgX + RHgX \rightleftharpoons R_2Hg + HgX_2 \tag{20.3}$$

The reaction is reversible and in fact the position of equilibrium is normally well to the left. The forward reaction, known as 'symmetrisation', is effected through the agency of an anionic or neutral ligand (L) when the inorganic product is the complex $HgX_2 . L_2$, or with a reducing agent when mercury is formed. Both methods can be regarded formally as displacing the equilibrium to the right by removing the mercury(II) salt, but the actual mechanisms involve interaction of the symmetrising agent with the organomercury salt (see below).

The reverse of symmetrisation (*syn*-proportionation) usually occurs spontaneously and constitutes a convenient way of preparing an organomercury salt when the diorganomercury compound is available. The cleavage of an alkyl group from R_2Hg (equation 20.4) also provides a method for conversion to RHgX, but this is rarely of preprative importance and is discussed in chapter 22 in the context of reactions of the mercury–carbon bond.

$$R_2Hg + X-Y \rightarrow RHgX + RY \tag{20.4}$$

Finally, organomercury salts can be transformed into other organomercury salts by substitution (equation 20.5; Z is mainly a metal or H) or addition (equation 20.6; Y = A–B–X) reactions of the Hg–X bond.

$$RHgX + Z-Y \rightarrow RHgY + Z-X \tag{20.5}$$

$$RHgX + A=B \rightarrow RHgY \tag{20.6}$$

20.1 Transformation of RHgX into RHgR[1]

Although Hilpert and Grüttner first reported the Grignard route to unsymmetrical organomercurials in 1915 (equation 20.7), its development was largely at the hands of Kharasch around 1930.

$$PhCH_2HgBr + PhMgBr \rightarrow PhCH_2HgPh + MgBr_2 \tag{20.7}$$

Contrary to earlier suggestions, Kharasch showed that both reagent combinations $(RHgBr + R^1MgBr)$ and $(R^1HgBr + RMgBr)$ can be employed, provided that an excess of the Grignard reagent is avoided and the reaction temperature is carefully controlled. A large number of examples are known[13,14] that include dialkyl-, alkyl(aryl)-, cycloalkyl(aryl)-, and diaryl-mercurials, and the scope also embraces alkenyl and alkynyl groups. In recent years the organolithium route has often been preferred (for example equations 20.8[280] and 19.11) and this has the advantage that the preparation can be carried out at lower temperatures.

$$PhHgCl + MeCCl_2Li \xrightarrow[\text{THF/Et}_2\text{O}]{-100\,^{\circ}\text{C}} PhHgCCl_2Me + LiCl \tag{20.8}$$

$$67 \text{ per cent}$$

The mercuration of sufficiently strong *C*-acids by organomercury bases (equation 20.2) has been covered in section 19.2 where the preparation of the Seyferth reagents (equations 19.69 and 19.70) was discussed and examples of mercuration involving pentafluorobenzene (equation 19.45), diazoalkanes (equation 19.59), pentachlorocyclopentadiene (equation 20.68) and acetylene (equation 19.74) were provided; R is always a simple group and in most examples is phenyl.

Less general methods of preparing mixed diorganomercurials include elimination and insertion reactions at the Hg–X bond. The formation of mixed diarylmercurials and alkyl(aryl)mercurials by thermal decarboxylation has already been mentioned (section 19.4). In the examples then given (equations 19.180 and 19.181) the organomercury carboxylates were prepared from metal carboxylates and organomercury halides, but as an alternative suitable carboxylic acids can be treated with organomercury hydroxides (equation 20.9[281]).

$$3\text{-F}.C_6H_4.HgOH \xrightarrow[-H_2O]{CF_3CO_2H} 3\text{-F}.C_6H_4.HgO_2C.CF_3$$

$$\xrightarrow[-CO_2]{60-70\,^\circ C/DME} 3\text{-F}.C_6H_4HgCF_3 \quad (20.9)$$
$$55 \text{ per cent}$$

Methylene can be inserted into the mercury–halogen bond of benzylmercury halides (equation 20.10; $X = Cl^{282}$ or I^{283}) and trichloromethylmercury bromide,[284] or the mercury–oxygen bond of arylmercury carboxylates (equation 20.11[285]) to afford stable diorganomercurials.

$$PhCH_2HgX + CH_2N_2 \rightarrow PhCH_2HgCH_2X + N_2 \quad (20.10)$$

$$ArHgO_2CAr^1 + CH_2N_2 \rightarrow ArHgCH_2O_2CAr^1 + N_2 \quad (20.11)$$

Similar attempts to make chloromethyl(aryl)mercurys result in formation of the symmetrical diorganomercury compounds (equation 20.12[282]).

$$2ArHgCl + 2CH_2N_2 \xrightarrow{-2N_2} [2ArHgCH_2Cl] \rightarrow Ar_2Hg + (ClCH_2)_2Hg$$
$$(20.12)$$

This kind of disproportionation is always a potential problem when unsymmetrical diorganomercurials are being prepared and it may be encountered no matter which method of synthesis is employed. Thus for example pentafluorophenyl(pentachlorophenyl)mercury could not be obtained by the organolithium route from either combination of reagents (equation 20.13[286]) and mercuration of hindered silicon-containing phenols afforded the symmetrical diarylmercurials (equation 20.14[87]).

$$2C_6Cl_5HgBr + 2C_6F_5Li \xrightarrow{\,-78\,^\circ C\,}$$
$$\longrightarrow (C_6F_5)_2Hg + (C_6Cl_5)_2Hg \quad (20.13)$$
$$2C_6F_5HgBr + 2C_6Cl_5Li \xrightarrow{\,-15\,^\circ C\,}$$

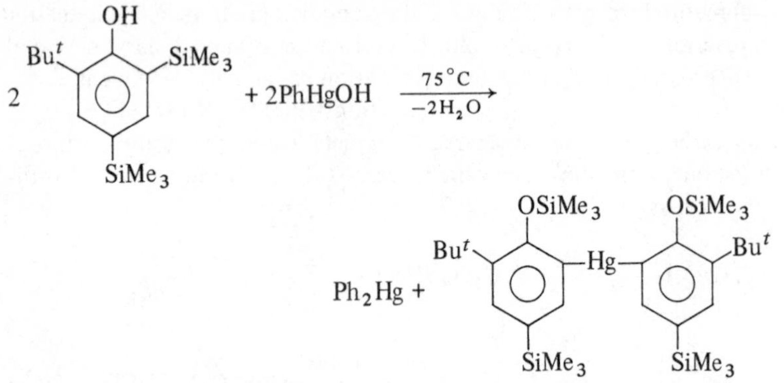

$$(20.14)$$

In any synthesis of a mixed diorganomercury compound the reaction temperature should be kept as low as possible to try to avoid disproportionation. Of course disproportionation will not occur in a homogeneous system if the unsymmetrical compound is thermodynamically stable with respect to the symmetrical mercurials. Indeed in some instances the reverse of disproportionation can provide a preparative route to $RHgR^1$ (for example equation 20.15[135]).

$$(PhCH_2CH_2)_2Hg + Hg(CCl_2.CO_2Me)_2 \rightarrow 2PhCH_2CH_2HgCCl_2.CO_2Me$$
$$(20.15)$$

Two other systems involving redistribution of organic groups between mercurials (equations 20.16 and 20.17) find restricted use in the preparation of mixed diorganomercury compounds.

$$R_2Hg + R^1HgX \rightarrow RHgR^1 + RHgX \qquad (20.16)$$

$$RHgX + R^1HgX \rightarrow RHgR^1 + HgX_2 \qquad (20.17)$$

Reaction 20.17, which is known as 'cosymmetrisation', is effected under similar conditions to those for symmetrisation (equation 20.3) so that formation of R_2Hg and R^1_2Hg by the latter route may compete with the desired reaction; symmetrisation is discussed in the next section.

20.2 Transformation of RHgX into R_2Hg (Symmetrisation)

20.2.1 By Complexing Agents

A number of anions that coordinate strongly to mercury(II) are effective in inducing symmetrisation. Recent examples employing the most common reagents, namely iodide,[72] thiocyanate,[71] and cyanide[287] are shown in equations 20.18, 20.19 and 20.20 respectively.

$$2\ 4\text{-MeO}.C_6F_4.HgO_2C.CF_3 \xrightarrow[\text{MeOH}]{\text{KI}} (4\text{-MeO}.C_6F_4)_2Hg \qquad (20.18)$$
$$88 \text{ per cent}$$

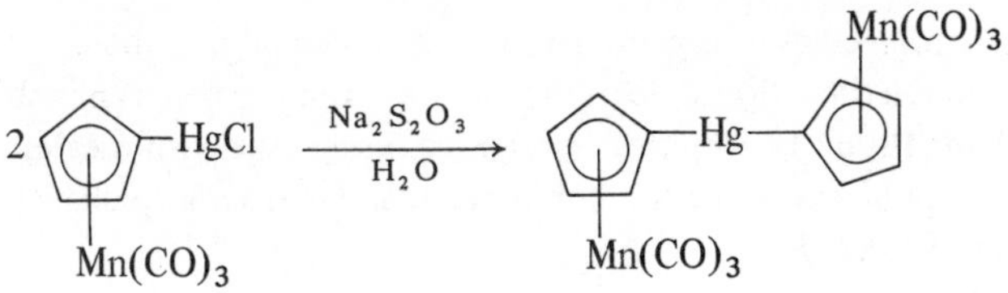

$$(20.19)$$

99 per cent

$$2 \, (NO_2)_3C \cdot CH_2CH_2HgCl \xrightarrow[\text{EtOH}]{\text{KCN}} \quad [(NO_2)_3C \cdot CH_2CH_2]_2Hg \quad (20.20)$$

90 per cent

A convenient modification of the usual method, which is suitable for preparing diarylmercurials, is to pass a solution of the arylmercury halide in chloroform through an alumina column that has been treated with sodium cyanide.[288] In fact 2-methylphenylmercury(II) halides undergo symmetrisation on columns of active basic alumina in the *absence* of added cyanide but the method appears to have a very limited range of applicability.[289] The reagent of choice for preparing all types of diarylmercurials, including heterocyclic and metal-complexed cyclo-pentadienide derivatives, is potassium iodide.

Neutral nitrogen- and phosphorus-centred ligands can also bring about sym-metrisation. Ammonia is the most versatile reagent and used in benzene or chloroform solution it is effective with many types of organomercury salts for which other symmetrising agents fail. For example addition products of alkynes and cyclic alkenes can be symmetrised whereas other ligands induce elimination; 2-carbonylalkylmercury salts can be symmetrised, yet are hydrolysed by aqueous potassium iodide. A modification of the diazo method of preparing arylmercury compounds (section 19.4) is to carry out the decomposition of the double salt in the presence of aqueous ammonia, when diarylmercurials are obtained directly (equation 20.21).

$$2 \, ArN_2 \cdot HgCl_3 + 6Cu \xrightarrow{\text{NH}_3} Ar_2Hg + Hg + 6CuCl + 2N_2 \quad (20.21)$$

Ammonia is generally unsuccessful in the symmetrisation of simple alkylmercury salts but nearly quantitative yields of dialkylmercurys can be obtained using polyethyleneimine.[290]

The mechanism of ammonia-induced symmetrisation has been investigated fairly extensively using mercurials of the type $PhCH(HgBr)CO_2R^1$. The over-all stoichiometry of the reaction is shown in equation 20.22 and the rate law that is obeyed is given in equation 20.23; $R = PhCH \cdot CO_2R^1$.

$$2 \, RHgBr + 2NH_3 \underset{}{\overset{\text{CHCl}_3}{\rightleftharpoons}} R_2Hg + HgBr_2 \cdot 2NH_3 \downarrow \quad (20.22)$$

$$\text{Rate} = k \, [RHgBr]^2 [NH_3]^2 \quad (20.23)$$

Jensen[15] has criticised the technique used by Reutov[291] to derive these results but he has confirmed the basic conclusions.[292]

Hence two molecules of organomercury bromide and two molecules of ammonia are involved in the transition state of the rate-determining step. A possible mechanism is one involving the 1:1 complex between organomercury bromide and ammonia as both electrophile and substrate in bimolecular electrophilic substitution (equations 20.24 and 20.25).

$$RHgBr + NH_3 \xrightleftharpoons{fast} RHgBr.NH_3 \qquad (20.24)$$

$$2\ RHgBr.NH_3 \xrightleftharpoons{slow} R_2Hg + HgBr_2.2NH_3 \qquad (20.25)$$

However, a more likely interpretation is that the 2:1 complex is formed and is attacked by uncomplexed, and therefore more electrophilic, organomercury bromide (equations 20.26 and 20.27).

$$RHgBr + 2NH_3 \xrightleftharpoons{fast} RHgBr.2NH_3 \qquad (20.26)$$

$$RHgBr.2NH_3 + RHgBr \xrightleftharpoons{slow} R_2Hg + HgBr_2.2NH_3 \qquad (20.27)$$

A third fact that indicates that the mechanism involves a bimolecular electrophilic substitution is that there is complete retention of configuration at the site of substitution.[293]

A four-centre transition state (**24**; figure 20.1) is usually proposed for the rate-determining step but there does not appear to be any evidence to favour this over the S_E2 (open) type. In an attempt to define the details of this transition state more clearly, the effect of substituents X and R^1 on the rate of symmetrisation of compounds $X.C_6H_4.CH(HgBr)CO_2R^1$ has been studied. However, it has been pointed out[15] that it is dangerous to draw conclusions about the rate-determining step (equation 20.27) from rate coefficients (k in equation 20.23) that include the equilibrium constant for the fast step (equation 20.26).

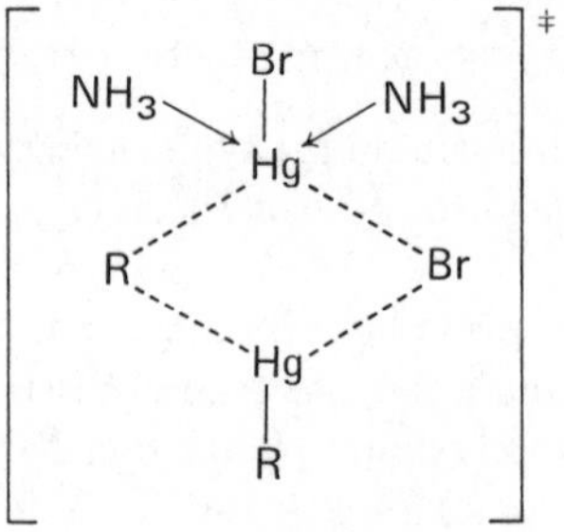

24

Figure 20.1

It is convenient to mention here some of the work that has been carried out in determining the mechanism of the reverse of symmetrisation (equation 20.28).

$$R_2Hg + HgX_2 \rightarrow 2RHgX \qquad (20.28)$$

Several studies employing systems capable of diastereoisomerism (for example R = *cis*-2-methoxycyclohexyl[294]) and systems capable of enantiomerism (for example R = (−)EtCHMe[295]) have established that *syn*-proportionation proceeds with retention of configuration at the site of substitution. Kinetic studies of the reaction of di-*sec*-butylmercury with mercury(II) salts show that it is first order in each reagent.[295]

Thus an $S_E 2$ mechanism is again implicated and on the grounds that $HgBr_3^-$ is less reactive than $HgBr_2$[295] it seems probable that nucleophilic assistance (cyclic transition state) is not very important. Most of these kinetic data relate to reactions in ethanol solution, but subsequent studies have established similar results in solvents such as dioxan,[296] benzene,[297] and pyridine,[298] and activation parameters have been evaluated. These results can have no bearing on the mechanism of symmetrisation since conditions are quite different.

Syn-proportionation continues to be utilised frequently in synthesis and usually involves merely mixing equimolar amounts of the reagents in a suitable solvent (for example equation 20.29[299]).

$$(PhCOCH_2)_2 Hg + HgI_2 \xrightarrow{THF} 2PhCOCH_2 HgI \qquad (20.29)$$

20.2.2 By Reducing Agents

Symmetrisation of many organomercury salts can be brought about by reduction, either electrolytically or with sodium stannite, hydrazine, or metals. Sodium stannite is the most versatile reagent, being effective for aryl-, alkyl-, and vinyl-mercurials; the symmetrisation of chloromercuriocyclopentadienyl(tricarbonyl)-manganese (analogous to equation 20.19) is a recent example.[230] Hydrazine has a similar scope but gives inferior yields with aliphatic systems. Of the reducing metals, copper has found the greatest application, particularly in the synthesis of diaryl-mercurials (for example equation 20.30[281]).

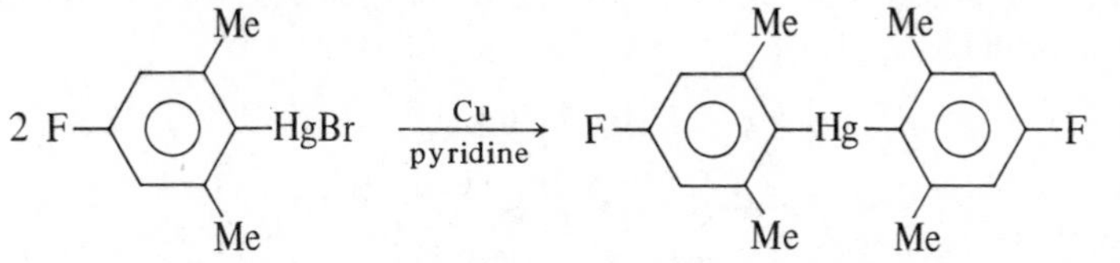

$$\text{(20.30)}$$

55 per cent

The use of zinc in the synthesis of dialkylmercury compounds via the organo-borane route[43] has already been mentioned (section 19.1, equation 19.25). Cadmium amalgam succeeds where all other reagents fail in the symmetrisation of perfluoroalkylmercury salts, and it has been used recently for fluorinated α-mercurioesters (equation 20.31[301]).

$$2 \; CF_3 CF(HgCl)CO_2 Et \xrightarrow{Cd/Hg} [CF_3 . CF(CO_2 Et)]_2 Hg \qquad (20.31)$$

Oxymercurials (section 19.3) pose a particular problem as regards symmetrisation. Both complexing agents and reducing agents tend to induce deoxymercuration and reducing agents also frequently bring about hydrogen-for-mercury substitution. Adducts of cyclic alkenes appear to be the easiest to symmetrise and an example employing hydrazine is shown in equation 20.32.[302]

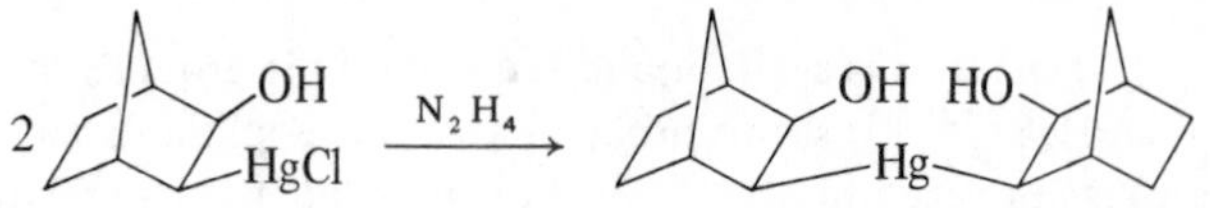

$$(20.32)$$

56 per cent

Symmetrisation of the simplest oxymercurial by an electrolytic method has been described (equation 20.33[303]), and the technique gives good yields of dialkyl-mercurials in general.

$$2\ HOCH_2CH_2HgCl \xrightarrow[NaOH]{2e} (HOCH_2CH_2)_2Hg + Hg + 2Cl^- \qquad (20.33)$$

There is little information about the mechanism of reductive symmetrisation but evidence exists to suggest that radicals are generated during the reaction.[15] Stereochemical studies of symmetrisation by electrolysis or by reaction with sodium stannite or hydrazine reveal extensive racemisation of *both* alkyl groups in the product. This is taken to indicate that radicals that can bring about race-misation in both starting materials and products are present; this appears to preclude any possibility of determining the stereochemistry of dialkylmercury formation. The presence of radicals is in turn taken to indicate that the reaction proceeds by a one-electron transfer (equation 20.34).

$$RHgX + e \rightarrow RH\dot{g} + X^- \qquad (20.34)$$

Three possible routes (equations 20.35–20.37) are proposed for the demise of organomercury(I) species. Of these the unimolecular process is responsible for generating the radicals that promote racemisation while the two bimolecular ones afford the product.

$$RH\dot{g} \rightarrow \dot{R} + Hg \qquad (20.35)$$

$$RH\dot{g} + H\dot{g}R \rightarrow [RHgHgR] \rightarrow RHgR + Hg \qquad (20.36)$$

$$RH\dot{g} + \dot{R} \rightarrow RHgR \qquad (20.37)$$

The balance between routes 20.35 and 20.37 depends on the nature of R but is also strongly influenced by the reaction conditions. Any stabilisation of organo-mercury(I) species should favour self combination and it appears that such stabilisa-tion can be provided by interaction with the surface of metallic mercury[304] or by solvation. Thus whereas there is only 30 per cent of symmetrisation when benzyl-mercury chloride reacts with tetramethylammonium borohydride in benzene, this rises to 70 per cent when the reaction is carried out in tetrahydrofuran.[305] It should be noted that for most organomercury compounds borohydride does not effect symmetrisation but brings about a smooth replacement of mercury by hydrogen and this accounts for its importance in synthesis (see chapter 23).

A notable exception to the rule that extensive racemisation accompanies reductive symmetrisation is that magnesium gives a high degree of retention (equation 20.38[306]).

$$2 \overset{*}{Et}CH(HgBr)Me \xrightarrow{Mg} (Et\overset{*}{C}HMe)_2Hg \xrightarrow{HgBr_2} 2 \overset{*}{Et}CH(HgBr)Me \quad (20.38)$$

$$[\alpha]_D^{25} -5.05° \qquad\qquad [\alpha]_D^{25} -7.57° \qquad\qquad [\alpha]_D^{25} -4.60°$$

It is suggested that this reaction involves a two-electron transfer (equation 21.39) and the absence of intermediate carbanions or Grignard reagents was indicated by an inability to trap such species with butanol or carbon dioxide.

$$RHgBr \xrightarrow{2e} RHg^- \xrightarrow[-Br^-]{RHgBr} [RHgHgR] \rightarrow R_2Hg + Hg \qquad (20.39)$$

The reaction is synthetically important in that it permits the synthesis of dialkyl-mercurials in which both alkyl substituents are optically active.

20.3 Transformation of RHgX into RHgY

The most common example of altering the inorganic group in an organomercury salt is the exchange of chloride for acetate (equation 20.40).

$$RHgOAc + MCl \rightarrow RHgCl + MOAc \qquad (20.40)$$

Because mercury(II) acetate is the most popular reagent in both mercuration (section 19.2) and oxymercuration (section 19.3), organomercury acetates are often the initial products of reactions generating the mercury–carbon bond. The anion exchange is readily effected with aqueous sodium or potassium chloride and is usually carried out because the organomercury chlorides are more amenable to purification than the corresponding acetates; many examples have been given in the aforementioned sections. Similar reactions occur with bromide and iodide but an excess of the reagent must be avoided or else symmetrisation may be induced.

Related transmetallation reactions (equation 20.41) can afford almost all the known types of RHgY.

$$RHgX + MY \rightarrow RHgY + MX \qquad (20.41)$$

The group X is usually a halogen and more often than not M is an alkali metal (for example equations 20.42[307] and 20.43[308]) although trialkyl derivatives of group IV metals have been used (equations 20.44[309] and 20.45[310]).

$$4\text{-MeO} . C_6H_4HgCl + NaOOCMe_2Ph \rightarrow 4\text{-MeO} . C_6H_4HgOOCMe_2Ph + NaCl$$

$$\qquad\qquad\qquad\qquad\qquad\qquad\qquad\qquad\qquad\qquad\qquad\qquad (20.42)$$

$$2\,^i\mathrm{PrHgBr} + \mathrm{Na_2S} \rightarrow (^i\mathrm{PrHg})_2\mathrm{S} + 2\mathrm{NaBr} \tag{20.43}$$

$$\mathrm{PhHgCl} + \mathrm{Bu_3GeSC_6H_4 \cdot 4\text{-}Me} \rightarrow \mathrm{PhHgSC_6H_4 \cdot 4\text{-}Me} + \mathrm{Bu_3GeCl} \tag{20.44}$$

$$\mathrm{MeHgCl} + \mathrm{Me_3SnMn(CO)_5} \rightarrow \mathrm{MeHgMn(CO)_5} + \mathrm{Me_3SnCl} \tag{20.45}$$

Where organomercury chlorides are available but carboxylates are required, they can be obtained by reaction with the appropriate silver salt (equation 20.46[281]).

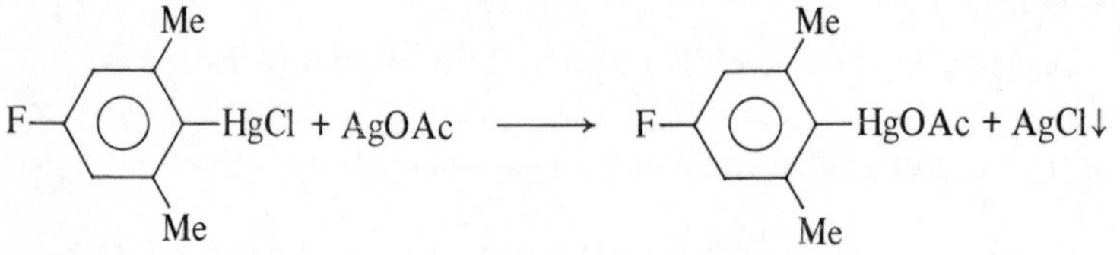

$$\tag{20.46}$$

Organomercury trimethylsilanes can be prepared by reaction 20.47 and the corresponding tin compound by subsequent exchange with trimethyltin methoxide (equation 20.48).[311]

$$\mathrm{Bu^t HgCl} + (\mathrm{Me_3Si})_2\mathrm{Hg} \rightarrow \mathrm{Bu^t HgSiMe_3} + \mathrm{Me_3SiCl} + \mathrm{Hg} \tag{20.47}$$

$$\mathrm{Bu^t HgSiMe_3} + \mathrm{Me_3} + \mathrm{Me_3SnOMe} \rightarrow \mathrm{Bu^t HgSnMe_3} + \mathrm{Me_3SiOMe} \tag{20.48}$$

Reactions 20.42–20.48 illustrate how varied the group Y can be. The most common type involves bonding to mercury through oxygen or nitrogen and the hydroxides (and related oxides), alkoxides, and amines are the most important of these because they provide an alternative starting point for the synthesis of many other organomercury compounds.

Phenylmercury(II) hydroxide is best prepared by heating under reflux a slurry of phenylmercury(II) acetate and methanol with 8 per cent aqueous sodium hydroxide[141] (equation 20.49).

$$\mathrm{PhHgOAc} + \mathrm{NaOH} \xrightarrow{\mathrm{MeOH/H_2O}} \mathrm{PhHgOH} + \mathrm{NaOAc} \tag{20.49}$$

The hydroxide is readily converted into the corresponding oxide (equation 20.50), and the existence of two discrete compounds was established by a combination of thermogravimetric analysis and infrared spectroscopy.[312]

$$2\,\mathrm{PhHgOH} \xrightarrow[\text{0.3--0.05 mm}]{100\,^\circ\mathrm{C}} (\mathrm{PhHg})_2\mathrm{O} + \mathrm{H_2O} \tag{20.50}$$

Methylmercury oxide[313,314] is a solid, melting at 139 °C, but the corresponding hydroxide does not appear to exist. The product from hydrolysing methylmercury bromide with methanolic potassium hydroxide has been described as a mixture of the oxide and the onium salt $(\mathrm{MeHg})_3\mathrm{O^+}^-\mathrm{OH}$ (melting point 88 °C).[313] However, treatment of the compound $\mathrm{MeHgN(SiMe_3)_2}$ (see below) with an equimolar amount of water at −70 °C affords a product that also melts at 88 °C but is said to have the composition $(\mathrm{MeHg})_2\mathrm{O} \cdot x\mathrm{H_2O}$.[314]

The organomercury silylamines are prepared by the route shown in equation 20.51 (R = Me, Et; X = Cl or Br).[314]

$$(Me_3Si)_2NH \xrightarrow[-BuH]{Bu\,Li/hexane} LiN(SiMe_3)_2 \xrightarrow[-LiX]{R\,HgX\,/toluene} RHgN(SiMe_3)_2$$

$$(20.51)$$

A similar method has been used to prepare organomercury alkoxides (for example equation 20.52[139]).

$$PrHgCl + NaOBu^t \xrightarrow{Et_2\,O} PrHgOBu^t + KCl \qquad (20.52)$$

These routes have the disadvantage that moisture-sensitive products must be separated from finely divided salts. This can be avoided for the alkoxides by using the alternative methods shown in equations 20.53 (R = Me or Et) and 20.54.[140]

$$(PhHg)_2O + (RO)_2CO \rightarrow 2\,PhHgOR + CO_2 \qquad (20.53)$$

$$PhHgOMe + ROH \rightarrow PhHgOR + MeOH \qquad (20.54)$$

We have seen that the mercuration of sufficiently strong carbon acids (HR^1) by the compounds PhHgOH, PhHgOR, or $RHgN(SiMe_3)_2$ is an important route to unsymmetrical diorganomercurials, $RHgR^1$ (section 20.1). Analogous reactions with other acids (equation 20.55) provide the main alternative to transmetallation for preparing compounds RHgY.

$$RHgX + HY \rightarrow RHgY + HX \qquad (20.55)$$

The range of products thus available from organomercury hydroxides is illustrated by reactions with N-acids (equations 20.56–20.61), O-acids (equations 20.9 and 20.62), and S-acids (equations 20.63 and 20.64). The ease of mercuration generally increases with increasing acidity of HY. Thus water must be removed as an azeotrope with benzene in the mercuration of aniline, but phenols and thiols react at room temperature in aqueous media.

$$n\,\text{'MeHgOH'} + NH_3 + HClO_4 \xrightarrow{\text{ref.}\atop315} [(MeHg)_n NH_{4-n}]^+ ClO_4^-$$

$$n = 1\text{-}4 \qquad (20.56)$$

$$PhHgOH + PhNH_2 \xrightarrow{\text{ref.}\atop86} PhHgNHPh \qquad (20.57)$$

$$PhHgOH + \;\;\; \xrightarrow{\text{ref.}\atop316} \qquad\qquad (20.58)$$

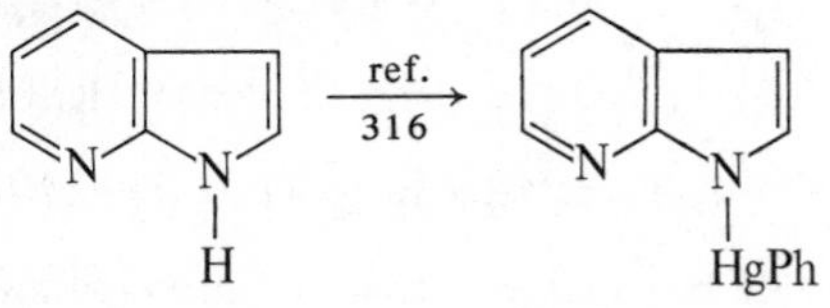

$$3\text{-F}.C_6H_4HgOH + MeCONH_2 \xrightarrow{\text{ref.}\atop281} 3\text{-F}.C_6H_4HgNH.COMe$$

$$(20.59)$$

$$\text{PhHgOH} + \text{PhSO}_2\text{NHC}_6\text{H}_4 . 2\text{-Me} \xrightarrow[317]{\text{ref.}} \text{PhHgN(2-Me} . \text{C}_6\text{H}_4)\text{SO}_2\text{Ph}$$

$$(20.60)$$

$$\text{PhHgOH} + 4\text{-NO}_2 . \text{C}_6\text{H}_4\text{NH} . \text{N} : \text{CHPh} \xrightarrow[318]{\text{ref.}} \text{PhHgN(4-NO}_2 . \text{C}_6\text{H}_4)\text{N} : \text{CHPh}$$

$$(20.61)$$

$$\text{'MeHgOH'} + 4\text{-NO}_2 . \text{C}_6\text{H}_4\text{OH} \xrightarrow[319]{\text{ref.}} \text{MeHgOC}_6\text{H}_4 . 4\text{-NO}_2 \qquad (20.62)$$

$$(20.63)$$

$$\text{PhHgOH} + \text{MeCSNHPh} \xrightarrow[320]{\text{ref.}} \text{PhHgSCMe} : \text{NPh} \qquad (20.64)$$

Understandably the moisture-sensitive organomercury bases have not been used as widely as the hydroxide, but they are very powerful mercurating agents. Diethylamine[321] and pyrrole[322] have been N-mercurated by phenylmercury(II) alkoxides, and hydrazoic acid,[314] cyanamide,[314] ammonia (equation 20.65[323]), and tris(dimethylamino)phosphine imine (equation 20.66[324]) by alkylmercury(II) bis-(trimethylsilyl)amines; the latter also bring about O-mercuration of nitroalkanes (equation 20.67[325]).

$$\text{MeHgN(SiMe}_3)_2 + \text{liquid NH}_3 \rightarrow (\text{MeHg})_3\text{N} + 3\text{HN(SiMe}_3)_2 \quad (20.65)$$

$$\text{MeHgN(SiMe}_3)_2 + \text{HN} : \text{P(NMe}_2)_3 \rightarrow \text{MeHgN} : \text{P(NMe}_2)_3 + \text{HN(SiMe}_3)_2$$

$$(20.66)$$

$$\text{EtHgN(SiMe}_3)_2 + \text{Me}_2\text{CH} . \text{NO}_2 \rightarrow \text{EtHgON(O)} : \text{CMe}_2 + \text{HN(SiMe}_3)_2$$

$$(20.67)$$

A few substitutions not involving metal or protic reagents are known, for example the formation of organomercury bromide from the corresponding acetate and bromine[184] and the reaction of N-phenylmercurioaniline with phthalic anhydride,[86] but these are of little preparative interest. New organomercury compounds can be prepared by addition reactions of oxides and alkoxides to carbon dioxide (equation 20.68[313]), carbon disulphide (equation 20.69[326]), and isocyanates[86] (equation 20.70; R = Me, Bu^t or HgPh).

$$(\text{MeHg})_2\text{O} + \text{CO}_2 \rightarrow (\text{MeHgO}_2)\text{CO} \qquad (20.68)$$

$$(\text{PrHg})_2\text{O} + \text{EtOH} + \text{CS}_2 \rightarrow \text{PrHgSCSOEt} \qquad (20.69)$$

$$\text{PhHgOR} + \text{R}^1\text{NCO} \rightarrow \text{PhHgNR}^1\text{COOR} \qquad (20.70)$$

This method should be capable of extension to organomercury amines (see reference 86) and to other multiply bonded acceptors, but has received little attention to date.

21 Structure of Organo-mercury Compounds

This is a short chapter, which is intended to gather together the results of studies in which internuclear parameters for organomercury compounds have been determined. It is mainly concerned, therefore, with work involving diffraction methods and hence with structures of mercurials in the vapour or solid phases. The reported mercury–carbon bond lengths vary from 0.182 to 0.234 nm, but most results lie in the ranges 0.210 ± 0.005 and 0.207 ± 0.003 nm for aliphatic and aromatic derivatives respectively. The angle subtended at mercury by the two R groups or the R- and X groups is very rarely outside the range 170–180° and the linear arrangement is often found in the solid as well as in the vapour state. There is spectroscopic evidence to suggest that the characteristic linear stereochemistry is carried over into solution.

The use of spectroscopic techniques to determine *qualitative* aspects of structure, that is, atomic sequences and stereochemical nature, is not discussed here but has been illustrated in section 19.2.2 in connection with the mode of bonding in mercurials obtained from 1,3-dicarbonyl compounds and cyclopentadiene derivatives, and in section 19.3.2 with regard to the stereochemistry of oxy-mercurials. n.m.r. spectroscopy is by far the most important technique in solution studies and a detailed summary[327] of its application in structural organo-mercury chemistry has made a timely appearance. The review is arranged according to type of organomercury compound and a lot of useful chemical shift (^{1}H, ^{13}C, ^{19}F, and ^{199}Hg) and coupling constant (^{1}H-^{199}Hg, ^{13}C-^{199}Hg, ^{19}F-^{199}Hg) data are presented, many of them conveniently in tabulated form; a complete coverage of the literature until the end of 1973 is provided. It seems likely that in the immediate future the biggest advances will derive from ^{13}C n.m.r. spectroscopic studies and accordingly we draw attention to some ^{13}C n.m.r. work[328–330] that has appeared since the review was published.

21.1 Vapour Phase

A recent study[331] of dimethylmercury by electron diffraction has provided information about the mercury–carbon bond that supercedes the earlier data quoted in Grdenić's review[332] of the structural chemistry of mercury. The new work gives $r_g(HgC) = 0.2083 \pm 0.0005$ nm, which corresponds to an r_α^0 HgC distance of 0.2080 ± 0.0005 nm, shorter by 0.0015 nm than the comparable r_z distance obtained by Raman spectroscopy (see reference 333 for the meaning of these different internuclear distances). The mercury–carbon bond length found for diphenylmercury by electron diffraction is 0.2093 ± 0.0005 nm[334] and, as with dimethylmercury, the CHgC arrangement is linear.

A linear disposition of ligands is maintained in methylmercury chloride, bromide, and iodide, and in phenylmercury bromide. The methylmercury halides have been studied by microwave spectroscopy and earlier work has been superceded by a more thorough investigation[335] in which the spectra of many more isotopic species have been measured. In particular the rotational constants obtained for fifteen species of methylmercury iodide invalidate data previously reported for this molecule. The internuclear distances found are presented in table 21.1.

Table 21.1

Compound	$r_s(HgC)$/nm	$r_s(HgX)$/nm
MeHgCl	0.2055	0.2283
MeHgBr	0.2072	0.2406
MeHgI	0.2077	0.2571

Electron-diffraction data for phenylmercury bromide afford bond lengths of 0.2068 ± 0.0002 and 0.2435 ± 0.0004 nm for Hg–C and Hg–Br respectively.[336]

We are aware of only one other electron-diffraction study and this shows that *cis*-2-chlorovinylmercury chloride is planar with a CHgCl angle of $168.5 \pm 1.5°$.[337] The distortion from linearity may arise from intramolecular coordination of the carbon-bonded chlorine to mercury but the separation of these atoms (0.327 nm) indicates that any such interaction must be weak. The internuclear distances of greatest interest are 0.214 ± 0.002 nm for Hg–C and 0.227 ± 0.001 nm for Hg–Cl.

21.2 Solid State

Recently there has been a considerable upsurge in X-ray diffraction work and half of the approximately thirty crystal structures that have now been reported have appeared in the literature in the last three years. We provide here what we believe to be a substantially complete list (at December 1974) of compounds for which internuclear parameters have been determined. This takes the form of four tables containing compound formulae with associated Hg–C and Hg–X distances together with the CHgX angle. All diorganomercurials appear in table 21.2 but the larger list

Table 21.2. X-Ray Crystallographic Data for R_2Hg

Compound	r(HgC)/nm	<CHgC/°	Ref.
PhHgPh†		180	339
4-Me . $C_6H_4HgC_6H_4$. 4-Me†	0.208(2)	180(1.4)	340
$C_6F_5HgC_6F_5$	0.209	176.2(1.2)	341
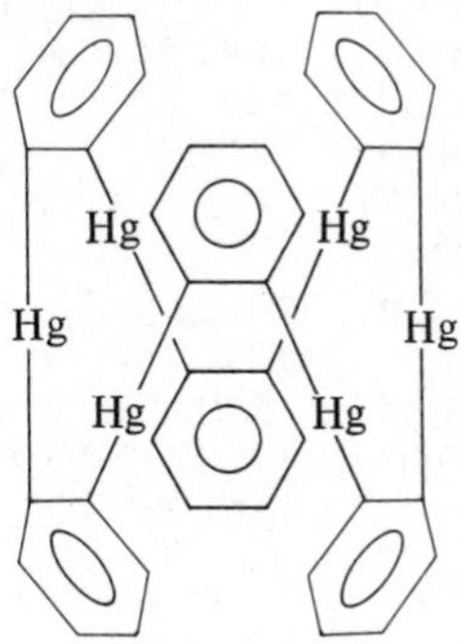 *(polycyclic structure with six Hg atoms)*	0.215(5)	180	342
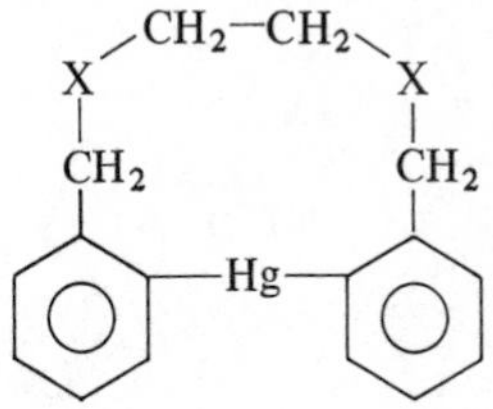 *(macrocyclic structure)* $X = O$	0.207(5) / 0.213(6)	173.6(2.1)	343
$X = NMe$	0.207(5) / 0.212(2)	172.3(0.9)	343
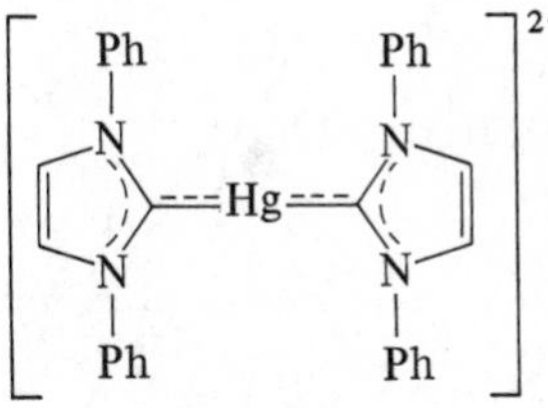 $\left[\text{(imidazolyl)}_2Hg\right]^{2+}$ 2 ClO_4^- ‡	0.206(1)	180	344
$[(Bu^tCO)_2CH]_2Hg$	0.213(3) / 0.218(3)	170(1)	105
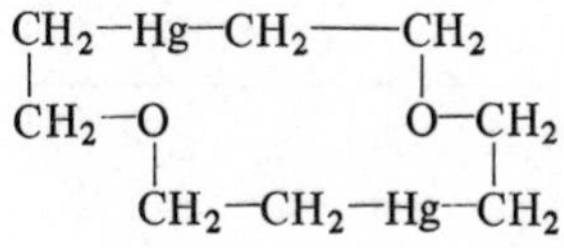 *(dioxa-dimercury ring structure)*	0.214	176	345

† Planar with mercury at a centre of symmetry.

‡ Mercury is at a centre of symmetry and the imidazolylmercury system is planar.

Table 21.3. X-ray Crystallographic Data for RHgX

Compound	r(HgC)/nm	r(HgX)/nm	<CHgX/°	Ref.
MeHgCl	0.206(3)	0.250(3)		346
MeHgCN[†]	0.208(2)	0.205(1)	180	347
MeHgN$_3$	0.227(11)	0.222(14)	173(4)	348
(MeHgOSiMe$_3$)$_4$	0.209(4)	0.211(2)	112(8)	349
MeHgSCMe$_2$CH(CO$_2$$^-$)NH$_3$$^+$.H$_2$O	0.198(6)	0.2376(14)	176(2)	350
	0.182(5)	0.2385(14)	179(1)	
MeHgSCMe$_2$CH(CO$_2$$^-$)NH$_2$$^+$HgMe	0.1885(7)[‡]	0.2320(2)[‡]	176.4[‡]	351
	0.2165(6)[§]	0.2216(5)[§]	168.1[§]	
MeHgSCH$_2$CH(CO$_2$$^-$)NH$_3$$^+$.H$_2$O	0.209(4)	0.2352(12)	180	352
MeHg$^+$NH$_2$CH(CO$_2$$^-$)CH$_2CH_2$SMe		0.206(4)	173(2)	352
CCl$_3$HgBr		0.236		353
		0.243		
trans-ClCH:CHHgCl	0.210(11)	0.230(2)	167(5)	354
PhCH$_2$HgSCPh$_3$	0.210	0.2363	179	355
(Bu^{t}CO)$_2$CHHgOAc	0.211(2)	0.210(1)	175(1)	78
cis-PhCOCH:CHHgCl	0.223(12)	0.2370(19)	173(2)	356
Et$_2$NCH$_2$CH$_2$HgCl	0.213(3)	0.236(1)	167.1(8)	231
cyclohexane, HgCl / """OMe	0.234	0.253	178	186
cyclohexane, HgCl / OMe	0.215	0.250	180	186
Me$_2$C(OMe)CMe$_2$CH$_2$HgI		0.260		357
octahydrobenzofuran with OH, Me, Me, HgCl	0.203(2)	0.213(1)	174(1)	358
R–C=C–HgX (B$_{10}$H$_{10}$ carborane), R = Me, X = Br		0.247		
R = Ph, X = Br		0.244		359
R = Ph, X = I		0.258		
C(HgOAc)$_4$	0.205(2)	0.210(8)		360
C(HgO$_2$C.CF$_3$)$_4$	0.2042(4)	0.2065(1)	175.3(0.3)	360

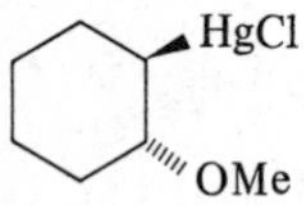

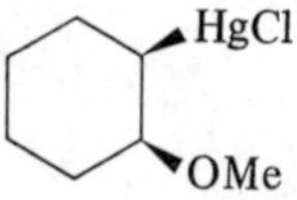

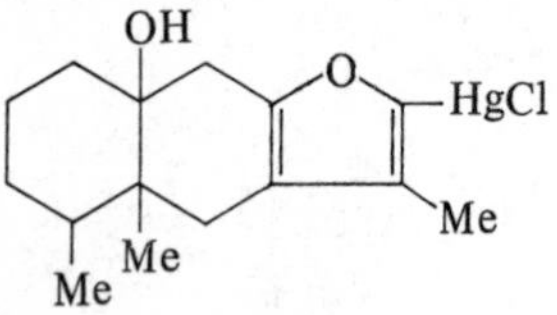

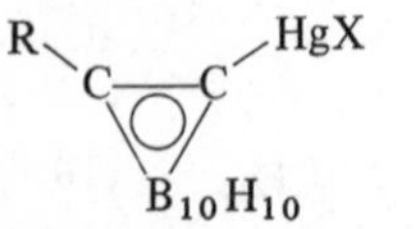

[†] Neutron diffraction data also obtained.

[‡] For the MeHgS group.

[§] For the MeHgN group.

Table 21.4. X-ray Crystallographic Data for ArHgX

Compound	$r(HgC)$/nm	$r(HgX)$/nm	$<C_1HgX$/°	Ref.
PhHgBr	0.210	0.243		361
PhHgOAc	0.192(6)	0.211(4)	170(2)	362
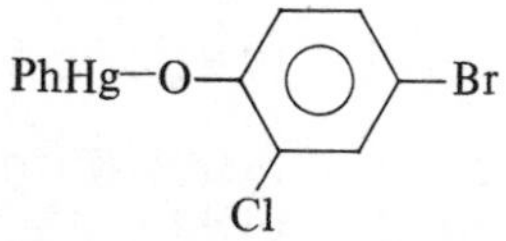 PhHg—O—(ring, Cl, Br)	0.207	0.201	175	363
PhHgSC$_6$H$_3$. 2,6-Me$_2$	0.197(6)	0.233	172(1)	338
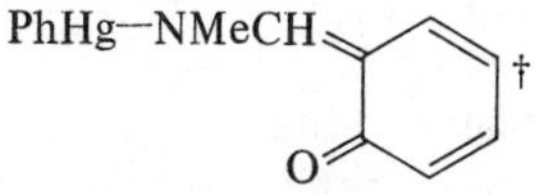 PhHg—NMeCH (ring, O) †	0.202(4)	0.215(3)	167(1)	364
4-EtO$_2$C . C$_6$H$_4$HgBr		0.242		365
(OH, Me ring)—Hg—N (O, ring, Me) ‡	0.207(4) / 0.201(5)	0.208(3) / 0.220(4)	174 / 176	366
1-C$_{10}$H$_7$HgI		0.261		367

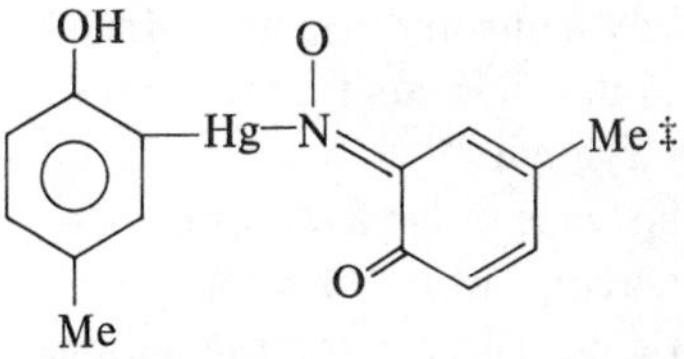

† The O . . . Hg distance is 0.244(3) nm, which indicates an appreciable contribution to the structure from the benzoid form, PhHgOC$_6$H$_4$. 2-CH : NMe.

‡ The C=O . . . Hg distance is 0.257(3) nm, which indicates an appreciable contribution to the structure from the benzoid form:

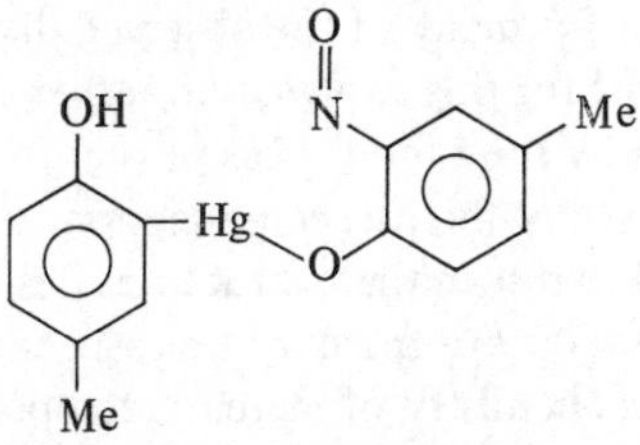

of organomercury salts is sub-divided into aliphatic (table 21.3) and aromatic (table 21.4) derivatives; table 21.5 contains details of organomercury complexes.

A problem in determining the structure of organomercury compounds by X-ray diffraction is that the mercury atom so dominates the scattering that location of the light atoms is difficult. A diffraction ripple near mercury can cause special problems in correctly locating the adjacent carbon so that the Hg—C distance may be particularly susceptible to error. Another important factor that must be allowed

Table 21.5. X-ray Crystallographic Data for Organomercury Complexes

Compound	r(HgC)/nm	r(HgX)/nm	<CHgX/°	Ref.
Ph$_2$Hg.(dmp)$_2$ [†]	0.210		180	368
Ph$_2$Hg.(tmp)$_2$ [‡]	0.210		180	368
	0.213			
[(C$_6$F$_5$)$_2$Hg]$_2$.Ph$_2$AsCH$_2$AsPh$_2$	0.207(4)		173(1.4)	369
	0.215(4)			
CCl$_3$HgCl.1,10-phenanthroline	0.212	0.230	155(2)	370
	0.207	0.232	145(2)	

[†] dmp = 2,9-dimethyl-1,10-phenanthroline.
[‡] tmp = 2,4,7,9-tetramethyl-1,10-phenanthroline.

for, especially if the crystal is not centrosymmetric, is the large anomalous-dispersion correction for mercury; anomalous dispersion can in fact be used to solve the structure. No attempt has been made here to assess the quality of each structure determination and the authors' estimated errors are reproduced in the tables without comment. Structures can be refined by neutron diffraction since the scattering factors for all atoms are then similar, but it appears that this has been carried out only in the case of methylmercury cyanide.[347]

Internuclear distances frequently indicate the eixstence of intra- and/or inter-molecular interactions between donor atoms and mercury, which raise the co-ordination number of mercury to three, four, five or six. These extra interactions are recognised by the usual criterion that the distance between the donor atom and mercury is less than the sum of their van der Waals radii, but there is consider-able uncertainty about the magnitude of the van der Waals radius for mercury. The upper limit of the range 0.150–0.173 nm[332] has often been taken in these diagnoses yet recent data[338] point to the smaller value of 0.156 nm. The co-ordinative interactions are generally weak but they (particularly the intramolecular variety) may be responsible for many of the observed distortions from a linear CHgX arrangement. Support for this explanation rather than one involving crystal-packing effects is provided by the vapour phase structure of *cis*-chlorovinylmercury chloride, which was discussed in the preceding section.

Although much of the interest in these structures lies in determining the co-ordination number and geometry of mercury this aspect will not be taken further here since the coordination chemistry of mercury compounds as a whole is discussed in another chapter.

22 Reactions of the Mercury–Carbon Bond

The metal–carbon bond in organomercury compounds is rather unreactive in comparison with that in organic compounds of other group II metals. Thus organomercurials are a poor source of nucleophilic carbon and usually do not react with compounds such as alcohols, aldehydes, ketones or alkyl halides. Nevertheless many of the principal reactions that they do undergo involve electrophilic substitution at carbon, the best-characterised examples being reactions with metal salts, acids, and halogens. In these substitutions the fully organic mercurials (R_2Hg) are generally more reactive than the corresponding organomercury salts (RHgX) since the presence of the X group attenuates the nucleophilicity of R. The Hg–C bond in aryl derivatives is more readily cleaved than its counterpart in aliphatic systems.

Within the classification of electrophilic substitution, several mechanisms can be envisaged even if we only consider reactions that occur at a saturated carbon centre. The various possibilities have been listed by Abraham[17] and the nomenclature recommended by him is adopted here. The S_E2 mechanism is found most commonly but it is difficult to distinguish between the 'open' (for example equation 22.1[371]) and 'cyclic' varieties. In the latter the nucleophilic part of the reactant assists Hg–C bond breaking by interacting, in the transition state, with the mercury atom of the substrate.

$$RHgX + Hg^*X_2 \rightleftharpoons \left[R \overset{\cdots HgX}{\underset{\cdots Hg^*X_2}{\Big<}} \right]^\ddagger \rightleftharpoons RHg^*X + HgX_2 \quad (22.1)$$

The criticism that the principle of microscopic reversibility is contravened by an open transition state for a symmetrical exchange has been shown to be unfounded.[372] Some recent calculations of steric effects lead to relative rate constants for the reaction 22.1 that are in close agreement with observed values provided that an open and not a cyclic transition state is assumed.[373]

Coordination of an external nucleophile to the mercury enhances the reactivity of the Hg–C bond towards electrophilic substitution. This effect has been established kinetically where the nucleophilic assistance is provided by added anions (for example equation 22.2; R = PhCH . CO_2Et[374]), but any such assistance by the solvent is not amenable to detection.

$$(-)RHgBr + 2Br^- \rightleftharpoons RHgBr_3{}^{2-} \xrightarrow{slow} \bar{R} + HgBr_3{}^- \rightarrow (\pm)RHgBr \quad (22.2)$$

The view has been expressed[375] that all S_E1 reactions require nucleophilic catalysis, and in agreement with this the few authenticated examples all involve reactions of organomercurials with relatively strong acceptor properties, in good donor solvents (equation 22.3; R = PhCH . CO_2Et,[373,376] $4\text{-}NO_2 . C_6H_4$,[377] or C_6F_5[378]).

$$RHgBr + Hg^*Br_2 \underset{DMSO}{\rightleftharpoons} RHg^*Br + HgBr_2 \quad (22.3)$$

Obviously coordination of solvent molecules to the organomercury substrate could also act to favour an open over a cyclic transition state in the bimolecular mechanism.

To illustrate these *general* remarks about electrophilic substitution of organomercurials, we have deliberately chosen examples involving the symmetrical one-alkyl exchange, for this neatly links mercury–carbon bond formation with mercury–carbon bond cleavage. The reaction is concerned with alkyl transfer both to and from mercury and it has been the subject of an intense mechanistic investigation.[15-17] The picture that emerges from this, and from the studies of *syn*-proportionation (section 20.2.1) and of alkyl transfers from other metals to mercury (section 19.1.2), provides a reasonable model for our view of the preparatively important reactions between mercury alkyls and metal salts (see below).

There are many reactions of organomercury compounds that do not involve electrophilic substitution at carbon. The idea that the electronegative group (X) in organomercury salts (RHgX) attenuates the nucleophilicity of R has already been mentioned, and the logical extension of this is that a strongly electron-withdrawing X group should render the alkyl group susceptible to *nucleophilic* attack. This is observed, for example, in the reactions by which mercury(II) salts bring about oxidation of alkenes (the Denigès and Treibs reactions).[379] Carbonium ion intermediates are sometimes implicated and their formation may be assisted by a redox reaction between the organomercurial and mercury(II) salt (equation 22.4, where X is typically ClO_4, NO_3, or $O_2C . CF_3$).

$$HgX_2 + XHgR \rightarrow X^- + Hg_2X_2 + R^+ \rightarrow products \quad (22.4)$$

The mercury–carbon bond can be cleaved homolytically and the concurrent operation of heterolytic and homolytic pathways for some reactions has complicated many mechanistic studies. In this connection particular mention should be made of halogenolysis, for which both S_E and free-radical chain mechanisms are known; the chain-propagating steps are shown in equations 22.5 and 22.6, where X is chlorine, bromine, or iodine.

$$\dot{R} + X_2 \rightarrow RX + \dot{X} \qquad (22.5)$$

$$\dot{X} + RHgX \rightarrow \dot{R} + HgX_2 \qquad (22.6)$$

Another reaction involving homolysis of the mercury–carbon bond that has received much attention in recent years, is the reduction of organomercury salts with alkaline sodium borohydride, and the pyrolysis or photolysis of organomercurials are classical routes to free radicals.

Thus carbanions, free radicals, and carbonium ions are all established intermediates in reactions of organomercurials and this illustrates the broad range of mechanisms encountered for mercury–carbon bond cleavage. Of course there are many reactions, some very important synthetically, for which mechanistic information is scanty or non-existent. Consequently we have not used mechanism as the basis of our approach to describing organomercury reactions. Rather we have adopted a product-based classification and the main part of our discussion is divided according to the type of substituent that has replaced mercury in the final product.

Finally in this introduction we must draw attention to Makarova's two-part treatise dealing with reactions of organomercury compounds and published in 1971.[380] This represents the most extensive review of the subject to date and a particularly valuable aspect of it is the inclusion of experimental details for reactions that are preparatively useful. Where possible we have tried to illustrate reactions with examples that have appeared since Makarova's review was completed.

22.1 Replacement of Mercury by other Metals (Organometallic Synthesis)

In synthesis the traditional application of organomercurials is in transferring their organic groups to other metals to prepare new organometallic compounds. Diorganomercurials (R_2Hg) are usually the reagents of choice but most exchanges will also occur with the corresponding organomercury salts. Reactions with metal alkyls (equations 22.7–22.9) and hydrides (equations 22.10–22.12) have been used in this context, but the most important systems involve transfers to free metals and their halides.

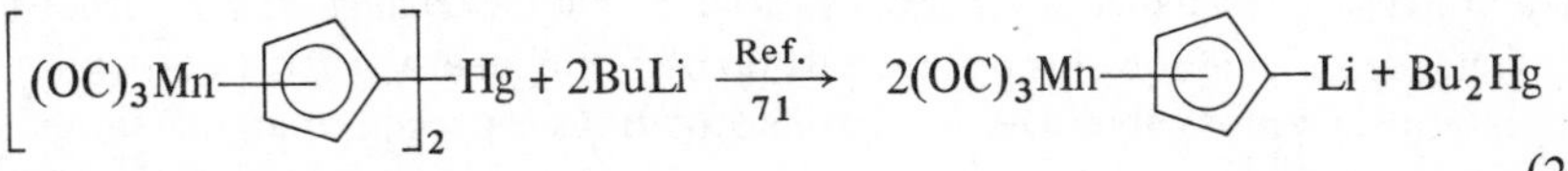

$$\left[(OC)_3Mn\!\!-\!\!\bigcirc\!\!-\right]_2 Hg + 2BuLi \xrightarrow[71]{Ref.} 2(OC)_3Mn\!\!-\!\!\bigcirc\!\!-Li + Bu_2Hg$$

$$(22.7)$$

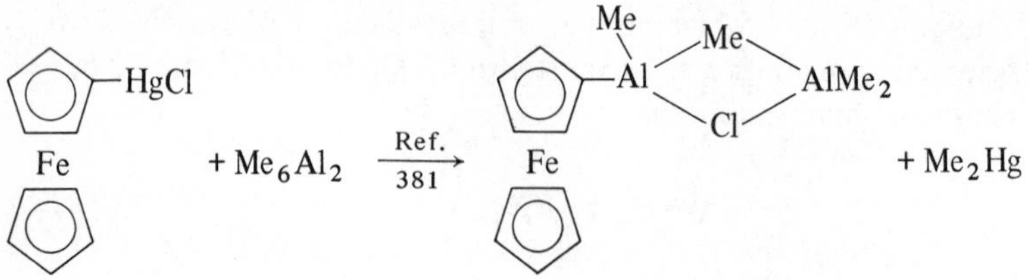

$$\text{Fc}-\text{HgCl} + \text{Me}_6\text{Al}_2 \xrightarrow{\text{Ref.}}{381} \text{Fc}-\overset{\overset{\text{Me}}{|}}{\text{Al}}\overset{\text{Me}}{\diagdown}\text{AlMe}_2 + \text{Me}_2\text{Hg} \qquad (22.8)$$

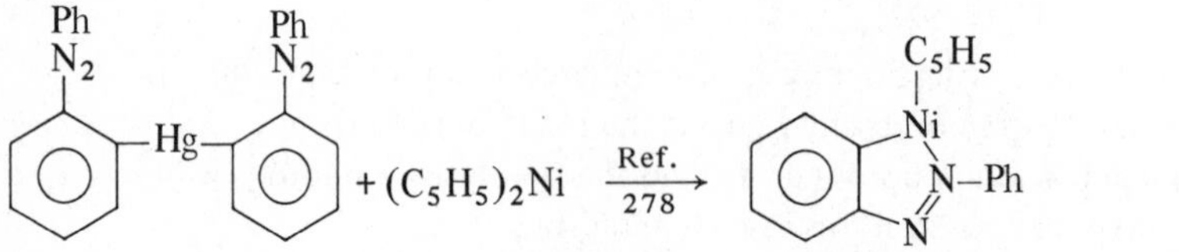

$$\text{Ph-N}_2\text{-}\underset{}{\bigcirc}\text{-Hg-}\underset{}{\bigcirc}\text{-N}_2\text{-Ph} + (\text{C}_5\text{H}_5)_2\text{Ni} \xrightarrow{\text{Ref.}}{278} \underset{}{\bigcirc}\overset{\text{C}_5\text{H}_5}{\underset{}{\text{Ni}}}\text{N-Ph} \qquad (22.9)$$

$$\text{Et}_2\text{Hg} + \text{EtBeH}.\text{NMe}_3 \xrightarrow{\text{Ref.}}{382} \text{Et}_2\text{Be}.\text{NMe}_3 + \text{Hg} + \text{EtH} \qquad (22.10)$$

$$\text{Ar}_2\text{Hg} + \text{BH}_3 \xrightarrow{\text{Ref.}}{383} [\text{ArBH}_2] \xrightarrow{\text{H}_2\text{O}} \text{ArB(OH)}_2 \qquad (22.11)$$

$$\text{Me}_2\text{Hg} + 2\text{GaH}_3.\text{NMe}_3 \xrightarrow{\text{Ref.}}{384} 2\,\text{MeGaH}_2.\text{NMe}_3 + \text{Hg} + \text{H}_2 \qquad (22.12)$$

22.1.1 By Metals

The reaction of diorganomercurials with metals of groups I, II, and III (equation 22.13; $n = 1$–3) is one of the classical routes to organometallic compounds. It was by this method that the first organic derivatives of a group III metal were prepared.

$$n/2\,\text{R}_2\text{Hg} + \text{M} \rightarrow \text{R}_n\text{M} + n/2\,\text{Hg} \qquad (22.13)$$

The reaction also proceeds with tin and bismuth, though better with their sodium alloys. A variety of simple alkyl- and aryl-metallics have been prepared in this way and a few examples of alkenyl and alkynyl exchanges are known. In some systems it has been shown that the alkenyl groups are transferred with retention of configuration (for example equations 22.14[385] and 22.15[386])

$$\textit{cis}\text{-ClCH}:\text{CHHgCl} + \text{Na/Sn} \rightarrow (\textit{cis}\text{-ClCH}:\text{CH})_3\text{SnCl} \qquad (22.14)$$

$$(\textit{trans}\text{-MeCH}:\text{CH})_2\text{Hg} + \text{Ga} \rightarrow (\textit{trans}\text{-MeCH}:\text{CH})_3\text{Ga} \qquad (22.15)$$

The reaction conditions are dictated mainly by the identity of the metal. Thus organic derivatives of group I metals are formed at, or just above, room temperature whereas transfers to other metals occur only with heating, often to quite high temperatures; aryl exchange is generally easier than alkyl exchange. Hydrocarbons and ethers are suitable solvents but the reactions are frequently carried out with neat reagents in sealed tubes. Dry anaerobic conditions (nitrogen or argon) should be used.

A very important feature of the reaction is that it provides organometallic compounds that are not contaminated with metal salts; this is particularly valuable when preparing metal organics for spectroscopic study (for example diallylmagnesium[387]). In synthetic applications the newly formed organometallic reagent is often used *in situ.*

Table 22.1. Preparation of Organometallic Compounds from Diorganomercurials and Metals

R in R_2Hg	Metal	Product	Ref.
CD_3	Li	CD_3Li	389
Me_3SiCH_2	Na, K, Rb, Cs	Me_3SiCH_2M	390
$Ar^\dagger$	Ca	Ar_2Ca	391
Ph_3C	Ca	$(Ph_3C)_2Ca$	391
$PhCH_2$	Al	$(PhCH_2)_3Al$	392
cyclo-C_3H_5	Ga	$(cyclo-C_3H_5)_3Ga$	393
C_6F_5	In	$(C_6F_5)_3In$	394

† Ar = Ph, 2-Me . C_6H_4, 3-Me . C_6H_4, 4-Me . C_6H_4, 4-MeO . C_6H_4, 2-thienyl, 1-naphthyl, and 1-indenyl.

Recent examples of the reaction are shown in table 22.1. A new route to organoplatinum compounds (equation 22.16; X = Cl or Br, n = 3 or 4) involves oxidation of the zero-valent metal, utilised in the form of a phosphine complex.[388]

$$RHgX + (Ph_3P)_nPt \xrightarrow[\text{R.T.}]{\text{benzene}} (Ph_3P)_2PtR(X) + Hg + (n-2)Ph_3P \quad (22.16)$$

R = Me, $MeO_2C . CH_2$, (4-MeO . $C_6H_4)_2C$: CH, 4-Me. C_6H_4, $C_5H_5FeC_5H_4$. This method should be capable of extension to other transition metals.

22.1.2 By Metal Halides

The preparation of organometallics by transfer of organic groups from mercury to metal halides (equations 22.17 and 22.18) is generally inferior to the analogous Grignard route because of the poor nucleophilicity of mercury-bonded carbon.

$$R_2Hg + MX_n \rightarrow RMX_{n-1} + RHgX \quad (22.17)$$

$$RHgX + MX_n \rightarrow RMX_{n-1} + HgX_2 \quad (22.18)$$

However, an advantage of the mercury method is that the transferable group can contain functional substituents that are incompatible with the Grignard synthesis.

The reaction has been used to prepare aryl, vinyl, and, with greater difficulty, alkyl derivatives of group III, IV, and V metals. Transmetallations with boron, thallium, and arsenic halides are typically carried out in refluxing benzene (for example equation 22.19^{395}), while reactions with phosphorus trichloride require more vigorous conditions.

$$3\text{-Me} . C_6H_4HgCl + BCl_3 \rightarrow 3\text{-Me} . C_6H_4BCl_2 + HgCl_2 \qquad (22.19)$$

Alkenyl groups are transferred with retention of configuration as shown in equation 22.20^{238}; this reaction also illustrates the point about functional groups and appears to be the only example involving an oxymercuration derivative in which the new organometallic compound has been isolated.

$$(trans\text{-MeC(OAc)}:CMe)_2Hg + 2TlCl_3 \rightarrow$$

$$2\ trans\text{-MeC(OAc)}:CMe . TlCl_2 + HgCl_2 \quad (22.20)$$

The most important application in group IV is to the formation of organolead compounds where the exchange is usually carried out with lead tetraacetate. Arylation of tin(II) (equation 22.21; for example Ar = $(OC)_3Mn . C_5H_4$ ^{396}) and germanium(II) halides occurs with oxidation of the metal.

$$Ar_2Hg + SnCl_2 \rightarrow Ar_2SnCl_2 + Hg \qquad (22.21)$$

In the reaction of α-carbonylalkylmercurials with organosilicon compounds the product of exchange is Si—O bonded. A mechanism involving nucleophilic attack on silicon by the carbonyl group accounts for this and for the predominant inversion of configuration that is observed (equation 22.22; Np = 1-naphthyl).^{397}

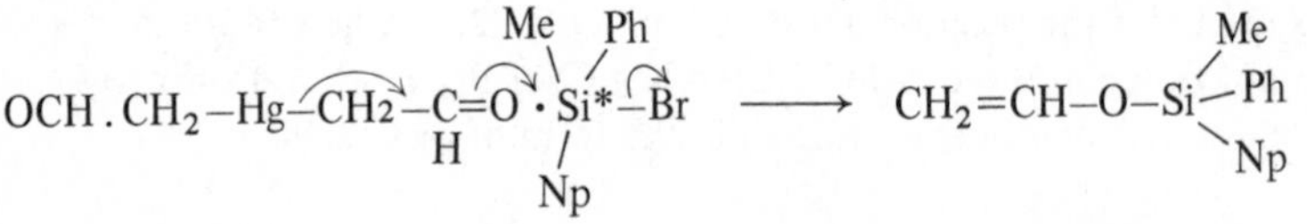

$$OCH . CH_2-Hg-CH_2-\underset{\underset{H}{|}}{C}=O \cdot \underset{\underset{Np}{|}}{\overset{Me\quad Ph}{Si*}}-Br \longrightarrow CH_2=CH-O-\underset{\diagdown Np}{\overset{Me}{\underset{}{Si}}}{\diagup Ph} \qquad (22.22)$$

An area of preparative importance that has grown up during the past few years concerns the transfer of organic groups to transition metals; examples are given in table 22.2.

22.2 Replacement of Mercury by Hydrogen

In general the mercury in organomercurials can be replaced by hydrogen through reaction with mineral acids or reducing agents. For those organomercury compounds prepared by mercuration (section 19.2), this merely represents a retrogression to the original organic substrate and hence holds no preparative interest unless a deuterium reagent is employed. The area where hydrogenodemercuration is most likely to be synthetically useful is in reactions with the mercurials formed by addition (section 19.3), for then it completes a sequence (equation 22.23) whereby protic reagents (R_nYH) are added regiospecifically to alkenes.

$$R^1R^2C:CHR^3 \xrightarrow[\text{addition}]{} R^1R^2C(YR_n)CH(HgX)R^3 \xrightarrow[\text{hydrogenodemercuration}]{}$$

$$R^1R^2C(YR_n)CH_2R^3 \qquad (22.23)$$

Table 22.2. Transfer of Organic Groups to Transition Metals

Organomercurial	Transition metal reagent	Product	Ref.
Me_2Hg	$NbCl_5$	$MeNbCl_4$	398
Me_2Hg	$TaCl_5$	$MeTaCl_4$	398
$2\text{-}PhN_2 . C_6H_4HgCl$	$Mn(CO)_5Cl$	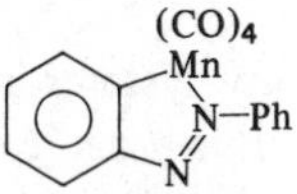 (CO)$_4$Mn-benzotriazole N–Ph complex	277
$CH_2:CH.CH_2HgCl$	Na_3RhCl_6	$[(\pi\text{-}C_3H_5)_2RhCl]_2$	399
$(C_5H_5)_2Hg$	$(Ph_3P)_2NiCl_2$	$\pi\text{-}C_5H_5(Ph_3P)NiCl$	400
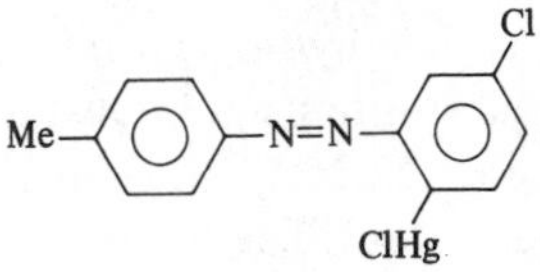 Me–C$_6$H$_4$–N=N–C$_6$H$_3$(Cl)(ClHg)	$PdCl_2$	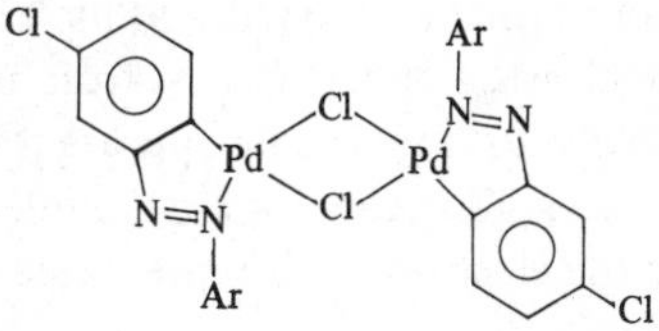 bis-Pd chloro-bridged dimer (Ar = 4-Me·C$_6$H$_4$)	277
Ph_2Hg	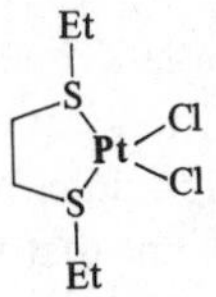 Et-S,S-Pt(Cl)(Cl) dithia complex	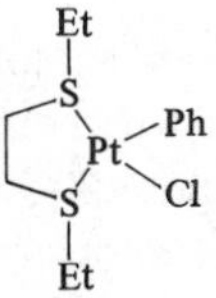 Et-S,S-Pt(Ph)(Cl) dithia complex	401

$(Ar = 4\text{-}Me \cdot C_6H_4)$

This particular substitution can be carried out successfully with certain reducing agents, notably sodium borohydride or sodium amalgam, and synthetic applications of the sequence are described in chapter 23. In contrast, these mercurials react with mineral acids to bring about a reversal of the original addition.

It follows that the acidolysis of organomercurials is not preparatively important. However, it provides a convenient model system for investigating acid cleavage of metal–carbon bonds in general, and consequently quite a lot of mechanistic work has been carried out in the area.[15-17] More recently there has been a surge of interest in the mechanism of reductive demercuration, catalysed in this case by the growing application of sequence 22.23 in organic synthesis. Quite different mechanistic pictures emerge of acidolysis and reduction, and this is the aspect we concentrate on in this section.

22.2.1 Acidolysis

Both alkyl groups can be removed from a diorganomercurial by mineral acids (equations 22.24 and 22.25) but more vigorous conditions are required for the second stage.

$$R_2Hg + HX \rightarrow RH + RHgX \qquad (22.24)$$

$$RHgX + HX \rightarrow RH + HgX_2 \qquad (22.25)$$

Acetic acid reacts with fully organic mercurials but not normally with organo-mercury salts.

Kinetic studies of many systems indicate that acidolysis is an electrophilic substitution, but within this broad category a range of transition states has been proposed. This is because the finer details of the mechanism are determined by the identity of the electrophile, which depends on the choice of acid and solvent system. The presence of species able to coordinate to mercury also influences the pathway as does, of course, the composition of the leaving organic group. The full range of mechanistic types that have been suggested is illustrated below.

The cleavage of aryl groups from mercury by reagents such as hydrogen chloride in partially aqueous media appears to resemble aromatic substitution and probably proceeds via formation of a σ-complex (for example equation 22.26;[402] for simplicity the electrophile is represented as an uncoordinated proton). The formation

$$ArHgCl + H^+ \underset{slow}{\rightleftharpoons} Ar\overset{+}{\underset{HgCl}{\diagup}}\overset{H}{} \underset{(Cl^-)}{\overset{fast}{\rightleftharpoons}} ArH + HgCl_2 \qquad (22.26)$$

of a saturated hydrocarbon by acidolysis of a dialkylmercurial is usually an S_E2 process but the question of whether the transition state is open or cyclic remains. It has been tentatively suggested[403] that acetic acid cleaves the Hg–C bond by a six-centred cyclic process (transition state **25**) but that when perchloric acid is present an open transition state (**26**) incorporating protonated acetic acid is involved (figure 22.1). An open and a cyclic transition state respectively have been proposed for the acidolysis of alkyl mercury iodides by aqueous perchloric or sulphuric acid[404] and the reaction of dialkylmercurials with hydrogen chloride in dimethyl-sulphoxide/dioxan, but the basis for the latter suggestion has been strongly criticised.[15]

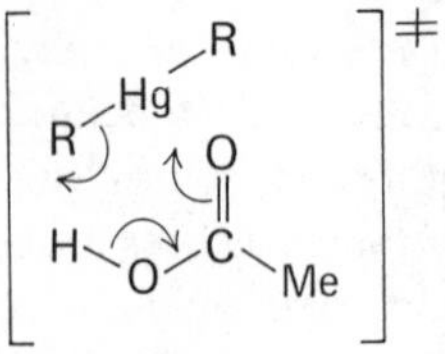

25 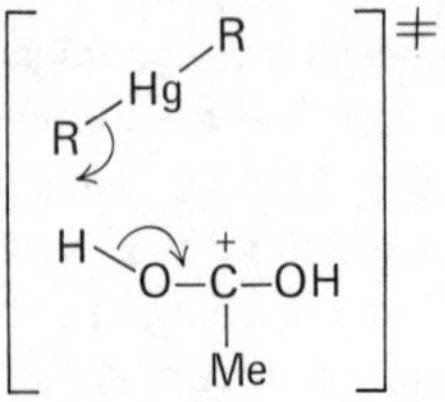   **26**

Figure 22.1

Where relatively stable carbanions can be formed and coordination to mercury to aid ionisation of the Hg–C bond is possible, a unimolecular mechanism can be

observed. The anion-catalysed S_E1 mechanism has been established for the acidolysis of 4-chloromercuriomethyl pyridinium (equations 22.27-22.29[405,406]) and
2-chloromercurio-1,3-dimethylbenzimidazolium[103] ions, and a solvent-induced
unimolecular mechanism has been proposed[407] for the reaction of a series of
diorganomercurials derived from strong C-acids with hydrogen chloride in
dimethylformamide.

$$HN^+\!\!\!\bigcirc\!\!\!-CH_2-HgCl + Cl^- \rightleftharpoons HN^+\!\!\!\bigcirc\!\!\!-CH_2-\bar{H}gCl_2 \quad (22.27)$$

$$HN^+\!\!\!\bigcirc\!\!\!-CH_2-\bar{H}gCl_2 \xrightarrow{slow} HN^+\!\!\!\bigcirc\!\!\!-\bar{C}H_2 + HgCl_2 \quad (22.28)$$

$$HN^+\!\!\!\bigcirc\!\!\!-\bar{C}H_2 + H^+ \longrightarrow HN^+\!\!\!\bigcirc\!\!\!-CH_3 \quad (22.29)$$

22.2.2 Reduction

In contrast to acidolysis, reductive demercuration occurs more readily with organomercury salts than with diorganomercurials. This suggests that the two reactions
proceed by fundamentally different mechanisms and there is mounting evidence
that most reductions involve a homolytic pathway. We have already seen
(section 20.2.2.) that reducing agents such as hydrazine and metals can bring
about symmetrisation of organomercury salts. The reaction is believed to proceed
by a one-electron transfer with generation of organomercury(I) species. If these
do not undergo self combination they can decompose unimolecularly to form
organic radicals which, by abstraction of hydrogen from suitable hydrogen donors,
can afford the hydrogenodemercuration product (equation 22.30).

$$RHgX + e \xrightarrow{-X^-} RH\dot{g} \xrightarrow{-Hg} R\cdot \xrightarrow{H\text{-donor}} RH \quad (22.30)$$

Diorganomercurials should be much less susceptible to reduction by this mechanism
since the unfavourable process of carbanion expulsion is required in the first step.

The mechanistic picture outlined above suggests that it should be possible to
bring about hydrogenodemercuration of diorganomercurials by generating the
organic radicals in a different way. This has been accomplished by pyrolysing or
photolysing the compounds in the presence of a hydrogen donor such as methanol
(equation 22.31; for a review of this subject see reference 408)

$$R_2Hg \xrightarrow[MeOH]{\Delta \text{ or } h\nu} 2RH + HCOH + Hg \quad (22.31)$$

In a recent example a high yield of benzene was obtained by photochemical decomposition of diphenylmercury in tetrahydrofuran.[409]

It should be remembered that hydrogen-for-mercury substitution is only
prepatively important when applied to the organomercurials obtained from

alkenes by oxymercuration and related reactions. Fortuitously these compounds often afford appreciable yields of hydrogenodemercuration product under conditions where symmetrisation dominates with other organomercurials, for example in reductions carried out with sodium amalgam (equation 22.32[410]), with hydrazine (equation 22.33[411]), or electrolytically (equation 22.34[412]).

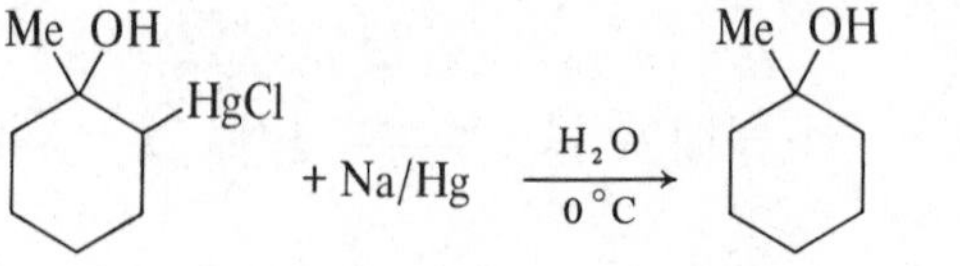

$$(22.32)$$

51 per cent

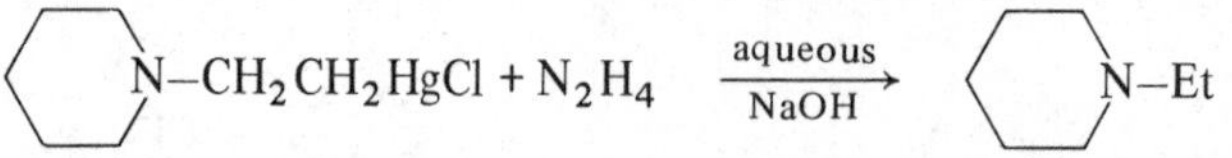

$$(22.33)$$

68 per cent

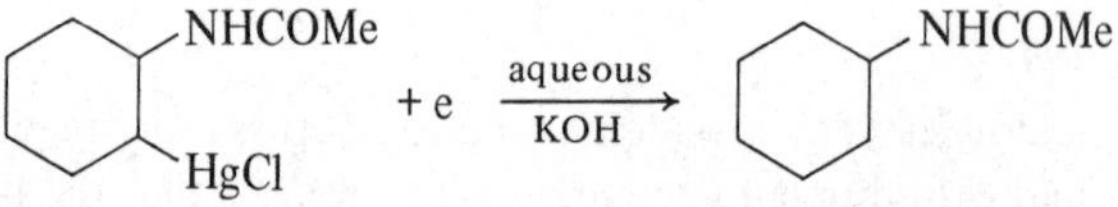

$$(22.34)$$

71 per cent

However, metal hydrides are the best reagents for effecting reductive demercuration, the process being rapid and usually remarkably free of side reactions. Among the hydrides that have been used are lithium borohydride, lithium aluminium hydride (equation 22.35[232]), and sodium trimethoxyborohydride (equation 22.36[413]).

$$PhNH . CH_2CH_2HgCl + LiAlH_4 \rightarrow PhNHEt \qquad (22.35)$$

$$MeCH(OMe)CH_2HgCl + NaB(OMe)_3H \rightarrow Me_2CH(OMe) \qquad (22.36)$$

By far the most popular reagent, however, is sodium borohydride in aqueous sodium hydroxide. This was developed in 1966 by Bordwell and Douglass[414] who established the stoichiometry to be that shown in equation 22.37, and who carried out some preliminary mechanistic investigations.

$$4RHgCl + 4OH^- + BH_4^- \rightarrow 4RH + 4Hg + 4Cl^- + H_2BO_3^- + H_2O \qquad (22.37)$$

Several observations combine to provide a powerful argument for the intermediacy of organic radicals in hydrogenodemercuration by sodium borohydride (and other metal hydrides). First the absence of rearrangement during reduction of 2-phenyl-2-methylpropylmercury bromide and 1,7,7-trimethylbicyclo[2.2.1]-heptyl-2-mercury bromide[415] rules out carbonium ion intermediates. Next the lack of deuterium incorporation into the product when the reagent is borohydride in D_2O,[414,416] and the high uptake of deuterium when using borodeuteride in H_2O[415,416] argue strongly against the intermediacy of carbanions. The nature of products obtained from certain systems also excludes a concerted mechanism, but is wholly consistent with the formation of alkyl radicals that are capable of undergoing rapid rearrangement. Thus the formation of the same ratio of products **32, 33**, and **34** from reduction of compounds **27** or **28** is explained by the equilibration of radicals **29, 30**, and **31** before hydrogen abstraction (figure 22.2).

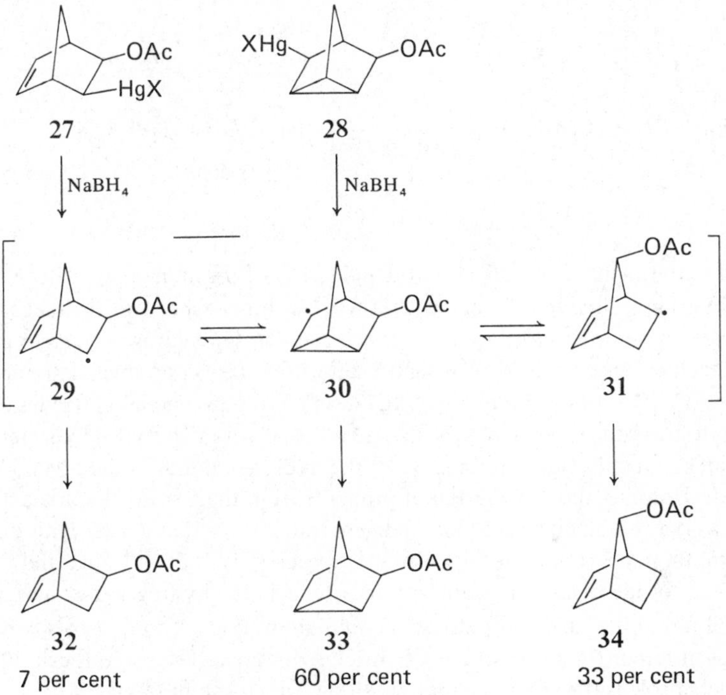

Figure 22.2

This system has been investigated independently by two groups (X = Br[415] and X = Cl[416,417]) who claim to have obtained essentially the same product distribution. However in one group's report[415] the percentages we have given for products **32** and **34** are transposed. Because of the similarity of the numbers and the correspondence of related data between the two groups, it seems beyond doubt that the results *are* in agreement but that one group has misidentified the minor products. A further indication of the participation of radicals **29–31** is the similarity, in both product distribution[415,417] and stereochemistry of deuterium incorporation,[417] between these reductions and those of the corresponding alkyl halides with organotin hydrides, which are known to proceed by a radical mechanism.

Trapping experiments have provided the strongest evidence for radicals. The 2-phenyl-2-methylpropyl radical has been scavenged with high efficiency by 2,2,6,6-tetramethylpiperidoxyl (equation 22.38[418]), and a variety of other radicals have been trapped with oxygen[418-420] to afford the corresponding alcohols in high yield (for example equation 22.39[419]).

$$\text{Ph}-\underset{\underset{\text{Me}}{|}}{\overset{\overset{\text{Me}}{|}}{\text{C}}}-\text{CH}_2\text{HgBr} + \overset{\bullet}{\text{O}}-\text{N}\underset{\text{Me Me}}{\overset{\text{Me Me}}{\bigcirc}} \xrightarrow[\text{DMF}]{\text{NaBH}_4} \text{Ph}-\underset{\underset{\text{Me}}{|}}{\overset{\overset{\text{Me}}{|}}{\text{C}}}-\text{CH}_2-\text{O}-\text{N}\underset{\text{Me Me}}{\overset{\text{Me Me}}{\bigcirc}} \qquad (22.38)$$

$$\text{Ph}_3\text{C}.\text{CH}_2\text{HgCl} + \text{NaBH}_4 \xrightarrow[\text{H}_2\text{O-OH}^-/\text{THF}]{\text{O}_2} \text{Ph}_2\text{C(OH)CH}_2\text{Ph} + \text{Ph}_3\text{C}.\text{CH}_2\text{OH}$$

$$ \text{58 per cent} \text{19 per cent}$$

$$ (22.39)$$

The intermediacy of radicals in the reduction of organomercury salts by sodium borohydride is therefore firmly established, but other aspects of the mechanism are less clear. In particular a variety of schemes for the hydrogen-transfer process have been considered. Small but significant differences in product ratios are observed[415,417] when alkylmercury halides (27 and 28) and alkyl halides (which generate the same system of radicals) are reduced with tributyltin hydride; this indicates that tributyltin hydride is *not* the hydrogen donor in the organomercury system. However the distribution of products from the organomercurials *is* the same as that obtained with sodium borohydride, or for that matter with diethyl-aluminium hydride or tributylphosphine copper(I) hydride.[415] This, and the discovery of identical deuterium isotope effects in the hydrogen-abstraction stage for sodium borohydride and lithium aluminium hydride,[420] point to a common hydrogen donor for all these metal-hydride reductions. The obvious candidate for such a role is the organomercury hydride, which can presumably be formed from the substrate by anion exchange of the type discussed in section 20.3. The trapping experiments described earlier exclude a cage process for hydrogen transfer and a free-radical chain mechanism with the propagation steps shown in equations 22.40 and 22.41 is probably involved.

$$\dot{\text{R}} + \text{HHgR} \rightarrow \text{RH} + \text{Hg}\dot{\text{R}} \qquad (22.40)$$

$$\text{RH}\dot{\text{g}} \rightarrow \dot{\text{R}} + \text{Hg} \qquad (22.41)$$

There appears to be no evidence regarding the way in which these chains are initiated.

In the practical applications of reductive demercuration (chapter 23) sodium borohydride is often used in a mixture of tetrahydrofuran and water, and this was the medium employed for it in most of the mechanistic studies, except for some of the trapping experiments where solvent dimethylformamide was used.[418] It should be noted, however, that extensive deoxymercuration of alkene addition products has been found[414,415,421] in some media (for example equation 22.42[421]).

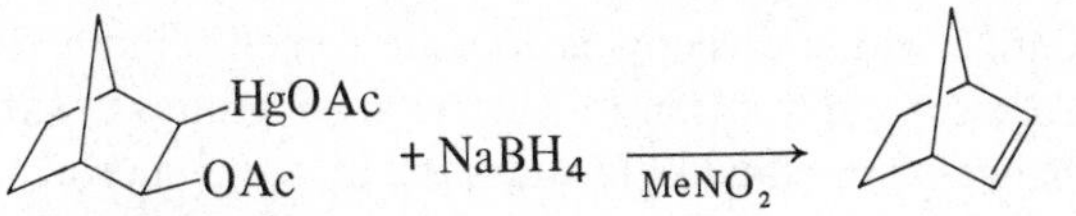

$$\text{(structure) } + \text{NaBH}_4 \xrightarrow[\text{MeNO}_2]{} \text{(structure)} \qquad (22.42)$$

22.3 Replacement of Mercury by Carbon Substituents

Any reaction that leads to the formation of a carbon–carbon bond is potentially important in synthesis. In the past, examples of such reactions in organomercury chemistry have been confined to alkylations of organic halides in which the structural requirements of the reactants are rather narrowly defined (equation 22.43), and to thermally or photolytically induced coupling processes (equation 22.44) which are attended by many side reactions.

$$R^1COCHR.HgX + Ar_3CY \rightarrow R^1COCHR.CAr_3 + YHgX \qquad (22.43)$$

$$R_2Hg \xrightarrow{\Delta \text{ or } h\nu} R-R + Hg + \text{other products} \qquad (22.44)$$

However, the situation appears to be changing with the discovery of reactions such as carbonylation[422] (for example equation 22.45[423]) and alkenylation (equation 22.46) that occur in the presence of transition metal compounds and that probably involve an initial transmetallation step.

$$2RHgX + Ni(CO)_4 \xrightarrow[\text{DMF}]{60-70^\circ C} R_2CO + NiX_2 + 2Hg + 3CO \qquad (22.45)$$

$$R_2Hg + 2R^1CH:CH_2 + 2PdX_2 \rightarrow 2R^1CH:CHR + HgX_2 + 2Pd + 2HX \qquad (22.46)$$

These kinds of processes may herald a new synthetic importance for the readily prepared but poorly reactive organomercurials in which they function as precursors for more labile organometallic reagents.

22.3.1 With Organic Halides

Organomercurials with simple organic groups are insufficiently nucleophilic to alkylate organic halides, but alkylations can be carried out by generating a reactive organometallic *in situ* through transmetallation (see section 22.1). A system of this type which, in some instances, brings about conversions with retention of configuration at the carbon originally bonded to mercury has recently been developed[424] (for example equation 22.47).

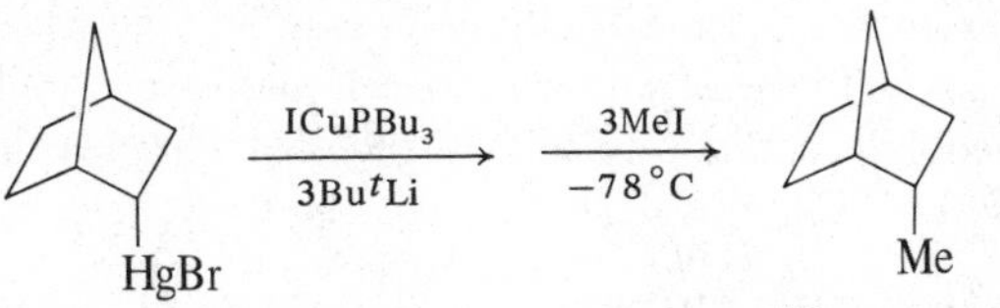

$$\text{(structure) HgBr} \xrightarrow[3Bu^tLi]{ICuPBu_3} \xrightarrow[-78^\circ C]{3MeI} \text{(structure) Me} \qquad (22.47)$$

70 per cent

Organomercurials with α-carbonyl groups are somewhat more reactive in nucleophilic processes and will transfer their organic groups to sufficiently electrophilic organic halides. Triarylmethyl halides fall into this category, especially when they are complexed with mercury(II) halides, $Ar_3C^{\delta+}\text{-----}{}^{\delta-}HgBr_3$. The reaction of such complexes with substituted α-carbethoxybenzylmercury bromides (equation 22.48) has been studied kinetically[425] and is concluded to follow an S_E2 (open) pathway.

$$4\text{-}X.C_6H_4CH(HgBr)CO_2Et + Ar_3CBr.HgBr_2 \rightarrow$$

$$4\text{-}X.C_6H_4CH(CAr_3)CO_2Et + 2HgBr_2 \quad (22.48)$$

In the absence of added mercury(II) bromide the reactants rapidly form a complex (**35**) which then decomposes to products with a first-order rate law. The pathway shown in equation 22.49 is therefore suggested and the effect of the substituent X indicates that fission of the Hg—C bond is the dominant feature in the transition state.[425]

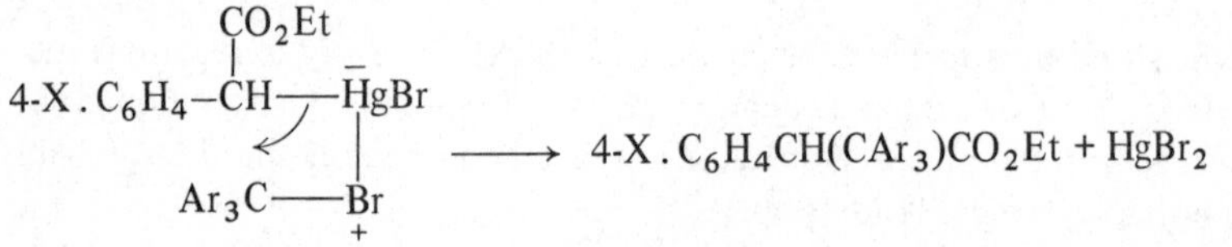

$$4\text{-}X.C_6H_4-\overset{\overset{\displaystyle CO_2Et}{|}}{CH}\underset{\underset{\displaystyle \underset{+}{Ar_3C\text{——}Br}}{|}}{\text{——}\overline{H}gBr} \longrightarrow 4\text{-}X.C_6H_4CH(CAr_3)CO_2Et + HgBr_2 \quad (22.49)$$

35

A similar mechanism may account for the ease with which bis(trinitromethyl)-mercury, which tends to behave like a mercury(II) salt (see below), alkylates methyl iodide (equation 22.50[426]).

$$[(NO_3)_3C]_2Hg + MeI \xrightarrow[\text{R.T.}]{H_2O} (NO_3)_3C\text{-}Me \quad (22.50)$$
$$71 \text{ per cent}$$

Organomercury compounds react with acyl halides (equation 22.51; for example Ar = 2-thienyl or 2-furyl[380]) more readily than with alkyl halides, but as a route to ketones the reaction cannot compete with the corresponding cadmium system.

$$Ar_2Hg + RCOCl \rightarrow ArCOR + ArHgCl \quad (22.51)$$

With acyl halides α-carbonylorganomercurials form C—O-rather than C—C-bonded products (equation 22.52[427])

$$(MeO_2C.CH_2COCH_2)_2Hg + MeCOBr \rightarrow MeCO_2C(CH_2.CO_2Me):CH_2 \quad (22.52)$$

These alkylations can be made easier by complexing the organic halide with a Lewis acid (for example equation 22.48; AlX_3 is often used with acyl halides) and coordination of a base to the organomercurial should have a similar effect. Anion catalysis of nucleophilic *addition* to suitable aldehydes has been reported[428] (equation 22.53; R = CCl_3 or C_6F_5).

$$MeCOCH_2HgBr + RCHO \xrightarrow[\text{DME}]{Br^-} \xrightarrow[\text{H}^+]{H_2O} MeCOCH_2.CH(OH)R \quad (22.53)$$

22.3.2 Coupling

Photolysis or pyrolysis of organomercurials affords organic radicals, which by hydrogen abstraction can give the hydrogenodemercuration product (section 22.2.2). As an alternative the radicals may couple (for example equation 22.54[429]).

$$(NC.CH_2CH_2)_2Hg \xrightarrow[\text{MeOH}]{h\nu} NC.CH_2CH_2.CH_2CH_2.CN \quad (22.54)$$
$$90 \text{ per cent}$$

The pyrolysis route is catalysed by metals such as silver or palladium and it can give rise to some novel organic products (for example equations 22.55[71] and 22.56[269]).

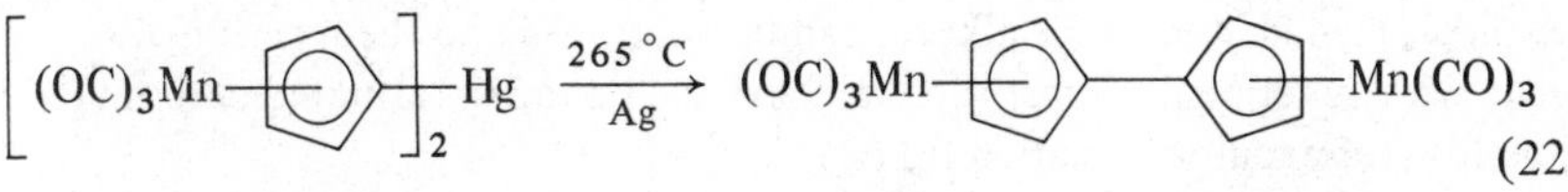

$$(22.55)$$

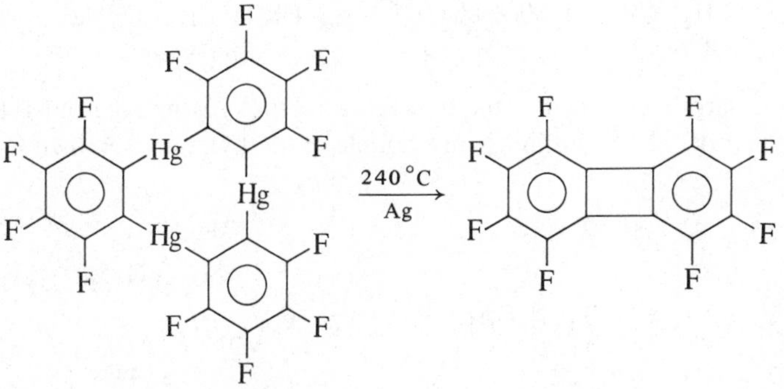

$$(22.56)$$

It has recently been discovered that a zerovalent palladium complex catalyses mercury extrusion from dipropenylmercury *at room temperature* (equation 22.57[430]); 2,4-hexadiene is obtained with high stereoselectivity.

$$(MeCH:CH)_2Hg \xrightarrow[\text{CH}_3\text{CN}]{(Ph_3P)_4Pd} MeCH:CH.CH:CHMe \quad (22.57)$$

In view of the known reaction of organomercury halides with the corresponding Pt^0 complex (see p. 213), it seems likely that this coupling proceeds via an organopalladium(II) intermediate. Such intermediates are certainly involved in reactions brought about by Pd(II) salts, namely biaryl formation (equation 22.58; Ar = 4-Me.C_6H_4[431]) and the transfer of organic groups from mercury to alkenes, which is described in the next section.

$$2Ar_2Hg + Pd(OAc)_2 \rightarrow Ar-Ar + 2ArHgOAc + Pd \quad (22.58)$$

22.3.3 With Alkenes

Since 1968 Heck[432] has developed a system whereby alkenes can be alkylated or arylated by a mixture of organomercurials and palladium(II) salts. The reaction is considered to proceed by three stages: (i) the organic group is transferred to palladium (equation 22.59), (ii) the Pd—C bond of the resulting organopalladium compound adds across the double bond of the alkene (equation 22.60), and (iii) an unstable palladium hydride is eliminated from the adduct (equation 22.61).

$$RHgX + PdX_2 \rightarrow [RPdX] + HgX_2 \tag{22.59}$$

$$[RPdX] + R^1CH:CH_2 \rightarrow [R^1CH(PdX)CH_2R] \tag{22.60}$$

$$[R^1CH(PdX)CH_2R] \rightarrow R^1CH:CHR + [HPdX] \rightarrow HX + Pd \tag{22.61}$$

The addition process is subject largely to steric control and the organic group generally becomes attached to the least substituted carbon atom of the double bond.

Terminal, medial, and cyclic alkenes can be employed and the reaction also proceeds with functionally substituted alkenes such as $\alpha\beta$-unsaturated carbonyl compounds (for example equation 22.62).

$$CH_2:CH.CO_2Me + Ph_2Hg \xrightarrow[\text{MeCN}]{\text{LiPdCl}_3} PhCH:CH.CO_2Me \tag{22.62}$$
$$\text{88 per cent}$$

Alkylation is largely confined to the transfer of methoxycarbonyl groups (for example equation 22.63), although an example of methylation is known (equation 22.64).

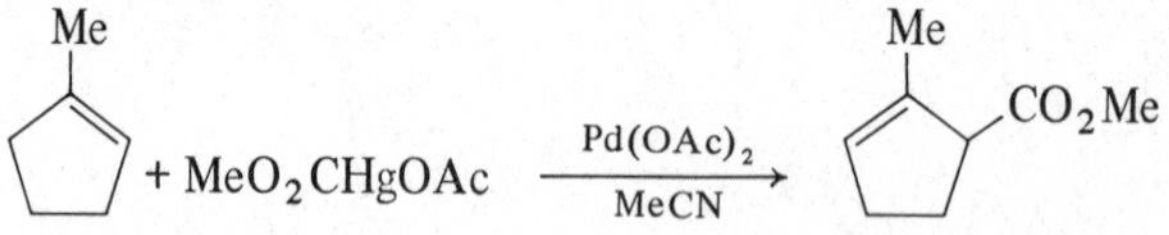

$$\tag{22.63}$$
$$\text{76 per cent}$$

$$PhCH:CH_2 + MeHgCl \xrightarrow[\text{MeOH}]{\text{Li}_2PdCl_4} trans\text{-}PhCH:CHMe \tag{22.64}$$
$$\text{75 per cent}$$

Some organopalladium species exhibit poor regioselectivity in the addition, which results in the formation of mixtures of isomeric products (for example equation 22.65; Ar = 4-MeO . C_6H_4). A chlorine atom suitably positioned in the starting alkene can be lost in preference to hydrogen (for example equation 22.66).

$$MeCH:CH_2 + ArHgOAc \xrightarrow[\text{MeCN}]{\text{Pd(OAc)}_2} ArCH:CHMe + ArCH_2.CH:CH_2 + ArCMe:CH_2$$

$$\qquad\qquad\qquad\qquad\quad 37 \text{ per cent} \qquad 3 \text{ per cent} \qquad 32 \text{ per cent}$$
$$\qquad\qquad\qquad\qquad\quad (trans)$$
$$\qquad\qquad\qquad\qquad\quad 2 \text{ per cent}$$
$$\qquad\qquad\qquad\qquad\quad (cis) \tag{22.65}$$

$$PhHgCl + CH_2:CH.CH_2Cl \xrightarrow{PdX_2} [PhCH_2CH(PdX)CH_2Cl] \xrightarrow{-PdXCl}$$

$$PhCH_2CH:CH_2 \tag{22.66}$$

Some diorganomercurials add to alkenes in their own right to form new organo-mercury compounds. Bis(trinitromethyl)mercury behaves like mercury(II) acetate towards alkenes (see section 19.3) except that there is a strong tendency for the trinitromethyl group to be incorporated into the adduct even in aqueous or alcoholic solution, and two moles of alkene are often taken up; a recent example is shown in equation 22.67 where $R = Me_2SiCH_2CH_2$.[433]

$$RCH:CH_2 + Hg[C(NO_3)_3]_2 \xrightarrow{H_2O} [(O_3N)_3CCHR.CH_2]_2Hg \quad (22.67)$$
$$94 \text{ per cent}$$

A more typical organomercurial has been shown to add across carbon–carbon double bonds that are conjugated with cyanide or alkoxycarbonyl groups (equation 22.68; X = H, Cl, Me and equation 22.69; R = CN or CO_2Et).[434]

$$4\text{-X}.C_6H_4CH:C(CN)_2 + Bu_2{}^tHg \rightarrow 4\text{-X}.C_6H_4CHBu^t.C(CN)_2HgBu^t$$
$$(22.68)$$

$$Me_2C:C(CN)R + Bu_2{}^tHg \rightarrow Me_2CBu^t.C(CN)R.HgBu^t \quad (22.69)$$

The reaction with a similar acetylene reveals that *cis*-addition takes place (equation 22.70).

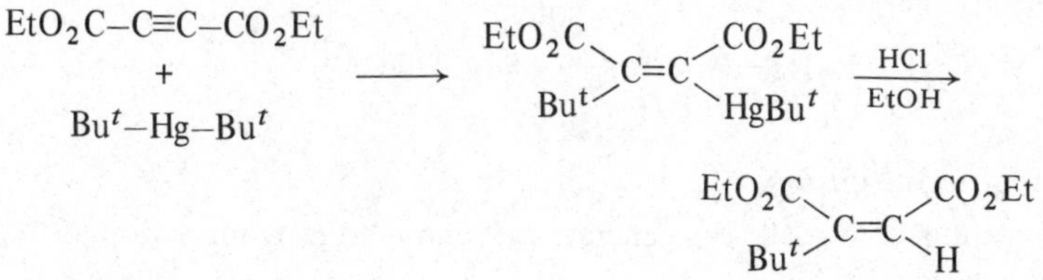

$$
\begin{array}{c}
EtO_2C-C\equiv C-CO_2Et \\
+ \\
Bu^t-Hg-Bu^t
\end{array}
\longrightarrow
\underset{Bu^t}{\overset{EtO_2C}{>}}C=C\underset{HgBu^t}{\overset{CO_2Et}{<}}
\xrightarrow[EtOH]{HCl}
$$

$$
\underset{Bu^t}{\overset{EtO_2C}{>}}C=C\underset{H}{\overset{CO_2Et}{<}}
\quad (22.70)
$$

When a trimethyl-silyl or -stannyl group replaces one of the tertiary butyls, a mercury-free product is obtained[435] (equation 22.71; M = Si or Sn).

$$PhCH:C(CN)_2 + Bu^tHgMMe_3 \rightarrow PhCHBu^t.C(CN):C:NMMe_3 \quad (22.71)$$

22.4 Replacement of Mercury by Oxygen Substituents

This is a relatively small area of organomercury chemistry but one growing in synthetic importance. Most reactions within this category can be rationalised in terms of one of two systems: (i) the generation and capture by oxygen of organic radicals and (ii) the formation and capture by oxygen-centred nucleo-philes of carbonium ions (included here are processes in which the nucleophile attacks before the carbonium ion is fully formed).

22.4.1 Homolytic Substitution

Dialkylmercurials (R_2Hg) that contain a secondary or tertiary carbon atom bonded to mercury (for example R = cyclo-C_6H_{11}, Pr^i, or Me_2CEt) are slowly oxidised by the atmosphere. Their reaction with oxygen in solution has been

examined by Razuvaev[436] and probably involves a variety of radical intermediates including the species $R^\cdot$ and $ROO^\cdot$. The organic products include
alcohols and ketones but complex mixtures are usually obtained and the
reaction is not preparatively useful.

Much more promising from the synthetic viewpoint are the reactions in which
alkyl radicals are generated from organomercury salts by treatment with sodium
borohydride. As mentioned in section 22.2.2, high yields of alcohols can be obtained by trapping these radicals with oxygen. By combining the reaction with
oxymercuration, vicinal oxygen substituents can be introduced into the molecule
(equation 22.72[418]).

$$\text{57 per cent} \qquad \text{38 per cent} \qquad (22.72)$$

Although α-carbonylalkylmercurials undergo an electrophilic substitution with
bromotriarylmethane (section 22.3.1), some simpler dialkylmercurials react by a
radical process and if oxygen is present, appreciable yields of organic peroxides
can be obtained (equation 22.73[437]).

$$R_2Hg + Ph_3CBr \xrightarrow{\;O_2\;} Ph_3COOR \qquad (22.73)$$

22.4.2 Heterolytic Substitution

In principle organomercury salts can generate carbonium ions by ionisation of the
Hg–X bond (equation 22.74).

$$RHgX \xrightarrow[-X^-]{} RHg^+ \xrightarrow[-Hg^\circ]{} R^+ \qquad (22.74)$$

It appears that salts of inorganic oxyacids are sufficiently polar to follow this route,
for they are susceptible to solvolysis under a variety of conditions and the pre-
dominance of rearranged products is indicative of carbonium ion intermediates
(for example equation 22.75[438]).

$$MeCH_2CH_2HgClO_4 + CF_3CO_2H \xrightarrow[8\,h]{reflux} Me_2CHO_2C\cdot CF_3 + Hg \quad (22.75)$$
$$\text{93 per cent}$$

Ionisation of more covalent organomercury salts such as acetates can be facilitated
by protonation (demercuration can be a troublesome side reaction in studies of
acidolysis) or interaction with a Lewis acid [for example equation 22.76; R =
$EtO_2CCH(COMe)$[161]].

$$RCH_2CH_2HgOAc \xrightarrow{BF_3/AcOH} RCH_2CH_2OAc + Hg \qquad (22.76)$$

Organomercury salts also form carbonium ions more readily in the presence of
mercury(II) salts, with which they can enter into a redox reaction. An example in

which a high degree of rearrangement is again found is shown in equation 22.77 where $X = O_2C.CF_3$.[9]

$$PhCH(OMe)CH(HgX)Ph + HgX_2 \xrightarrow[CH_2Cl_2]{MeOH} Ph_2CH.CH(OMe)_2 + Hg_2X_2 \quad (22.77)$$
$$88 \text{ per cent}$$

The oxidation of alkenes by mercury(II) salts (for a review of this subject see reference 379) involves intermediate organomercury compounds that undergo oxydemercurations similar to those we have been describing. The Treibs reaction, in which alkenes are converted into allylic esters by mercury(II) acetate in hot acetic acid, proceeds by rate-determining allylic mercuration[118] followed by product-forming oxydemercuration.[439] Both steps may occur with rearrangement and among the possible mechanisms are the S_E2' (cyclic) processes shown in equations 22.78 and 22.79.

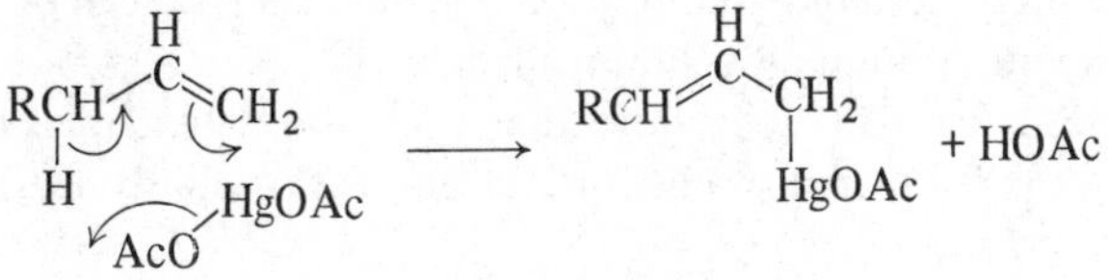

$$+ \text{HOAc} \quad (22.78)$$

$$+ \text{Hg}_2(\text{OAc})_2 \quad (22.79)$$

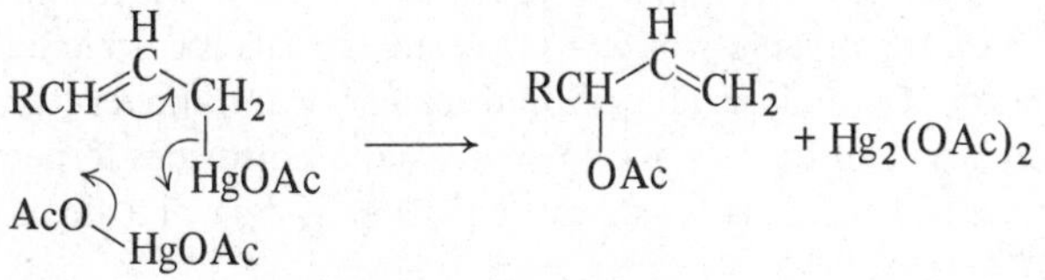

The demercuration is catalysed not only by mercury(II) acetate, but by lead(IV), thallium(III), and palladium(II) acetates also. The lead- and thallium-promoted reactions involve transmetallation followed by demetallation of the new allyl-metallic acetate (for example equation 23.80[439]).

$$MeCH:CHCH_2HgOAc + Pb(OAc)_4 \rightarrow MeCH[Pb(OAc)_3]CH:CH_2 \rightleftharpoons$$
trans
$$MeCH:CHCH_2Pb(OAc)_3 \rightarrow MeCH:CH.CH_2OAc + MeCH(OAc)CH;CH_2$$
$$51 \text{ per cent } (cis) \qquad 45 \text{ per cent}$$
$$4 \text{ per cent } (trans)$$
$$(22.80)$$

In the Denigès reaction terminal alkenes are converted into $\alpha\beta$-unsaturated carbonyl compounds by reaction with an acidic aqueous solution of mercury(II) sulphate (equation 22.81).

$$RCH_2CH:CH_2 \xrightarrow[H_2SO_4]{HgSO_4} RCOCH:CH_2 \quad (22.81)$$

The corresponding allyl alcohols, $RCH(OH)CH:CH_2$, are formed initially in this system,[379] presumably by a mechanism analogous to that in the Treibs reaction. Medial and cyclic alkenes afford *saturated* carbonyl compounds under the

Denigès conditions. These arise from demercuration of the hydroxymercuration product accompanied by a stabilising 1,2-nucleophilic rearrangement (for example equations 22.82[440] and 22.83[178]).

$$\text{MeCH:CHMe} \xrightarrow[\text{Hg}^{2+}]{\text{H}_2\text{O}} \quad \text{Me}-\overset{\text{H}}{\underset{\text{O}\ \text{Hg}^+}{\text{C}}}\text{CH}-\text{Me} \xrightarrow{\text{Hg}^{2+}} \text{H}^+ + \text{Me}-\underset{\text{O}}{\overset{\parallel}{\text{C}}}-\text{CH}_2\text{Me} + \text{Hg}_2{}^{2+}$$

(22.82)

(22.83)

A similar demercuration of β-hydroxyalkylmercury *acetates* can be induced by transmetallation with lithium tetrachloropalladate.[441]

Strictly speaking the oxidations proceeding via hydroxymercurials involve hydrogen- or carbon-for-mercury substitutions but it is obviously sensible to include them in this section.

Reagents containing electrophilic oxygen are not numerous but the attack of such species on the organic group of organomercurials is a potential route for oxydemercuration. Ozonolysis falls into this category and the reaction with primary, secondary, and tertiary alkylmercury halides provides respectable yields of carboxylic acids, ketones (for example equation 22.84), and alcohols respectively.[442]

$$\text{Me}_2\text{CH}\,.\,\text{HgBr} + \text{O}_3 \xrightarrow[\text{CH}_2\text{Cl}_2]{10\,^{\circ}\text{C}} \text{Me}_2\text{CO} + \text{AcOH}$$
$$\qquad\qquad\qquad\qquad\qquad 79 \text{ per cent} \quad 19 \text{ per cent}$$

(22.84)

The transfer of aryl groups to boron by reaction of arylmercurials with diborane was mentioned in section 22.1. All the various functionalising deborations developed by Brown[35] to accompany hydroboration are now potentially applicable to *aryl*boranes. Thus oxidation with alkaline hydrogen peroxide can be carried out *in situ* and gives good yields of phenols (equation 22.85; for example Y = H, Cl, Me, MeO).[11]

$$4-\text{Y}.\text{C}_6\text{H}_4\text{HgX} \xrightarrow[\text{THF}]{\text{B}_2\text{H}_6} [4-\text{Y}.\text{C}_6\text{H}_4\text{B}{\small\diagdown}] \xrightarrow[\text{HO}^-]{\text{H}_2\text{O}_2} 4-\text{Y}.\text{C}_6\text{H}_4\text{OH}$$
$$\qquad\qquad\qquad\qquad\qquad\qquad\qquad\qquad 65\text{–}99 \text{ per cent}$$

(22.85)

This is yet another example of a synthetic application of organomercurials that involves an initial transmetallation (compare section 22.3).

The transfer of alkyl groups in the reverse sense, that is, from boron to mercury (section 19.1), forms part of a new preparative sequence by which the anti-Markovnikov esterification of terminal alkenes can be achieved (equation 22.86[443])

$$\text{EtCH:CH}_2 \xrightarrow[\text{THF}]{\text{BH}_3} \xrightarrow{\text{Hg(OAc)}_2} \xrightarrow{\text{I}_2} \text{BuOAc}$$
$$\qquad\qquad\qquad\qquad\qquad\qquad 81 \text{ per cent}$$

(22.86)

The procedure actually involves iododemercuration, but the alkyl iodides thus formed are converted into acetates under the reaction conditions. Halogeno-demercuration is discussed in the next section.

22.5 Replacement of Mercury by Halogens

All kinds of diorganomercurials and organomercury salts react with chlorine, bromine, or iodine under a variety of generally mild conditions to yield alkyl halides (equation 22.87 and 22.88).

$$R_2Hg + X_2 \rightarrow RX + RHgX \tag{22.87}$$

$$RHgX + X_2 \rightarrow RX + HgX_2 \tag{22.88}$$

By analogy with acidolysis (section 22.2.1) it is reasonable to expect the fully organic mercurials to be more reactive than the organomercury salts in *electrophilic* halogenodemercuration, but homolytic displacements may occur at comparable rates. However, an equimolar mixture of R_2Hg and X_2 always obeys the stoichiometry of equation 22.87 since if any HgX_2 is produced, it rapidly consumes an equivalent amount of diorganomercurial to generate organomercury salt (see section 20.2.1).

A variety of other reagents have been used to bring about halogenodemercuration. These include Br_3^-, I_3^-, and ICl (see the discussion of mechanism below), hypobromous acid, *N*-bromosuccinimide (for example equation 22.89[444]), and copper(II) halides, which are particularly useful in the synthesis of halogenoferrocenes (equation 22.90[445]).

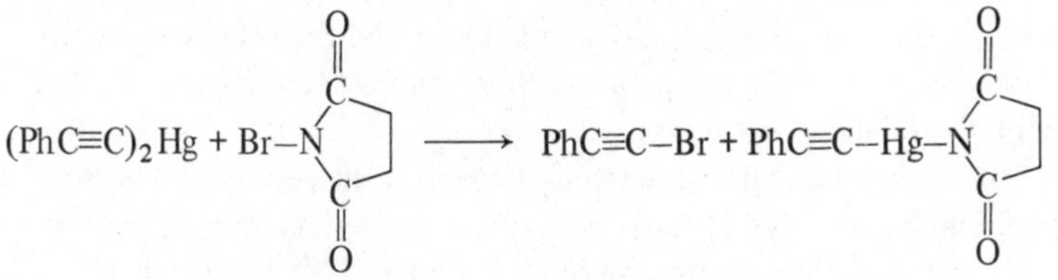

$$(PhC \equiv C)_2 Hg + Br-N \cdots \longrightarrow PhC \equiv C-Br + PhC \equiv C-Hg-N \cdots \tag{22.89}$$

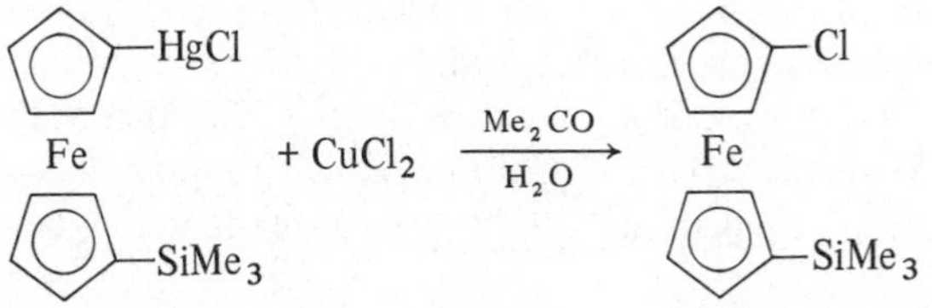

80 per cent

Two main aspects of halogenodemercuration will be discussed. Firstly we shall summarise the evidence for the existence of electrophilic and free-radical pathways, emphasising the conditions under which each is favoured, and secondly we shall draw attention to the main applications of the reaction.

22.5.1 Mechanism

In non-polar deoxygenated solvents halogenodemercuration proceeds by a non-stereospecific free-radical mechanism, but in polar solvents in the presence of air an electrophilic process with retention of configuration assumes importance. Complete suppression of the radical pathway can be difficult, but some reactions with chlorine, bromine, or iodine chloride in pyridine[15] and with trihalide ions (X_3^-) in various polar solvents appear to proceed exclusively by the heterolytic mechanism. A number of kinetic and stereochemical studies provide the experimental basis for these generalisations.

The diagnosis of a rate law that is first order in organomercurial and first order in halogenating agent signifies an $S_E 2$ mechanism. Such kinetics have been observed, for example, in the reaction of benzylmercury chloride with iodine plus cadmium iodide, or bromine plus ammonium bromide (equation 22.91; $X = I^{[446,447]}$ or $Br^{[448]}$), and in the iodinolysis of dimethylmercury when carried out in carbon tetrachloride in the dark.[449]

$$PhCH_2HgCl + X_2 \xrightarrow{X^-} PhCH_2X + XHgCl \qquad (22.91)$$

The rates of these reactions are unaffected by the presence or absence of air in the reaction vessel or by the intensity of incident light.

In contrast the iodinolysis of alkylmercury iodides in dioxan[450] and the brominolyses of benzylmercury chloride[451,452] and *sec*-butylmercury bromide[453] in carbon tetrachloride are strongly retarded by oxygen, have rates that are affected by illumination and rate laws that are independent of the concentration of organomercurial. It is believed that these reactions proceed by the free-radical route. If methanol is added to the carbon tetrachloride in concentrations comparable with that of the bromine, the reactions become second order and the rates insensitive to light. This indicates a switch to the polar mechanism, favoured perhaps by the formation of a methanol–bromine complex.

The conclusions from kinetic data are strengthened by the results of stereochemical studies. The brominolyses of *cis*- and *trans*-4-methylcyclohexylmercury bromide and of optically active *sec*-butylmercury bromide[454,455] have been investigated under a wide range of conditions. In CCl_4 or CS_2 under nitrogen there is complete loss of configuration, while in pyridine retention strongly dominates and can be total. More recent work has shown that the brominolyses of a *primary* alkylmercury halide (equation 22.92[56]) and of alkenylmercury bromides (for example equation 22.93[456]) also proceed with retention of configuration in pyridine.

$$\textit{threo-}PhCHD \cdot CHD \cdot HgCl + Br_2 \xrightarrow[\text{pyridine}]{0\,^{\circ}C} \textit{threo-}PhCHD \cdot CHD \cdot Br \quad (22.92)$$

$$\textit{cis-}MeCH{:}CH \cdot HgBr + Br_2 \xrightarrow[\text{pyridine}]{} \textit{cis-}MeCH{:}CHBr \qquad (22.93)$$

Complete loss of configuration is consistent with a free-radical chain mechanism involving the propagation steps shown in equations 22.94 and 22.95, and retention is the usual outcome of an S_E2 reaction.

$$\dot{X} + RHgX \rightarrow \dot{R} + HgX_2 \qquad (22.94)$$

$$\dot{R} + X_2 \rightarrow RX + \dot{X} \qquad (22.95)$$

If lack of stereospecificity is therefore taken to signify that a halogenodemercuration has proceeded partially by the homolytic route, Jensen's results[454,455,15] indicate that there is a substantial radical component in most polar solvents and under conditions where a heterolytic mechanism is diagnosed from kinetics. However, the stereochemical criterion must be applied with great caution for it is possible to envisage a stereospecific radical process (for example equation 22.96; this is particularly likely when $X = Cl$[15]) and a non-stereospecific polar mechanism.

$$\dot{X} + RHgX \rightarrow RX + \dot{Hg}X \qquad (22.96)$$

A nucleophilically assisted unimolecular mechanism $[S_E1(N)]$[375] has been proposed for the iododemercuration (I_2 in dimethylformamide or I_3^- in benzene) of R_2Hg[457] and $RHgX$[458,459] when R carries strongly electron-withdrawing substituents, for example $R = C_6F_5$, $(CF_3)_3C$, or $PhCH(CO_2Et)$; the Hammett ρ value of $+2.3$ for the reaction with compounds $4\text{-}X.C_6H_4CH(HgBr)CO_2Et$ is consistent with an $S_E1(I_3^-)$ mechanism.[375] The intermediacy of a *free* carbanion is not implied by the $S_E1(N)$ nomenclature but has been considered a possibility in the brominolysis of the compound $PhCH(OOBu^t)\text{-}CH(HgBr)CO_2Me$, which proceeds in pyridine with complete loss of configuration.[184]

Pyridine probably favours a heterolytic mechanism because by complexing with the reagents it enhances both the electrophilicity of the halogen and the nucleophilicity of the organomercurial. On the other hand, while halide ions activate the Hg–C bond by coordination to mercury, they convert halogens into trihalide ions which, being negatively charged, are *weaker* electrophiles. It should be noted that although a rate dependence on $[X_3^-]$ is established, a mechanism involving the independent, simultaneous action of X_2 and X^- is kinetically indistinguishable from one involving attack by the X_3^- unit. Cyclic transition states have usually been proposed for reactions proceeding by the S_E2 mechanism but experimental evidence for the internal nucleophilic assistance is lacking.

It is clear that much more work on the mechanistic aspect of halogenodemercuration is required, but that this will be made difficult by the extreme sensitivity of mechanism to reaction conditions, which arises from the availability of alternative pathways with similar energy requirements. These problems can be exacerbated by choice of organomercury substrate as illustrated by recent work on the brominolysis of *cis*- and *trans*-propenyl- and but-2-enylmercury bromide.[456] Predominant *inversion* of configuration is found for the reaction in carbon disulphide and is explained in terms of cleavage by an addition–elimination sequence with preferred *trans*-stereochemistry in each step (for example equation 22.97).

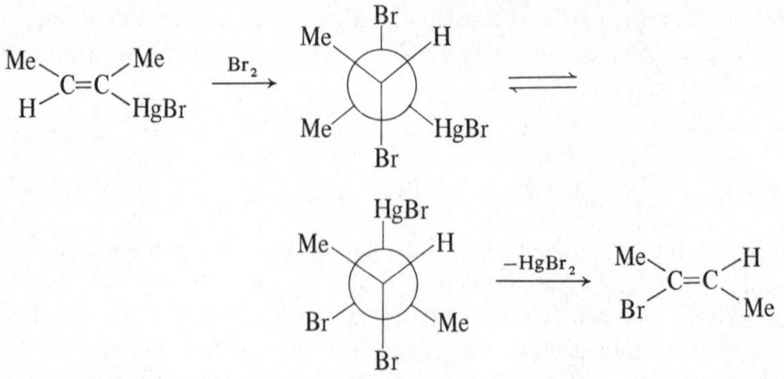

Me₂C=C(Me)(H)... (chemical structures for equation 22.97)

$$
\begin{array}{c}
\underset{H}{\overset{Me}{}}C=C\underset{HgBr}{\overset{Me}{}} \xrightarrow{Br_2} \quad\text{(Newman projection: Br, Me, H top; Me, HgBr, Br)} \quad \rightleftharpoons
\end{array}
$$

$$
\text{(Newman projection: HgBr, Me, H; Br, Me, Br)} \xrightarrow{-HgBr_2} \underset{Br}{\overset{Me}{}}C=C\underset{Me}{\overset{H}{}} \qquad (22.97)
$$

22.5.2 Applications

Halogenodemercuration has been used mainly as a method for determining the position and stereochemistry of the mercury in organomercurials by converting them into known alkyl halides. With the development of modern structural probes such as n.m.r. spectroscopy there has been a decline in this application of the reaction, but it is still occasionally useful, for example in assigning configurations to organomercury derivatives of tricyclo[2.2.1.0^{2,6}]heptane,[57,460] In stereochemical determinations of this kind it is assumed that halogen in pyridine at low temperature cleaves the Hg—C bond with retention.

Currently there is interest in exploiting the preparative potential of combining halogenolysis with mercuration and oxymercuration. Thus good yields of polybromoarenes have been obtained by the sequence shown in equation 22.98, where X = $O_2C.CF_3$ and Y = NO_2, $CONH_2$ or CO_2H.[66]

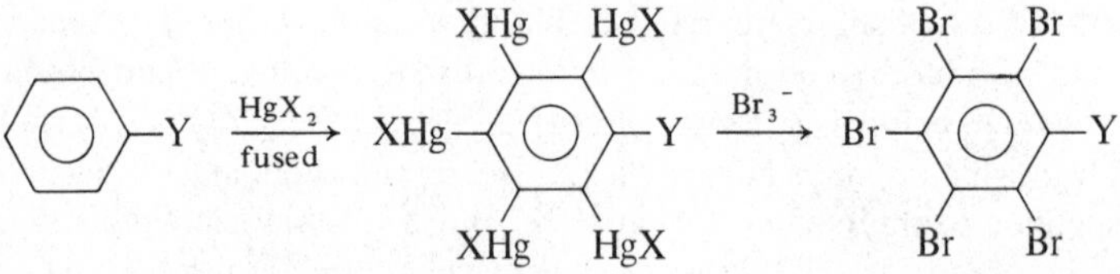

$$
\langle\text{benzene}\rangle\!-\!Y \xrightarrow[\text{fused}]{HgX_2} XHg\!-\!\langle\text{arene, XHg/HgX}\rangle\!-\!Y \xrightarrow{Br_3^-} Br\!-\!\langle\text{arene, Br}\rangle\!-\!Y \qquad (22.98)
$$

56–64 per cent

Many new β-halogenoalkyl peroxides have been prepared by combining halogenolysis with peroxymercuration (section 19.3.3). Three main types can be defined and are shown in equation 22.99,[461] 22.100 (R^1 = OMe or Ph; X = Cl, Br, or I),[184] and 22.101 (R^1 = OMe or Me; X = Br or I),[184] where the yields quoted are for the halogenodemercuration step in isolation.

$$
R^1R^2C\!:\!CHR^3 \xrightarrow[\text{Bu}^t\text{OOH}]{Hg(O_2C.CF_3)_2} \xrightarrow[]{KBr} \xrightarrow[CH_2Cl_2]{Br_2} R^1R^2C(OOBu^t)CHBr.R^3
$$

80–90 per cent

$$\qquad (22.99)$$

$$\text{RCH}:\text{CH}\cdot\text{COR}^1 \xrightarrow[\text{Bu}^t\text{OOH}]{\text{Hg(OAc)}_2} \xrightarrow{\text{KY}} \xrightarrow[\substack{\text{MeOH or} \\ \text{CH}_2\text{Cl}_2}]{\text{X}_2} \text{RCH(OOBu}^t\text{)CXH}\cdot\text{COR}^1$$
$$75\text{–}85 \text{ per cent}$$

$$(22.100)$$

$$\text{CH}_2:\text{CMe}\cdot\text{COR}^1 \xrightarrow[\text{Bu}^t\text{OOH}]{\text{Hg(OAc)}_2} \xrightarrow{\text{KBr}} \xrightarrow[\text{MeOH}]{\text{X}_2} \text{XCH}_2\cdot\text{CMe(OOBu}^t\text{)COR}^1$$
$$55 \text{ per cent}$$

$$(22.101)$$

Organomercury chlorides must be used when preparing chloroalkyl peroxides (equation 22.100; Y = X = Cl) otherwise mixed halogens, from which other alkyl halides can be obtained, are generated. Where medial alkenes are used, a mixture of diastereoisomeric products is obtained. The sequence is very attractive preparatively being versatile, easy to carry out, and affording high yields of products that require little purification.

It seems likely that the development of synthetic applications based on sequences of this type will become the main aspect of halogenodemercuration in the immediate future.

23 Organomercury Compounds in Organic Synthesis

In describing the reactions that are characteristic of the mercury–carbon bond we have been at pains to emphasise their established or potential value in organic synthesis. We now single out for special attention the two areas where activity in this context is greatest: (i) the introduction of oxygen or nitrogen substituents through the sequence oxy- or aminomercuration and hydrogenodemercuration with sodium borohydride, and (ii) the use of α-halogenoalkylmercurials (Seyferth reagents) to transfer divalent carbon fragments to suitable substrates. The mechanism of oxy- and aminomercuration has been described in section 19.3.2 and that of hydrogenodemercuration in section 22.2.2; the preparation of the Seyferth reagents was described in section 19.2.2.

23.1 Oxy- or Aminomercuration with Hydrogenodemercuration

The combination of oxy- or aminomercuration with sodium borohydride reduction provides a convenient method for preparing alcohols, ethers, alkyl peroxides, and amines (equation 23.1), and oxygen- or nitrogen-containing heterocyclic compounds (equation 23.2; the positions of attachment of Y and HgX may be reversed).

$$R^1R^2C{=}CHR^3 \xrightarrow[\text{HYR}_n]{\text{HgX}_2} R^1R^2C(YR_n)CH(HgX)R^3 \xrightarrow[^-\text{OH}]{\text{NaBH}_4} R^1R^2C(YR_n)CH_2R^3$$

$$(23.1)$$

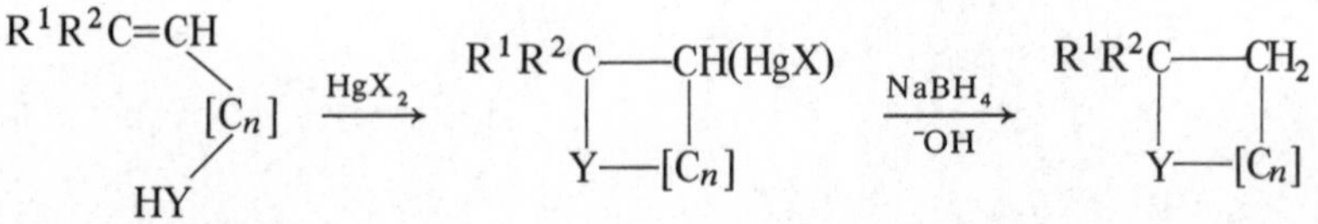

$$\begin{array}{c} R^1R^2C{=}CH \\ \diagdown \\ \qquad [C_n] \\ \diagup \\ HY \end{array} \xrightarrow{\text{HgX}_2} \begin{array}{c} R^1R^2C{-}\!\!-\!\!CH(HgX) \\ | \qquad\quad | \\ Y{-}\!\!-[C_n] \end{array} \xrightarrow[^-\text{OH}]{\text{NaBH}_4} \begin{array}{c} R^1R^2C{-}\!\!-\!\!CH_2 \\ | \qquad\quad | \\ Y{-}\!\!-[C_n] \end{array}$$

$$(23.2)$$

Both steps proceed rapidly under mild conditions and the reduction can usually be carried out *in situ*. The basic characteristics of the addition stage, namely that it is highly regio- and stereo-selective and is very rarely accompanied by skeletal rearrangements contribute greatly to the preparative value of the sequence.

The use of oxymercuration and reduction as a route to alcohols and ethers was first proposed by Henbest and Nicholls[207] in 1959 but was transformed into a practical reality by H. C. Brown and his coworkers some eight years later. Brown's contribution was to show that Bordwell and Douglass's clean hydrogenodemercuration of oxymercurials by alkaline sodium borohydride[414] can be carried out *in situ*, and he developed a one-pot, high-yield synthesis of alcohols.[462] Hydration of the alkene occurs in the Markovnikov sense (for example equations 23.3 and 23.4) so that the mercury route is complementary to the hydroboration–oxidation sequence developed earlier by the same school.

$$PhCH : CH_2 \xrightarrow[H_2O/THF]{Hg(OAc)_2} \xrightarrow[^-OH]{NaBH_4} PhCH(OH)Me \qquad (23.3)$$
$$96 \text{ per cent}$$

$$Me_2C : CHMe \xrightarrow[H_2O/THF]{Hg(OAc)_2} \xrightarrow[^-OH]{NaBH_4} Me_2C(OH)CH_2Me \qquad (23.4)$$
$$95 \text{ per cent}$$

In the bicyclic alkene series the reaction is highly stereoselective[463,464] and is complementary to the Grignard synthesis from related ketones since epimers are produced from the two routes.

Brown's alcohol synthesis kindled interest in the preparative value of oxymercuration and initiated a considerable amount of research aimed at exploiting the combination of mercury-induced addition with borohydride reduction. Thus within two years, communications appeared describing applications of the system to the synthesis of ethers[216] (equation 23.5; R = Me, Et, or Pr^i), secondary alkyl peroxides[465] (equation 23.6), amides[159,466] (equation 23.7), and secondary or tertiary alkyl azides[467] (equation 23.8).

$$R^1R^2C : CH_2 \xrightarrow[ROH]{Hg(OAc)_2} \xrightarrow[^-OH]{NaBH_4} R^1R^2C(OR)Me \qquad (23.5)$$
$$90\text{--}100 \text{ per cent}$$

$$RCH : CH_2 \xrightarrow[Bu^tOOH/CH_2Cl_2]{Hg(OAc)_2} \xrightarrow[^-OH]{NaBH_4} RCH(OOBu^t)Me \qquad (23.6)$$
$$60\text{--}70 \text{ per cent}$$

$$RCH : CH_2 \xrightarrow[MeCN]{Hg(NO_3)_2} \xrightarrow[^-OH]{NaBH_4} RCH(NHCOMe)Me \qquad (23.7)$$
$$50\text{--}92 \text{ per cent}$$

$$R^1R^2C : CH_2 \xrightarrow[H_2O/THF/NaN_3]{Hg(OAc)_2} \xrightarrow[^-OH]{NaBH_4} R^1R^2C(N_3)Me \qquad (23.8)$$
$$50\text{--}88 \text{ per cent}$$

The hydroxymercuration–reduction sequence has become a standard procedure in organic synthesis that has been applied to a wide range of acyclic, cyclic, polycyclic, and functionally substituted alkenes.[14] The versatility of the method is illustrated by the preparation of alcohols from terpenes (for example equation

23.9[468]), flavenes (for example equation 23.10[469]), and *endo*-dicyclopentadiene (equation 23.11[470]).

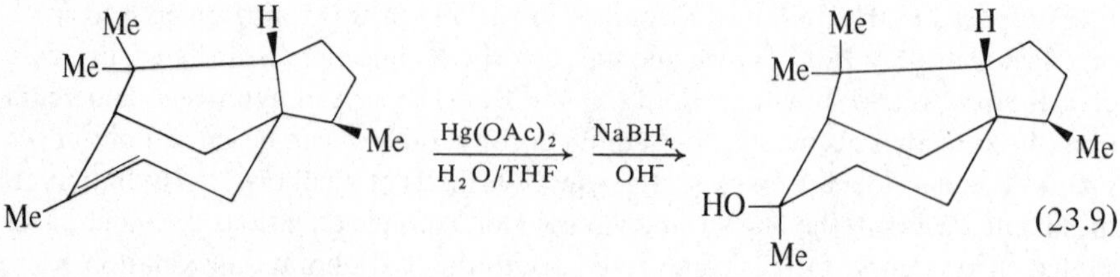

(23.9)

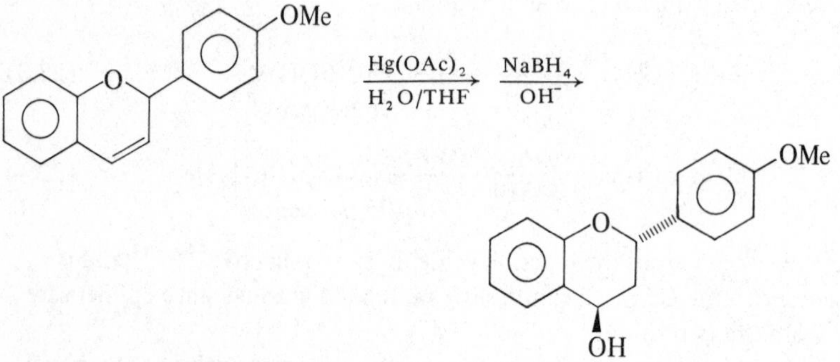

(23.10)

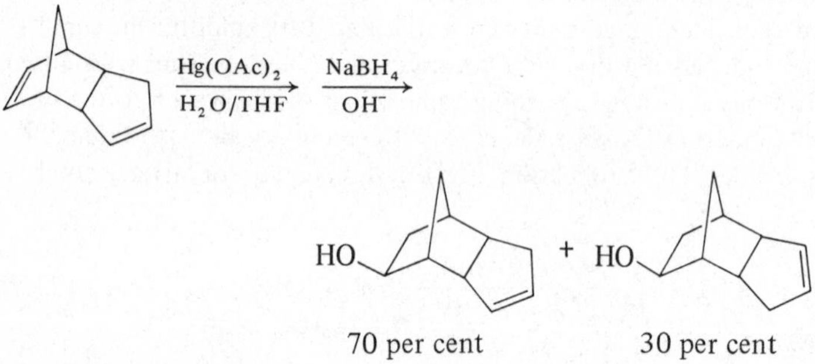

70 per cent   30 per cent

(23.11)

Whereas mercury attacks only the norbornene-type double bond in *endo*-dicyclopentadiene, hydroboration–oxidation affords alcohols arising mainly from hydration of the cyclopentene ring. The asymmetric synthesis of secondary alcohols using mercury(II) tartrate (p. 164), and the synthesis of ketones from medial alkynes (p. 181) are other synthetic aspects of hydroxymercuration-reduction that have already been mentioned.

The synthesis of secondary alkyl peroxides[225] is particularly valuable because the conventional route is lengthy and affords poor yields due to decomposition of the product under reaction conditions. The peroxymercuration method has been extended to provide a generalised and much-improved synthesis of β-methoxycarbonylalkyl peroxides (equation 23.12) and a route to the previously unknown α-carbonylalkyl peroxides (equation 23.13; R = OMe or Me).[471]

$$R^1R^2C:CH.CO_2Me \xrightarrow[Bu^tOOH/CH_2Cl_2]{Hg(OAc)_2} \xrightarrow[^-OH]{NaBH_4} R^1R^2C(OOBu^t)CH_2CO_2Me$$
$$85\text{--}95 \text{ per cent}$$

$$(23.12)$$

$$CH_2:CMe.COR \xrightarrow[Bu^tOOH/CH_2Cl_2]{Hg(OAc)_2} \xrightarrow[^-OH]{NaBH_4} Me_2C(OOBu^t)COR \quad (23.13)$$
$$45\text{--}85 \text{ per cent}$$

One of the main developments in applications of alkoxy- and aminomercuration has been the synthesis of heterocyclic compounds. The examples in equations 23.14 to 23.20 are chosen to illustrate the scope. In the oxygen series, derivatives of hydrogenated furan and pyran rings, including polycyclic systems, have been prepared through reactions with dienes [equations 23.14 (n = 2 or 3) and 23.15[221]] or unsaturated alcohols [equations 23.16 (R = H or Me)[472] and 23.17 (R = H or OH)[473]].

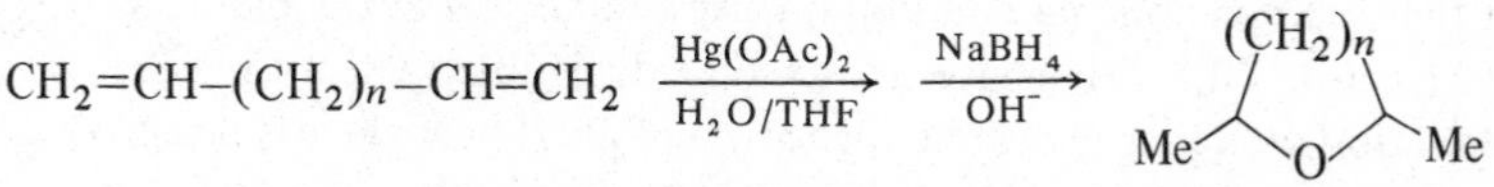

$$CH_2=CH-(CH_2)_n-CH=CH_2 \xrightarrow[H_2O/THF]{Hg(OAc)_2} \xrightarrow[OH^-]{NaBH_4}$$

$$(23.14)$$

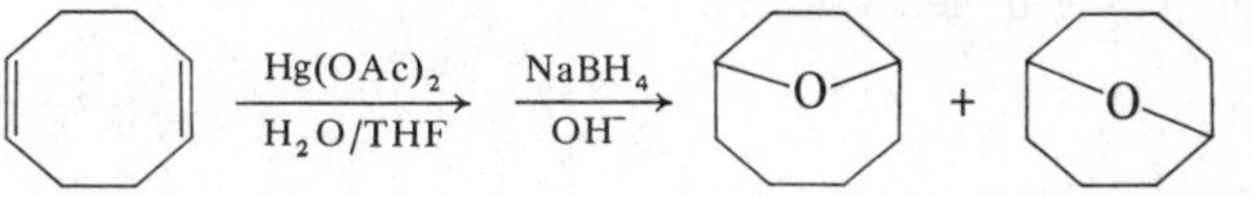

$$75 \text{ per cent} \qquad 25 \text{ per cent}$$

$$(23.15)$$

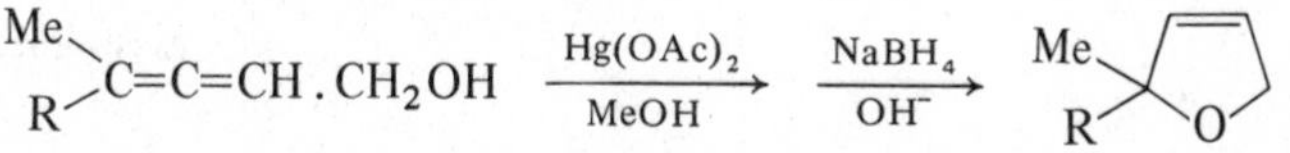

$$(23.16)$$

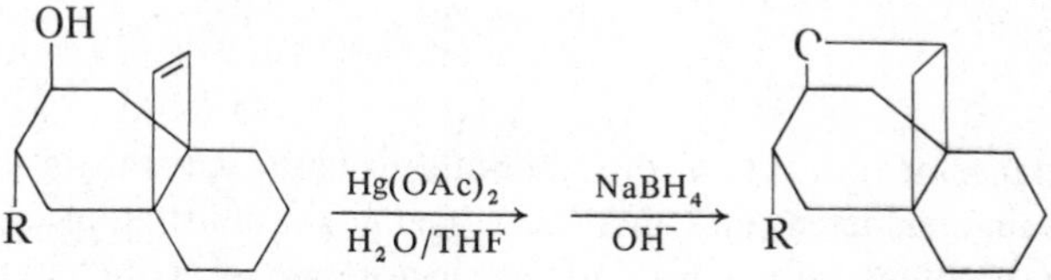

$$(23.17)$$

Likewise, 1,2-dioxacyclo-pentanes and -hexanes have been obtained in high yield via the reaction of dienes with hydrogen peroxide and mercury(II) nitrate.[175] A similar range of five- and six-membered rings is formed in the nitrogen series (for example equations 23.18[474] and 23.19[475]).

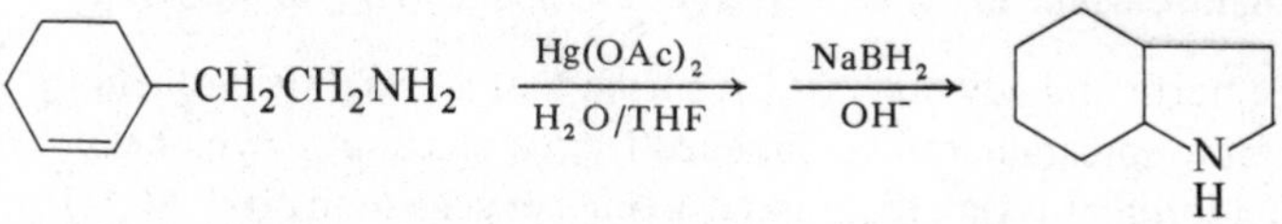

$$(23.18)$$

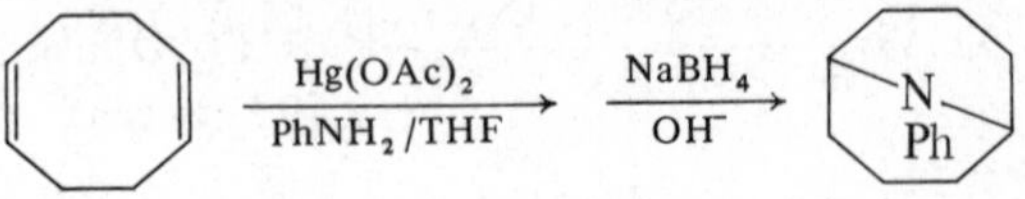

$$(23.19)$$

Finally the versatility of the method is illustrated by the synthesis of a spiro-heterocycle in which both oxy- and aminomercuration participate (equation 23.20[474]).

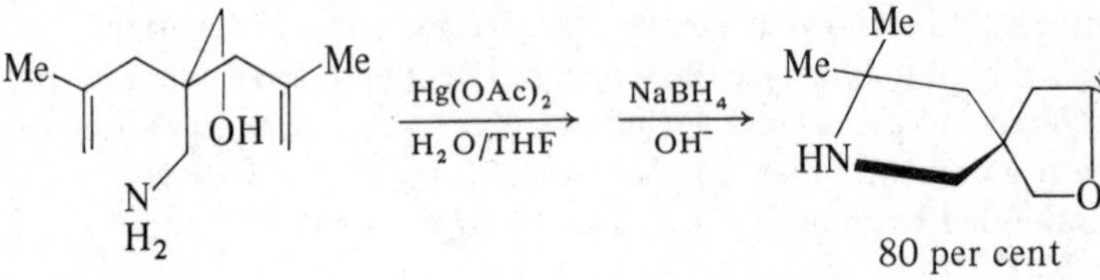

$$(23.20)$$

80 per cent

The fact that hydrogenodemercuration involves homolysis of the mercury-carbon bond (section 22.2.2) can give rise to some difficulties when using the addition–reduction sequence. Thus the formation of alkyl peroxides is usually accompanied by epoxidation because the intermediate radicals can cyclise rather than abstract hydrogen (equation 23.21[476]).

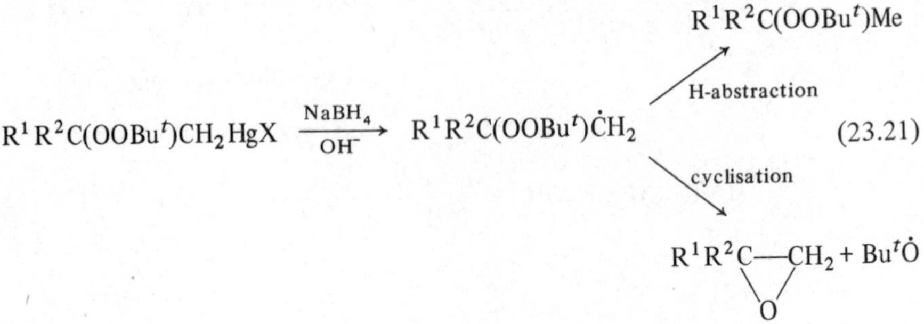

$$(23.21)$$

The rearrangement of intermediate radicals may also account for the mixtures of five- and six-membered rings obtained on reduction of intramolecular amino-mercuration products, although a polar mechanism was suggested by the original workers.[237] Reductions with sodium borodeuteride generally lead to non-stereospecific replacement of mercury, but it appears that stereospecific deuterio-demercuration can be achieved by using sodium amalgam in alkaline deuterium oxide as the reducing system.[477]

23.2 Divalent Carbon Transfer Reactions of Halogenoalkylmercurials

In 1962 Seyferth and coworkers[478] reported that an essentially quantative yield of 7,7-dichloronorcarane can be obtained from a mixture of cyclohexene and phenyl(trichloromethyl)mercury in refluxing benzene (equation 23.22).

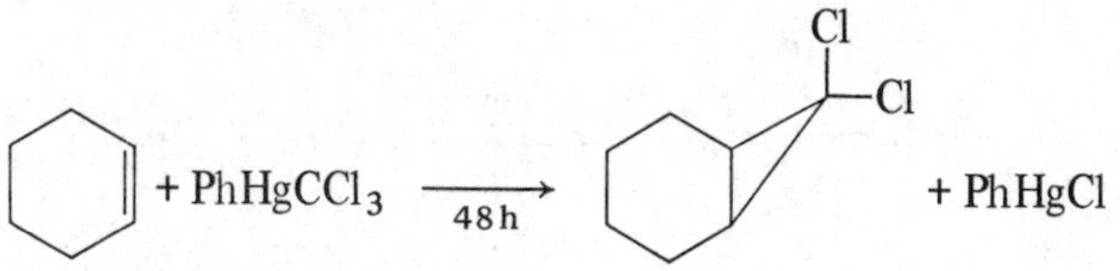

$$\text{[cyclohexene]} + PhHgCCl_3 \xrightarrow[48h]{} \text{[bicyclic dichloro product]} + PhHgCl \qquad (23.22)$$

Since then the ability of halogenoalkylmercurials to transfer divalent carbon fragments to suitable organic substrates has been investigated in great depth, principally by the Seyferth school. A wide range of organomercury reagents (RHgCXYZ) capable of transferring various carbene fragments (CXY) to a variety of multiply and singly bonded carbenophiles has been discovered, and well over 100 papers concerned with this development have now appeared in the literature. Fortunately Seyferth[479] has recently reviewed the use of phenyl(tri-halogenomethyl)mercurials in the preparation of *gem*-dihalogenocyclopropanes from alkenes, which is the reaction most likely to be utilised by synthetic organic chemists.

Before the advent of the organomercury reagents, the transfer of dihalogeno-carbenes to alkenes was effected with haloforms and strong base, or via the decarboxylation of alkali metal salts of trihalogenoacetic acids. The first of these procedures cannot be used for alkenes containing base-sensitive functional groups, and since trihalogenomethyl carbanions are the carbene precursors in both systems, difficulties can arise with alkenes that readily trap carbanions. These problems do not arise with trihalogenomethylmercurials (for example equation 23.23; vinyl acetate is a $^-CCl_3$ trap) and they have the added advantage of readily transferring dihalogenomethylene to a number of alkenes that are poorly reactive towards the other carbene sources (for example equation 23.24).

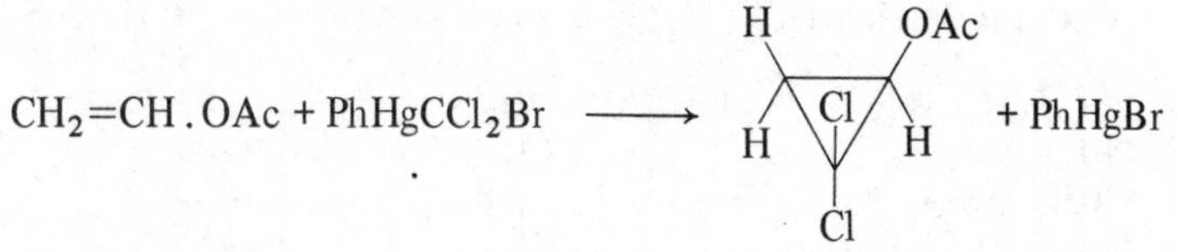

$$CH_2=CH.OAc + PhHgCCl_2Br \longrightarrow \text{[dichlorocyclopropane, H, OAc, H, H, Cl, Cl]} + PhHgBr \qquad (23.23)$$

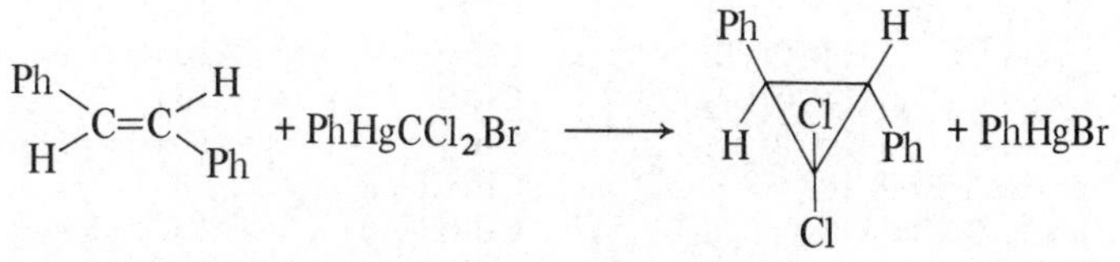

$$\underset{H}{\overset{Ph}{>}}C=C\underset{Ph}{\overset{H}{<}} + PhHgCCl_2Br \longrightarrow \text{[dichlorocyclopropane, Ph, H, H, Ph, Cl, Cl]} + PhHgBr \qquad (23.24)$$

That trihalogenomethyl carbanions are not captured by recognised traps is an important observation from the point of view of determining the mechanism of carbene transfer from organomercurials. The pathway outlined in equations 23.25 and 23.26 has been proposed for the PhHgCCl$_2$Br–alkene reaction on the basis of kinetic studies[480,481] and comparisons with the non-organomercury routes.[482]

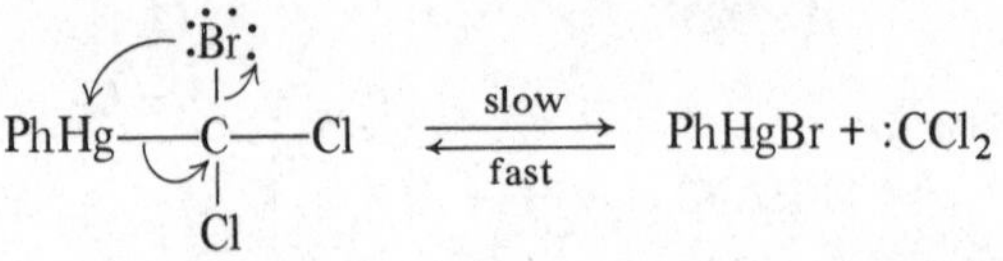

$$\text{PhHg-C-Cl} \xrightleftharpoons[\text{fast}]{\text{slow}} \text{PhHgBr} + :CCl_2 \qquad (23.25)$$

$$:CCl_2 + \underset{/\backslash}{\overset{\backslash/}{\underset{C}{\overset{C}{\parallel}}}} \xrightarrow{\text{fast}} \underset{/\backslash}{\overset{\backslash/}{\underset{C}{\overset{C}{\triangleright}}}}CCl_2 \qquad (23.26)$$

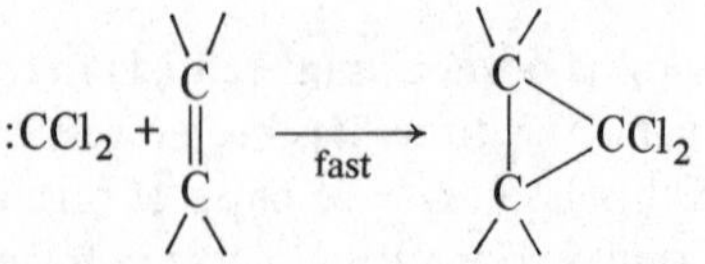

A similar mechanism is believed to hold for the reaction of compounds $PhHgCCl_3$, $PhHgCClBr_2$, $PhHgCBr_3$, and $PhHgCCl_2F$ with alkenes and with some other carbenophiles such as organosilicon hydrides. However methylenation of alkenes with bis(bromomethyl)mercury and CCl_2 transfers to dipolar multiple bonds such as C=N, C=O, and C=S probably do not involve free carbenes as intermediates.

In the remainder of this section we shall elaborate the scope of organomercury carbene-transfer chemistry with particular emphasis on the mercury aspect. The range of divalent carbon fragments that can be generated from organomercurials is listed in table 23.1. For some carbenes, a number of organomercury precursors are available. Thus in addition to the reagents listed in the table, the systems $(BrCH_2)_2Hg$, ICH_2HgI, $Ph_2Hg/BrCH_2HgBr$, and $Ph_2Hg/(ICH_2)_2Hg$ have been

Table 23.1. Halogenoalkylmercurials as Divalent Carbon Transfer Reagents

Organomercurial	Carbene transferred	Ref.
$PhCH_2HgCH_2I$	CH_2	283, 483
$PhHgCHClBr$	$CHCl$	484
$PhHgCHBr_2$	$CHBr$	484
$PhHgCF_3/NaI$	CF_2	485
$PhHgCFCl_2$	$CFCl$	486
$PhHgCFBr_2$	$CFBr$	137
cyclo-$C_6H_{11}HgCCl_2Br$	CCl_2	131
cyclo-$C_6H_{11}HgCClBr_2$	$CClBr$	131
cyclo-$C_6H_{11}HgCBr_3$	CBr_2	131
$PhHgCFBr.CF_3$	$CF(CF_3)$	487
$PhHgCClBr.CF_3$	$CCl(CF_3)$	132, 133
$PhHgCCl_2Ph$	$CClPh$	488
$Ph_2Hg/(Me_3SiCCl_2)_2Hg$	$CCl(SiMe_3)$	489
$Ph_2Hg/(Me_3SiCBr_2)_2Hg$	$CBr(SiMe_3)$	489
$PhHgCFCl.CO_2Me$	$CF(CO_2Me)$	490
$PhHgCFBr.CO_2Et$	$CF(CO_2Et)$	490
$PhHgCClBr.CO_2Me$	$CCl(CO_2Me)$	133
$PhHgCBr_2.CO_2Me$	$CBr(CO_2Me)$	133

used to effect methylene transfer, and the compounds $PhHgCCl_3$, $PhHgCCl_2Br$, and $PhHgCCl_2I$ have been used to generate dichlorocarbene. Carbenoid reactivity is obviously the main factor that influences choice of reagent, but storage properties and ease of preparation must also be taken into account. Thus although the compound $PhHgCCl_2I$ is a highly reactive source of dichlorocarbene, it has a short shelf life, which considerably diminishes its practical value. In this context some of the reagents (for example $PhHgCCl_2Br$ and $PhHgCBr_3$) are now commercially available.

The question of reactivity deserves further comment. The ease of carbene extrusion from the compound $RHgCXYZ$ in accordance with equation 23.27 increases along the series $Z = F < Cl < Br < I$, and the compound $PhHgCCl_2Br$, for example, affords *only* dichlorocarbene.

$$RHgCXYZ \xrightarrow{\Delta} [:CXY] + RHgZ \qquad (23.27)$$

Carbene generation is catalysed by sodium iodide but the mechanism then changes to one involving an intermediate carbanion from which halide-ion loss is less selective. This procedure is used to promote CF_2-transfer from phenyl(trifluoromethyl)mercury (equation 23.28), a process that does not take place under conditions of simple thermolysis.[485]

$$PhHgCF_3 + I^- \xrightarrow[-PhHgI]{80\,^\circ C} [:CF_3^-] \xrightarrow[-F^-]{} [:CF_2] \qquad (23.28)$$

In the phenylmercury series, transfers from the compounds $PhHgCCl_nBr_{3-n}$ ($n = 0$–2) to alkenes occur rapidly at 80 °C and typically are carried out in refluxing benzene. However, when one of the halogens is replaced by a polyatomic substituent (CF_3, Ph, $SiMe_3$, or CO_2R), long reaction times at temperatures of 120–155 °C are required and the sealed-tube technique is often adopted.

The influence of R in $RHgCXYZ$ on the ease of carbene transfer is largely unexplored but the indications are that it can be considerable. Notably the replacement of phenyl by alkyl groups leads to markedly more reactive carbenoids. Thus benzyl(iodomethyl)mercury is a much more reactive methylene-transfer reagent than the equivalent phenyl compound.[283] The cyclohexyl series, $C_6H_{11}HgCCl_nBr_{3-n}$ ($n = 0$–2), will efficiently transfer dihalogenocarbenes to alkenes in two to three days *at room temperature*[131] while reaction times of fifteen to eighteen days are needed to achieve comparable yields with the corresponding phenylmercury compounds. Furthermore the cyclohexyl(halogenomethyl)mercurials are stable in the solid state at room temperature and work-up procedures remain simple because cyclohexylmercury halides; like their phenyl counterparts, are poorly soluble in common organic solvents. The development of reagents that have an acceptable reactivity at room temperature, yet are easily stored, should greatly enhance the value of mercury carbenoids in organic synthesis.

We have seen (table 23.1) the divalent carbon fragments that are available from organomercury sources. With a knowledge of the organic substrates to which these may be transferred, the full range of synthetic applications becomes apparent. It

must be emphasised, however, that the *established* scope is much smaller than
the potential implied by taking all combinations of carbenoids and substrates.
Alkenes (particularly cyclohexene) and triethylsilane have been used as standard
carbenophiles for evaluating the carbenoid activity of new halogenoalkylmercurials
(equations 23.29 and 23.30), while the range of substrates outlined below is
primarily a list of CCl_2-acceptors.

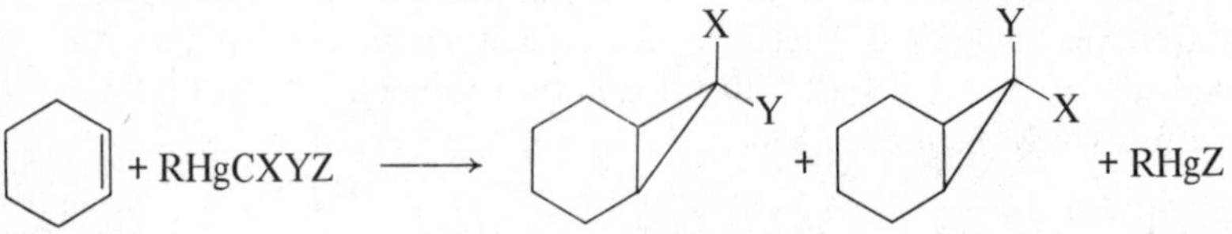

$$\text{(23.29)}$$

$$Et_3SiH + RHgCXYZ \rightarrow Et_3SiCXYH + RHgZ \qquad \text{(23.30)}$$

The most important carbenophiles are multiply bonded species from which
three-membered rings are obtained (equation 23.31; W = C, N, O, or S).

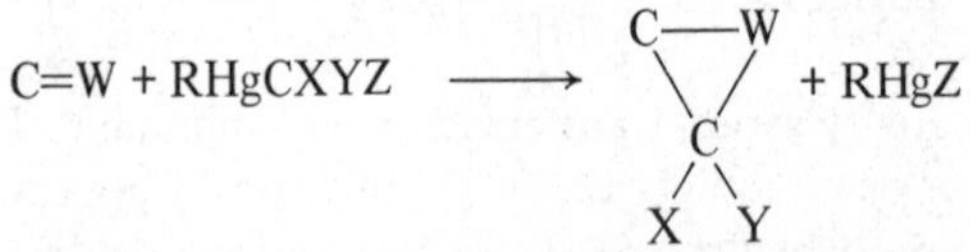

$$\text{(23.31)}$$

A considerable range of acyclic, cyclic, polycyclic, and functionally substituted alkenes participate in the reaction to afford cyclopropane derivatives;[14,479]
additions to medial alkenes occur stereospecifically with retention (for example
equation 23.24). Much of the recent interest has been in extending the reaction
to heteronuclear double bonds, and aziridines (for example equation 23.32[491]),
oxiranes (for example equation 23.33[492]) and thiiranes (for example equation
23.34[493]) have been prepared by this route.

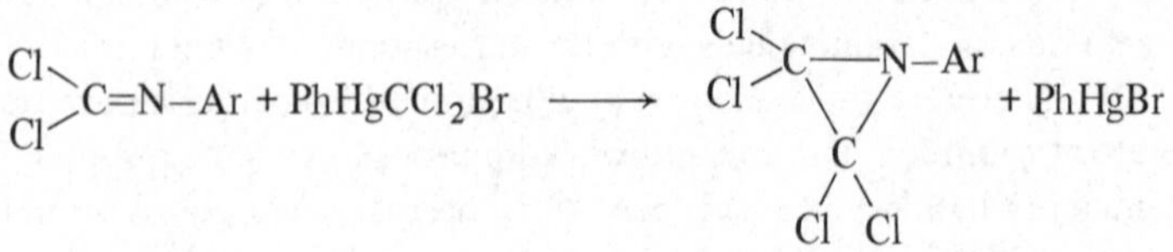

$$\text{(23.32)}$$

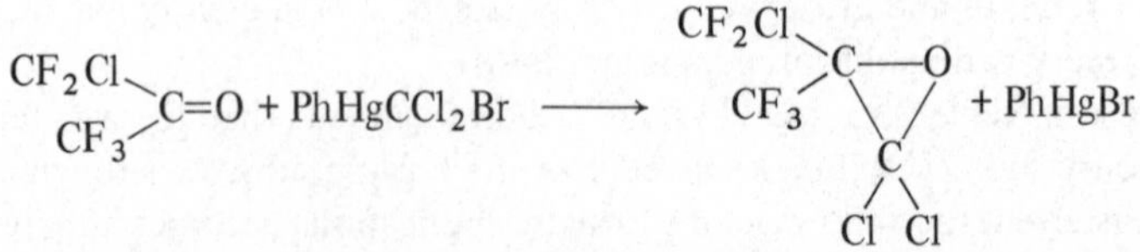

$$\text{(23.33)}$$

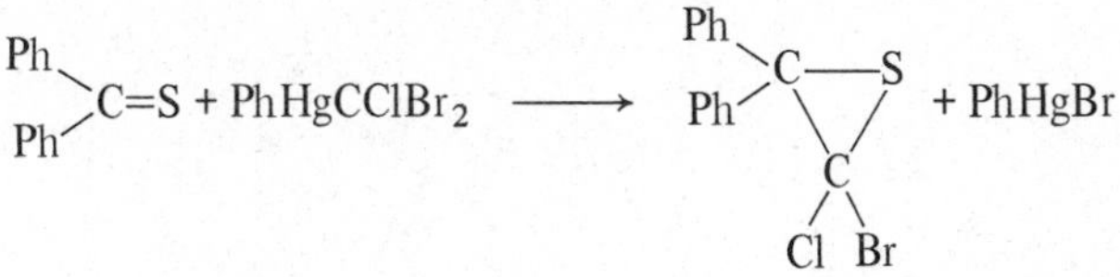

$$\underset{Ph}{\overset{Ph}{>}}C=S + PhHgCClBr_2 \longrightarrow \underset{Ph}{\overset{Ph}{>}}\underset{\underset{Cl \ Br}{C}}{\overset{C-S}{\triangle}} + PhHgBr \quad (23.34)$$

Carbenes do not normally undergo 1,4-addition to conjugated double bonds, but products obtained from N=C–C=O and N=N–C=O systems have been rationalised in terms of such a pathway (for example equation 23.35[494]).

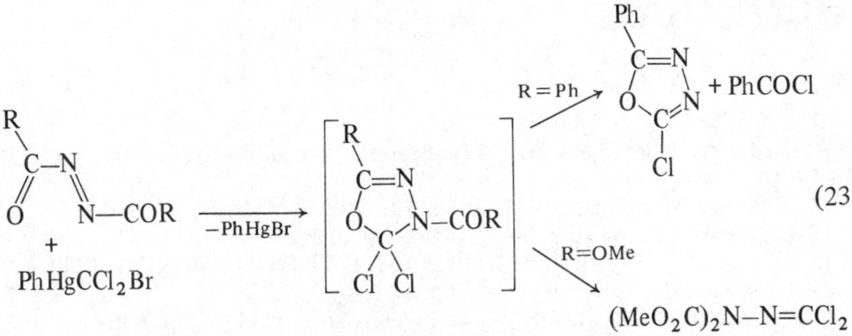

$$\underset{PhHgCCl_2Br}{\overset{\underset{O}{\overset{R}{>}}C-N}{\underset{N-COR}{|}}} \xrightarrow{-PhHgBr} \left[\underset{\overset{|}{\underset{Cl \ Cl}{C}}}{\overset{\underset{O}{\overset{R}{>}}C=N}{\underset{N-COR}{|}}} \right] \quad (23.35)$$

R = Ph → (+ PhCOCl)

R = OMe → $(MeO_2C)_2N–N=CCl_2$

Phenyl(halogenomethyl)mercurials also insert dihalogenocarbene into single bonds, notably benzylic C–H bonds (with predominant retention[495]), C–H bonds β to group IV metals,[496] Si–H bonds (with predominant retention[497]), Si–C bonds in silacyclobutanes[498] and Sn–Sn bonds.[499] Halogenomethylmercurials containing an amide function decompose in refluxing bromobenzene via intramolecular insertion of the carbene centre into an α C–H bond of the piperidino substituent (equation 23.36[134]).

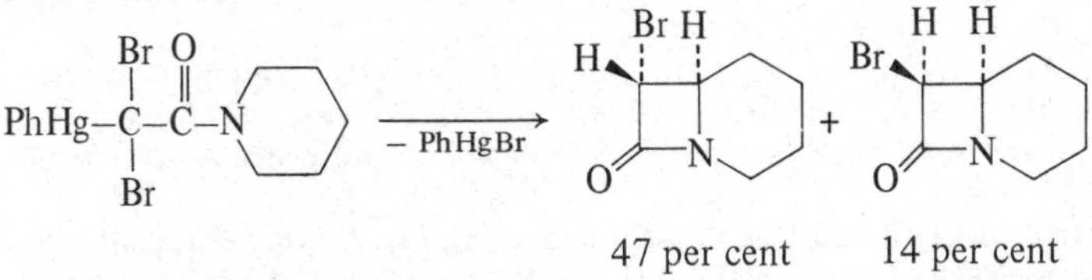

$$\underset{\overset{|}{Br}}{\overset{\overset{Br}{|}}{PhHg-C}}\overset{\overset{O}{\parallel}}{-C}-N\bigcirc \xrightarrow{-PhHgBr} \quad (23.36)$$

47 per cent   14 per cent

A final application worthy of mention is the use of phenyl(trihalogenomethyl)mercurials in preparing alkenyl halides via a Wittig reaction in which the ylid is generated *in situ* (equation 23.37[500])

$$PhHgCXYBr + Ph_3P \xrightarrow[-PhHgBr]{} [Ph_3P:CXY] \xrightarrow[-Ph_3PO]{R^1R^2C=O} R^1R^2C:CXY \quad (23.37)$$

References

1. V. Peruzzo, G. Plazzogna and G. Tagliavini, *J. organometallic Chem.*, 24 (1970) p. 347.
2. D. Dodd and M. D. Johnson, *J. chem. Soc. Perkin II* (1974) p. 219.
3. M. L. Bullpitt and W. Kitching, *J. organometallic Chem.*, 46 (1972) p. 21.
4. K. P. Butin, V. V. Strelets, A. N. Kashin, I. P. Beletskaya and O. A. Reutov, *J. organometallic Chem.*, 64 (1974) p. 181.
5. G. A. Olah and P. R. Clifford, *J. Am. chem. Soc.*, 95 (1973) p. 6067.
6. P. V. Roling, J. L. Dill and M. D. Rausch, *J. organometallic Chem.*, 69 (1974) p. C33.
7. L. G. Yudin, A. N. Kost and A. I. Pavlyuchenko, *Khim. Geterotsikl Soedin* (1971), 1517; *Chem. Abstr.*, 77 (1972) p. 34663y.
8. R. F. Kovar and M. D. Rausch, *J. organometallic Chem.*, 35 (1972) p. 351.
9. A. J. Bloodworth and I. M. Griffin, *J. organometallic Chem.*, 66 (1974) p. C1; *J. chem. Soc. Perkin I*, (1975) p. 195.
10. J. Roussel, J. J. Périé, J. P. Laval and A. Lattes, *Tetrahedron*, 28 (1972) p. 701.
11. S. W. Breuer, M. J. Leatham and F. G. Thorpe, *Chem. Comm.* (1971) p. 1475.
12. F. C. Whitmore (1921), *Organic Compounds of Mercury* (Chemical Catalog Company Inc., New York).
13. L. G. Makarova and A. N. Nesmeyanov, *Methods of Elemento-Organic Chemistry*, vol. 4, *The Organic Compounds of Mercury*, ed. A. N. Nesmeyanov and K. A. Kocheshkov (North-Holland Press, Amsterdam, 1967).
14. H. Straub, K. P. Zeller and H. Leditschke (1974). *Houben-Weyl. Methoden der Organischen Chemie* Band XIII/2b *Metallorganische Verbindungen:* Hg, Georg. Thieme Verlag, Stuttgart.
15. F. R. Jensen and B. Rickborn, *Electrophilic Substitution of Organomercurials* (McGraw-Hill, New York, 1968).
16. O. A. Reutov and I. P. Beletskaya, *Reaction Mechanisms of Organometallic Compounds* (North-Holland Press, Amsterdam, 1968).
17. M. H. Abraham, *Comprehensive Chemical Kinetics*, vol. 12, *Electrophilic Substitution at a Saturated Carbon Atom*, ed. C. H. Bamford and C. F. H. Tipper (Elsevier Scientific Publishing Co., Amsterdam, 1973).
18. E. Samuel and M. D. Rausch, *J. organometallic Chem.*, 37 (1972) p. 29.
19. G. Minghetti, F. Bonati and M. Massobrio, *Chem. Commun.* (1973) p. 260.
20. V. S. Petrosyan, A. B. Permin, S. G. Sacharov and O. A. Reutov, *J. organometallic Chem.*, 65 (1974) p. C7.

21. K. H. Slotta and K. R. Jacobi, *J. prakt. chem.*, 120 (1928) p. 249.
22. C. S. Marvel and V. L. Gould, *J. Am. chem. Soc.*, 44 (1922) p. 153.
23. C. S. Marvel and H. O. Calvery, *J. Am. chem. Soc.*, 45 (1923) p. 820.
24. M. Gaudemar, *Compt. rend.*, 254 (1962) p. 1100.
25. B. Floris and G. Ortaggi, *J. organometallic Chem.*, 50 (1973) p. 33.
26. A. N. Nesmeyanov, D. N. Kravtsov, B. A. Faingar and L. I. Petrovskaya, *Izv. Akad. Nauk SSSR, Ser. Khim.* (1968) p. 534.
27. K. Issler and H. P. Abicht, *J. prakt. chem.*, 312 (1970) p. 456.
28. M. W. Buxton, R. H. Mobbs and D. E. M. Wotton, *J. Fluorine Chem.*, 1 (1971) p. 179.
29. D. W. Slocum, C. A. Jennings, T. R. Englemann, B. W. Rockett and C. R. Hauser, *J. org. Chem.*, 36 (1971) p. 377.
30. L. I. Zakharkin, V. N. Kalinin and E. G. Rys, *Zhur. Obshch. Khim.*, 43 (1973) p. 847.
31. E. Frankland, *Annalen*, 85 (1853) p. 365.
32. I. P. Beletskaya, G. A. Artamkina, E. A. Shevlyagina and O. A. Reutov, *Zhur. Obshch. Khim.*, 34 (1964) p. 321.
33. A. Michaelis and P. Becker, *Chem. Ber.*, 15 (1882) p. 180.
34. D. S. Matteson and P. B. Tripathy, *J. organometallic Chem.*, 69 (1974) p. 53.
35. H. C. Brown, *Boranes in Organic Chemistry* (Cornell Univ. Press, New York, 1972).
36. R. C. Larock and H. C. Brown, *J. Am. chem. Soc.*, 92 (1970) p. 2467.
37. R. C. Larock and H. C. Brown, *J. organometallic Chem.*, 26 (1971) p. 35.
38. R. C. Larock, *J. organometallic Chem.*, 67 (1974) p. 353.
39. R. C. Larock, *J. organometallic Chem.*, 72 (1974) p. 35.
40. R. C. Larock, *J. organometallic Chem.*, 61 (1973) p. 27.
41. R. C. Larock and H. C. Brown, *J. organometallic Chem.*, 36 (1972) p. 1.
42. R. C. Larock, S. K. Gupta and H. C. Brown, *J. Am. chem. Soc.*, 94 (1972) p. 4371.
43. J. D. Buhler and H. C. Brown, *J. organometallic Chem.*, 40 (1972) p. 265.
44. (a) M. H. Abraham and T. R. Spalding, *J. chem. Soc. A* (1969) p. 399.
    (b) M. H. Abraham and G. F. Johnston, *J. chem. Soc. A* (1970) p. 193.
45. B. J. Gregory and C. K. Ingold, *J. chem. Soc. B* (1969) p. 276.
46. (a) R. G. Coombes and M. D. Johnson, *J. chem. Soc. A* (1966) p. 1805.
    (b) D. Dodd, M. D. Johnson and D. Vamplew, *J. chem. Soc. B* (1971) p. 1841.
47. D. Dodd, M. D. Johnson and N. Winterton, *J. chem. Soc. A* (1971) p. 910.
48. D. Dodd and M. D. Johnson, *J. chem. Soc. B* (1971) p. 662.
49. E. H. Bartlett and M. D. Johnson, *J. chem. Soc. A* (1970) p. 517.
50. D. S. Matteson, R. A. Bowie and G. Srivastava, *J. organometallic Chem.*, 16 (1969) p. 33.
51. A. N. Nesmeyanov, A. E. Borisov and N. V. Novikova, *Izv. Akad. Nauk. SSSR, Otdel. Khim. Nauk.* (1959) p. 1216.
52. F. R. Jensen and K. L. Nakamaye, *J. Am. chem. Soc.*, 88 (1966) p. 3437.
53. D. S. Matteson and R. A. Bowie, *J. Am. chem. Soc.*, 87 (1965) p. 2587.
54. A. N. Nesmeyanov, A. E. Borisov, N. V. Novikova and E. I. Fedin, *J. Organometallic Chem.*, 15 (1968) p. 279.
55. (a) G. M. Whitesides and D. J. Boschetto, *J. Am. chem. Soc.*, 93 (1971) p. 1529. (b) P. L. Bock and G. M. Whitesides, *J. Am. chem. Soc.*, 96 (1974) p. 2826.
56. D. Slack and M. C. Baird, *Chem. Commun.* (1974) p. 701.
57. E. Vedejs and M. F. Saloman, *J. org. Chem.*, 37 (1972) p. 2075.
58. M. Tada and H. Ogawa, *Tetrahedron Lett.* (1973) p. 2639.

59. H. L. Fritz, J. H. Espenson, D. A. Williams and G. A. Molander, *J. Am. chem. Soc.*, 96 (1974) p. 2378.
60. M. H. Abraham and D. F. Dadjour, *J. chem. Soc. Perkin II* (1974) p. 233.
61. R. M. G. Roberts, *J. organometallic Chem.*, 12 (1968) p. 97.
62. R. M. G. Roberts, *J. organometallic Chem.*, 18 (1969) p. 307.
63. H. Hashimoto and Y. Morimoto, *J. organometallic Chem.*, 8 (1967) p. 271.
64. J. R. Chipperfield, G. D. France and D. E. Webster, *J. chem. Soc. Perkin II* (1972) p. 405.
65. A. A. Morton and H. P. Penner, *J. Am. chem. Soc.*, 73 (1951) p. 3300.
66. G. B. Deacon and G. J. Farquharson, *J. organometallic Chem.*, 67 (1974) p. C1.
67. J. Lorberth, F. Schmock and G. Lange, *J. organometallic Chem.*, 54 (1973) p. 23.
68. B. A. Arbuzov and E. G. Kataev, *Doklady Akad. Nauk SSSR*, 96 (1954) p. 983.
69. H. V. A. Briscoe, J. B. Peel and G. W. Young, *J. chem. Soc.* (1929) p. 2589.
70. M. D. Rausch, L. P. Kleman, A. Siegel, R. F. Kovar and T. H. Gund, *Synth. inorg. Metal. Chem.*, 3 (1973) p. 193.
71. R. F. Kovar and M. D. Rausch, *J. org. Chem.*, 38 (1973) p. 1918.
72. H. B. Albrecht and G. B. Deacon, *J. organometallic Chem.*, 57 (1973) p. 77.
73. H. B. Albrecht, G. B. Deacon and M. J. Tailby, *J. organometallic Chem.*, 70 (1974) p. 313.
74. W. Kitching and P. R. Wells, *Austral. J. Chem.*, 20 (1967) p. 2029 and earlier references.
75. W. Kitching, *Organometallic Chem. Rev.*, 3 (1968) p. 35.
76. R. J. Bertino and G. B. Deacon, *J. organometallic Chem.*, 67 (1974) p. C61.
77. R. H. Fish, *J. Am. chem. Soc.*, 96 (1974). p. 6664.
78. R. Allmann and H. Musso, *Chem. Ber.*, 106 (1973) p. 3001.
79. R. Taylor, *Comprehensive Chemical Kinetics*, vol. 13, *Reactions of Aromatic Compounds*, ed. C. H. Bamford and C. F. H. Tipper (Elsevier Publishing Co., Amsterdam, 1972) p. 186.
80. C. Perrin and F. H. Westheimer, *J. Am. chem. Soc.*, 85 (1963) p. 2773.
81. H. C. Brown and R. A. Wirkkala, *J. Am. chem. Soc.*, 88 (1966) pp. 1447, 1453, 1456.
82. A. J. Kresge, M. Dubeck and H. C. Brown, *J. org. Chem.*, 32 (1967) p. 745.
83. V. I. Sokolov, V. V. Bashilov and O. A. Reutov, *Doklady Akad. Nauk. SSSR*, 197 (1971) p. 101.
84. A. J. Kresge and J. F. Brennan, *J. org. Chem.*, 32 (1967) p. 752.
85. E. Cherbuliez and M. Mori, *Helv. Chim. Acta*, 28 (1945) p. 17.
86. A. J. Bloodworth and J. Serlin, *J. chem. Soc. Perkin I* (1973) p. 261.
87. G. A. Razuvaev, N. S. Vasileiskaya and N. N. Vavilina, *J. organometallic Chem.*, 80 (1974) p. 19.
88. G. B. Deacon and F. B. Taylor, *Inorg. nucl. Chem. Letters* (1969) p. 477.
89. G. B. Deacon and F. B. Taylor, *Austral. J. Chem.*, 21 (1968) p. 2675.
90. G. K. I. Magomedov, V. G. Syrkin and A. S. Frenkel, *Zhur. Obshch. Khim*, 42 (1972) p. 2450.
91. G. A. Razuvaev, A. N. Artemov, G. G. Petukhov, N. I. Sirotkin and N. A. Pukhnarevich, *Izv. Akad. Nauk. SSSR Ser. Khim.* (1973) p. 1172.
92. V. A. Nefedov, *Zhur. Obshch. Khim.*, 36 (1966) p. 1954.
93. M. Rosenblum, N. M. Brawn, D. Clappenelli and J. Tancrede, *J. organometallic Chem.*, 24 (1970) p. 469.
94. M. D. Rausch, E. O. Fischer and H. Grabert, *J. Am. chem. Soc.*, 82 (1960) p. 76.

95. A. N. Nesmeyanov, E. N. Kolobova, K. N. Anisimov and L. I. Baryshnikov, *Izv. Akad. Nauk SSSR Ser. Khim.* (1964) p. 1135.
96. M. D. Rausch and R. A. Gennetti, *J. org. Chem.*, 35 (1970) p. 3888.
97. G. Amiet, K. Nicholas and R. Pettit, *Chem. Commun.* (1970) p. 161.
98. H. Gilman and G. F. Wright, *J. Am. chem. Soc.*, 55 (1933) p. 3302.
99. M. W. Swaney, M. J. Skeeters and R. Norris Shreve, *Ind. Eng. Chem.*, 32 (1940) p. 360.
100. G. W. Kirby and S. W. Shah, *Chem. Commun.* (1965) p. 381.
101. L. G. Yudin, A. I. Pavlyuchenko and A. N. Kost, *Zhur. Obshch. Khim.*, 39 (1969) p. 2784.
102. H. J. Schonherr and H. W. Wanzlick, *Chem. Ber.*, 103 (1970) p. 1037.
103. C. J. Cooksey, D. Dodd and M. D. Johnson, *J. chem. Soc. B* (1971) p. 1380.
104. W. Schrauth and W. Schoeller, *Chem. Ber.*, 41 (1908) p. 2089.
105. R. Allmann, K. Flatau and H. Musso, *Chem. Ber.*, 105 (1972) p. 3067.
106. R. H. Fish, R. E. Lundin and W. F. Haddon, *Tetrahedron Letters* (1972) p. 921.
107. F. Bonati and G. Minghetti, *J. organometallic Chem.*, 22 (1970) p. 5.
108. K. S. Patel and B. N. Mankad, *Curr. Sci.*, 30 (1961) p. 335.
109. T. Do Minh, O. P. Strausz and H. E. Gunning, *Tetrahedron Letters* (1968) p. 5237.
110. J. Lorberth, *J. organometallic Chem.*, 27 (1971) p. 303.
111. S. J. Valenty and P. S. Skell, *J. org. Chem.*, 38 (1973) p. 3937.
112. U. Schöllkopf and P. Markusch, *Annalen*, 753 (1971) p. 143.
113. M. Regitz, A. Liedhegener, U. Eckstein, M. Martin and W. Anschutz, *Annalen*, 748 (1971) p. 207.
114. D. Seyferth, R. S. Marmor and P. Hilbert, *J. org. Chem.*, 36 (1971) p. 1379.
115. L. V. Okhlobystina, G. Ya. Legin and A. A. Fainzilberg, *Izv. Akad. Nauk. SSSR Ser. Khim* (1969) p. 708.
116. D. Seyferth and R. S. Marmor, *J. organometallic Chem.*, 59 (1973) p. 231.
117. V. M. Neplynev, R. G. Dubenko and P. S. Pelkis, *Zhur. Org. Khim.*, 6 (1970) p. 2113.
118. Z. Rappoport, S. Winstein and W. G. Young, *J. Am. chem. Soc.*, 94 (1972) p. 2320.
119. J. Lorberth, *J. organometallic Chem.*, 19 (1969) p. 189.
120. S. Lenzer, *Austral. J. Chem.*, 22 (1969) p. 1303.
121. G. W. Watt and L. J. Baye, *J. inorg. nucl. Chem.*, 26 (1964) p. 1531.
122. G. Wulfsberg, R. West and V. N. M. Rao, *J. Am. chem. Soc.*, 95 (1973) p. 8658.
123. F. A. Cotton and T. J. Marks, *J. Am. chem. Soc.*, 91 (1969) p. 7281.
124. E. Maslowsky and K. Nakamoto, *Inorg. Chem.*, 8 (1969) p. 1108.
125. J. Mink, L. Bursics and G. Végh, *J. organometallic Chem.*, 34 (1972) p. C4.
126. P. West, M. C. Woodville and M. D. Rausch, *J. Am. chem. Soc.*, 91 (1969) p. 5649.
127. A. J. Campbell, C. A. Fyfe, R. G. Goel, E. Maslowsky and C. V. Senoff, *J. Am. chem. Soc.*, 94 (1972) p. 8387.
128. C. H. Campbell and M. L. H. Green, *J. chem. Soc. A* (1971) p. 3282.
129. B. Floris, G. Illuminati and G. Ortaggi, *Chem. Commun.* (1969) p. 492.
130. D. Seyferth and R. L. Lambert, *J. organometallic Chem.*, 16 (1969) p. 21.
131. D. Seyferth and C. K. Haas, *J. organometallic Chem.*, 46 (1972) p. C33.
132. D. Seyferth and D. C. Mueller, *J. Am. chem. Soc.*, 93 (1971) p. 3714.
133. D. Seyferth, R. A. Woodruff, D. C. Mueller and R. L. Lambert, *J. organometallic Chem.*, 43 (1972) p. 55.
134. N. G. Johansson, *Acta Chem. Scand.*, 27 (1973) p. 1417.

135. D. Seyferth and R. A. Woodruff, *J. organometallic Chem.*, 71 (1974) p. 335.
136. D. Seyferth and H. D. Simmons, *J. organometallic Chem.*, 6 (1966) p. 306.
137. D. Seyferth and S. P. Hopper, *J. organometallic Chem.*, 51 (1973) p. 77.
138. M. Fedarynski and M. Makosza, *J. organometallic Chem.*, 51 (1973) p. 89.
139. G. A. Razuvaev, V. I. Shcherbakov and S. F. Zhiltsov, *Izv. Akad. Nauk SSSR Ser. Khim.* (1968) p. 2803.
140. A. J. Bloodworth, *J. chem. Soc. C* (1970) p. 2051.
141. J. R. Johnson and W. L. McEwen, *J. Am. chem. Soc.*, 48 (1926) p. 469.
142. R. J. Spahr, R. R. Vogt and J. A. Nieuwland, *J. Am. chem. Soc.*, 55 (1933) p. 3728.
143. F. G. Kleiner and W. P. Neumann, *Annalen*, 716 (1968) p. 19.
144. G. Eglinton and W. McCrae, *J. chem. Soc.* (1963) p. 2295.
145. R. E. Dessy, W. L. Budde and C. Woodruff, *J. Am. chem. Soc.*, 84 (1962) p. 1172.
146. J. M. Coxon, M. P. Hartshorn and A. J. Lewis, *Tetrahedron*, 26 (1970) p. 3755.
147. R. G. Smith, H. E. Ensley and H. E. Smith, *J. org. Chem.*, 37 (1972) p. 4430.
148. W. Kitching, *Organometallic Chem. Rev. A*, 3 (1968) p. 61; *Organometallic Reactions*, 3 (1972) p. 319.
149. J. Chatt, *Chem. Rev.*, 48 (1951) p. 1.
150. C. W. Whitehead and J. J. Traverso, *J. Am. chem. Soc.*, 80 (1958) p. 2182.
151. R. M. Carlson and A. H. Funk, *Tetrahedron Letters* (1971) p. 3661.
152. T. Sugita, Y. Yamasaki, O. Itoh and K. Ichikawa, *Bull. Chem. Soc. Japan*, 47 (1974) p. 1945.
153. H. C. Brown and P. J. Geoghegan, *J. org. Chem.*, 37 (1972) p. 1937.
154. R. D. Bach and R. F. Richter, *Tetrahedron Letters* (1973) p. 4099.
155. H. Arzoumanian and J. Metzger, *J. organometallic Chem.*, 57 (1973) p. C1.
156. G. B. Bachman and M. L. Whitehouse, *J. org. Chem.*, 32 (1967) p. 2303.
157. V. I. Sokolov and O. A. Reutov, *Izv. Akad. Nauk SSSR, Ser. Khim.* (1967) p. 1632.
158. D. Chow, J. H. Robsen and G. F. Wright, *Canad. J. Chem.*, 43 (1965) p. 312.
159. J. Beger and D. Vogel, *J. prakt. Chem.*, 311 (1969) p. 737.
160. K. Ichikawa, S. Fukushima, H. Ouchi and M. Tsuchida, *J. Am. chem. Soc.*, 80 (1958) p. 6005.
161. K. Ichikawa, H. Ouchi and S. Fukushima, *J. org. Chem.*, 24 (1959) p. 1129.
162. K. Ichikawa, S. Fukushima, H. Ouchi and M. Tsuchida, *J. Am. chem. Soc.*, 81 (1959) p. 3401.
163. R. D. Bach, R. N. Brummel and R. F. Richter, *Tetrahedron Letters* (1971) p. 2879.
164. M. Matsuo and Y. Saito, *J. organometallic Chem.*, 27 (1971) p. C41.
165. K. Ichikawa, K. Fujita and H. Ouchi, *J. Am. chem. Soc.*, 81 (1959) p. 5316.
166. K. Ichikawa, O. Itoh, T. Kawamura, M. Fujiwara and T. Ueno, *J. org. Chem.*, 31 (1966) p. 447.
167. A. G. Brook and G. F. Wright, *Canad. J. Res.*, 28 (1950) p. 623.
168. H. C. Brown and M.-H. Rei, *Chem. Commun.*, 96 (1969) p. 1296.
169. L. E. Overman, *J. Am. chem. Soc.*, 96 (1974) p. 597.
170. J. G. Traynham, G. R. Franzen, G. A. Knesel and D. J. Northington, *J. org. Chem.*, 32 (1967) p. 3285.
171. F. J. McQuillin and D. G. Parker, *J. chem. Soc. Perkin I* (1974) p. 809.
172. M. Julia and J.-D. Fourneron, *Tetrahedron Letters* (1973) p. 3429.
173. M. T. Mustafaeva, V. A. Smit, A. V. Semenovskii and V. F. Kucherov, *Izv. Akad. Nauk SSSR, Ser. Khim.* (1973) p. 1151.

174. K. C. Pande and S. Winstein, *Tetrahedron Letters* (1964) p. 3393.
175. A. J. Bloodworth and M. E. Loveitt, *Chem. Commun.* (1976) p. 94 and unpublished work.
176. M. C. Cabaleiro, A. D. Ayala and M. D. Johnson, *J. chem. Soc. Perkin II* (1973) p. 1207.
177. J. Halpern and H. B. Tinker, *J. Am. chem. Soc.*, 89 (1967) p. 6427.
178. P. Abley, J. E. Byrd and J. Halpern, *J. Am. chem. Soc.*, 95 (1973) p. 2591.
179. J. Romeyn and G. F. Wright, *J. Am. chem. Soc.*, 69 (1947) p. 697.
180. A. K. Chaudhuri, K. L. Mallik and M. N. Das, *Tetrahedron*, 19 (1963) p. 1981.
181. R. D. Bach and R. F. Richter, *J. org. Chem.*, 38 (1973) p. 3442.
182. I. V. Bodrikov, V. R. Kartashov, L. I. Koval'ova and N. S. Zefirov, *J. organometallic Chem.*, 82 (1974) p. C23.
183. J.-E. Bäckvall and B. Åkermark, *J. organometallic Chem.*, 78 (1974) p. 177.
184. A. J. Bloodworth and I. M. Griffin, *J. chem. Soc. Perkin I* (1974) p. 688.
185. M. M. Krevoy, L. C. Schaleger and J. C. Ware, *Trans. Faraday Soc.*, 58 (1962) p. 2433.
186. A. G. Brook and G. F. Wright, *Acta Cryst.*, 4 (1951) p. 50.
187. M. M. Anderson and P. M. Henry, *Chem. and Ind.* (1961) p. 2053.
188. S. Wolfe and P. G. C. Campbell, *Canad. J. Chem.*, 43 (1965) p. 1184.
189. H. C. Brown, M.-H. Rei and K.-T. Liu, *J. Am. chem. Soc.*, 92 (1970) p. 1760.
190. A. J. Bloodworth and I. M. Griffin, *J. chem. Soc. Perkin II* (1975) p. 531.
191. T. Ibusuki and Y. Saito, *J. organometallic Chem.*, 56 (1973) p. 103.
192. T. Ibusuki and Y. Saito, *Org. Mag. Res.*, 6 (1974) p. 436.
193. W. L. Waters, T. G. Traylor and A. Factor, *J. org. Chem.*, 38 (1973) p. 2306 and references therein.
194. T. G. Traylor, *J. Am. Chem. Soc.*, 86 (1964) p. 244.
195. M. Sakai, *Tetrahedron Letters* (1973) p. 347.
196. V. I. Sokolov, L. L. Troitskaya and O. A. Reutov, unpublished work quoted in reference 193.
197. R. D. Bach and R. F. Richter, *J. Am. chem. Soc.*, 93 (1972) p. 4747.
198. A. Factor and T. G. Traylor, *J. org. Chem.*, 33 (1968) p. 2615.
199. T. Sasaki, K. Kanematsu, A. Kondo and Y. Nishitani, *J. org. Chem.*, 39 (1974) p. 3569.
200. E. Schmitz, A. Rieche and O. Brede, *J. prakt. Chem.*, 312 (1970) p. 30.
201. V. I. Sokolov, *Izv. Akad. Nauk SSSR, Ser. Khim.*, (1972) p. 1089.
202. W. Schöller, W. Schrauth and W. Esser, *Chem. Ber.*, 46 (1913) p. 2864.
203. H. Bilka, G. Collin, C. Duschek, W. Höbold, R. Höhn, W. Pritzkow, H. Schmidt and D. Schnurpfeil, *J. prakt. Chem.*, 311 (1969) p. 1037.
204. E. Müller, *Chem. Ber.*, 106 (1973) p. 3920.
205. G. Schroeder, U. Prange, B. Putze, J. Thio and J. F. M. Oth, *Chem. Ber.*, 104 (1971) p. 3406.
206. N. S. Zefirov and L. G. Gurvich, *J. organometallic Chem.*, 81 (1974) p. 309.
207. H. B. Henbest and B. Nichols, *J. chem. Soc.* (1959) p. 227.
208. W. L. Waters and E. F. Kiefer, *J. Am. chem. Soc.*, 89 (1967) p. 6261.
209. R. K. Sharma, B. A. Shoulders and P. D. Gardner, *J. org. Chem.*, 32 (1967) p. 241.
210. W. S. Linn, W. L. Waters and M. C. Caserio, *J. Am. chem. Soc.*, 92 (1970) p. 4018.
211. A. J. Bloodworth and R. J. Bunce, *J. chem. Soc. C* (1971) p. 1453.
212. A. J. Bloodworth and R. J. Bunce, *J. organometallic Chem.*, 60 (1973) p. 11.
213. G. Spengler, M. Wilderotter and T. Trommer, *Brennstoff-Chem.*, 45 (1964) p. 182.

214. R. A. Alexander, N. C. Baenziger, C. Carpender and J. R. Doyle, *J. Am. chem. Soc.*, 82 (1960) p. 535.
215. R. D. Bach, *J. Am. chem. Soc.*, 91 (1969) p. 1771.
216. H. C. Brown and M.-H. Rei, *J. Am. chem. Soc.*, 91 (1969) p. 5646.
217. A. Factor and T. G. Traylor, *J. org. Chem.*, 33 (1968) p. 2607.
218. L. A. Paquette and P. C. Storm, *J. org. Chem.*, 35 (1970) p. 3390.
219. M. F. Grundon, D. Stewart and W. E. Watts, *Chem. Commun.* (1973) p. 573.
220. R. S. Bly, R. K. Bly, O. A. Bedenbaugh and O. R. Vail, *J. Am. chem. Soc.*, 89 (1967) p. 881.
221. H. C. Brown, P. J. Geoghegan, J. T. Kurek and G. L. Lynch, *Organometal. chem. Synth.*, 1 (1970) p. 7.
222. A. G. Brook, A. Rodgman and G. F. Wright, *J. org. Chem.*, 17 (1952) p. 988.
223. A. N. Nesmeyanov and I. F. Lutsenko, *Izv. Akad. Nauk SSSR, Otdel. Khim. Nauk* (1943) p. 296.
224. R. K. Summerbell and E. S. Polbacki, *J. org. Chem.*, 27 (1962) p. 2074.
225. D. H. Ballard and A. J. Bloodworth, *J. chem. Soc. C* (1971) p. 945.
226. E. Schmitz and O. Brede, *J. prakt. Chem.*, 312 (1970) p. 43.
227. R. L. Rowland, W. L. Perry and H. L. Friedman, *J. Am. chem. Soc.*, 73 (1951) p. 1040.
228. O. A. El Seoud, A. T. do Amaral, M. M. Campos and L. do Amaral, *J. org. Chem.*, 39 (1974) p. 1915.
229. R. Kh. Freidlina and N. S. Kochetkova, *Izv. Akad. Nauk SSSR, Otdel Khim. Nauk* (1945) p. 128.
230. N. S. Kochetkova, R. Kh. Freidlina and A. N. Nesmeyanov, *Izv. Akad. Nauk SSSR, Otdel Khim. Nauk* (1947) p. 347.
231. K. Toman and G. G. Hess, *J. organometallic Chem.*, 49 (1973) p. 133.
232. A. Lattes and J. J. Périé, *Compt. Rend. C* 262 (1966) p. 1591.
233. J. J. Périé and A. Lattes, *Bull. Chim. Soc. France* (1970) p. 583.
234. G. Wendt, B. V. Shetty and W. F. Bruce, *J. Am. chem. Soc.*, 81 (1959) p. 423.
235. H. K. Hall, J. P. Schaefer and R. J. Spanggord, *J. org. Chem.*, 37 (1972) p. 3069.
236. J. J. Périé, J.-P. Laval, J' Roussel and A. Lattes, *Tetrahedron Letters* (1971) p. 4399.
237. J. J. Périé, J.-P. Laval, J. Roussel and A. Lattes, *Tetrahedron*, 28 (1972) p. 675.
238. A. N. Nesmeyanov, A. E. Borisov and V. D. Vil'chevskaya, *Izv. Akad. Nauk SSSR, Otdel Khim. Nauk* (1954) p. 1008.
239. G. Drefahl, G. Heublein and A. Wintzer, *Angew. Chem.*, 70 (1958) p. 166.
240. A. N. Nesmeyanov, A. E. Borisov, I. S. Savel'eva and M. A. Osipova, *Izv. Akad. Nauk SSSR, Otdel. Khim. Nauk* (1961) p. 1249.
241. G. Chandra, D. Devaprabhakara and M. S. Muthana, *Curr. Sci.*, 40 (1971) p. 400.
242. W. W. Middleton, A. W. Barrett and J. H. Seager, *J. Am. chem. Soc.*, 52 (1930) p. 4405.
243. P. F. Hudrlik and A. M. Hudrlik, *J. org. Chem.*, 38 (1973) p. 4254.
244. A. N. Nesmeyanov and R. Kh. Freidlina, *Doklady Akad. Nauk SSSR*, 24 (1940) p. 59.
245. A. N. Nesmeyanov and N. K. Kochetkova, *Izv. Akad. Nauk SSSR, Otdel Khim. Nauk* (1949) p. 76.
246. A. Fabrycy and Z. Wichert, *Rocz. Chem.*, 42 (1968) p. 35.
247. H. A. Staab and J. Ipaktschi, *Chem. Ber.*, 104 (1971) p. 1170.

248. R. Ya. Levina and B. M. Gladstein, *Doklady Akad. Nauk SSSR*, 71 (1950) p. 65.
249. B. A. Arbusov, V. V. Ratuev, Z. G. Isaeva and E. K. Kazakova, *Izv. Akad. Nauk SSSR, Ser. Khim.* (1972) p. 385.
250. Yu. S. Shabarov, S. N. Burenko and T. S. Shulman, *Zhur. Obshch. Khim.*, 42 (1972) p. 1310.
251. F. R. Jensen, D. B. Patterson and S. E. Dinizo, *Tetrahedron Letters* (1974) p. 1315.
252. C. H. De Puy and R. H. McGirk, *J. Am. chem. Soc.*, 95 (1973) p. 2366; 96 (1974) p. 1121.
253. A. De Boer and C. H. De Puy, *J. Am. chem. Soc.*, 92 (1970) p. 4008.
254. V. I. Sokolov, N. B. Rodina and O. A. Reutov, *Zhur. Org. Khim.*, 3 (1967) p. 2089.
255. E. Müller, *Tetrahedron Letters* (1973) p. 1203.
256. T. Schirafuji and H. Nozaki, *Tetrahedron*, 29 (1973) p. 77.
257. F. F. Gadallah, A. A. Cantu and R. M. Elofson, *J. org. Chem.*, 38 (1973) p. 2386.
258. F. R. Jensen and L. H. Gale, *J. Am. chem. Soc.*, 82 (1960) p. 145.
259. J. Grimshaw and J. J. Ramsey, *J. chem. Soc. B* (1968) p. 60.
260. O. A. Reutov and O. A. Ptitsyna, *Organometallic Reactions*, 4 (1972) p. 73.
261. Yu. A. Ol'dekop, N. A. Maier and Yu. D. But'ko, *Zhur. Obshch. Khim.*, 41 (1971) p. 2253.
262. Yu. A. Ol'dekop, N. A. Maier, Yu. D. But'ko and M. S. Mindel, *Zhur. Obshch. Khim.*, 41 (1971) p. 828.
263. Yu. A. O'dekop, N. A. Maier, Yu. D. But'ko and M. S. Mindel, *Zhur. Obshch. Khim.*, 41 (1971) p. 1066.
264. Yu. A. Ol'dekop, N. A. Maier and Yu. D. But'ko, *Zhur. Obshch. Khim.*, 41 (1971) p. 2047.
265. Yu. A. Ol'dekop, N. A. Maier and A. A. Erdman, *Vestsi Akad. Navuk Belarus. SSR, Ser. Khim. Navuk.* (1971) p. 74; *Chem. Abstr.*, 74 (1971) p. 124585s.
266. G. B. Deacon, *Organometallic Chem. Rev. A*, 5 (1970) p. 355.
267. J. E. Connett, A. G. Davies, G. B. Deacon and J. H. S. Green, *J. chem. Soc. C* (1966) p. 106.
268. G. B. Deacon and P. W. Felder, *J. chem. Soc. C* (1967) p. 2313.
269. P. Sartori and A. Golloch, *Chem.. Ber.*, 101 (1967) p. 2004.
270. B. A. Tertov, A. V. Koblik and P. P. Onishchenko, *Zhur. Obshch. Khim.*, 44 (1974) p. 628.
271. D. Seyferth, S. P. Hopper and G. J. Murphy, *J. organometallic Chem.*, 46 (1972) p. 201.
272. I. L. Knunyants, Y. F. Komissorov, B. L. Dyatkin and L. T. Lantseva, *Izv. Akad. Nauk SSSR, Ser. Khim.* (1973) p. 943.
273. J. E. Connett and G. B. Deacon, *J. chem. Soc. C* (1966) p. 1058.
274. G. B. Deacon and P. W. Felder, *Austral. J. Chem.*, 23 (1970) p. 1359.
275. R. J. Bertino, P. G. Cookson, G. B. Deacon and I. K. Johnson, *J. Fluorine Chem.*, 3 (1973/74) p. 122.
276. P. G. Cookson and G. B. Deacon, *Austral. J. Chem.*, 24 (1971) p. 1599.
277. R. J. Cross and N. H. Tennent, *J. organometallic Chem.*, 61 (1973) p. 33; 72 (1974) p. 21.
278. P. G. Cookson and G. B. Deacon, *Austral. J. Chem.*, 26 (1973) p. 541.
279. P. G. Cookson and G. B. Deacon, *Austral. J. Chem.*, 26 (1973) p. 1893.
280. D. Seyferth and D. C. Mueller, *J. organometallic Chem.*, 28 (1971) p. 325.

281. D. N. Kravtsov, B. A. Kvasov, L. S. Golovchenko and E. I. Fedin, *J. organo-metallic Chem.*, 36 (1972) p. 227.
282. L. Hellerman and M. D. Newman, *J. Am. chem. Soc.*, 54 (1932) p. 2859.
283. R. Scheffold and U. Michel, *Angew. Chem.*, 84 (1972) p. 160.
284. R. Kh. Freidlina and F. K. Velichko, *Izv. Akad. Nauk SSSR Otdel. Khim. Nauk* (1959) p. 1225.
285. P. Pfeiffer, R. Schulze-Bentrop, K. H. La Roche and E. Schmitz, *Chem. Ber.*, 85 (1952) p. 232.
286. R. D. Chambers and D. J. Spring, *J. organometallic Chem.*, 31 (1971) p. C13.
287. A. K. Prokofev and O. Yu. Okhlobystin, *J. organometallic Chem.*, 36 (1972) p. 239.
288. P. V. Roling, S. B. Roling and M. D. Rausch, *Synth. inorg. metalorg. Chem.*, 1 (1971) p. 97.
289. R. J. Cross and C. M. Jenkins, *J. organometallic Chem.*, 56 (1973) p. 125.
290. R. C. Wade and D. Seyferth, *J. organometallic Chem.*, 22 (1970) p. 265.
291. O. A. Reutov, I. P. Beletskaya and R. E. Mardaleishirli, *Zhur. Fiz. Khim.*, 33 (1959) pp. 152, 1962.
292. F. R. Jensen, B. Rickborn and J. J. Miller, *J. Am. chem. Soc.*, 88 (1966) p. 340.
293. A. N. Nesmeyanov, O. A. Reutov and S. S. Poddubnaya, *Izv. Akad. Nauk SSSR Otdel. Khim. Nauk* (1953) p. 850.
294. G. F. Wright, *Canad. J. Chem.*, 30 (1952) p. 268.
295. H. B. Charman, E. D. Hughes and C. K. Ingold, *J. chem. Soc.*, (1959) p. 2530.
296. R. E. Dessy and Y. K. Lee, *J. Am. chem. Soc.*, 82 (1960) p. 689.
297. R. E. Dessy, Y. K. Lee and J.-Y. Kim, *J. Am. chem. Soc.*, 83 (1961) p. 1163.
298. V. S. Petrosyan, S. M. Sakembaeva and O. A. Reutov, *Izv. Akad. Nauk SSSR Ser. Khim.* (1973) p. 1403.
299. H. O. House, R. A. Auerbach, M. Gall and N. P. Peet, *J. org. Chem.*, 38 (1973) p. 514.
300. A. N. Nesmeyanov, T. P. Tolstaya and V. V. Korotkov, *Doklady Akad. Nauk SSSR*, 209 (1973) p. 1113.
301. V. R. Policschuk, L. S. German and I. L. Knunyants, *Izv. Akad. Nauk SSSR Ser. Khim.* (1971) p. 2024.
302. T. G. Traylor and A. W. Baker, *J. Am. chem. Soc.*, 85 (1963) p. 2746.
303. I. A. Esikova, O. N. Temkin, A. P. Tomilov, G. P. Pavlikova and R. M. Flid, *Elektrochimiya*, 6 (1970) p. 743.
304. R. Benesch and R. E. Benesch, *J. phys. Chem.*, 57 (1952) p. 648.
305. M. L. Bullpitt and W. Kitching, *J. organometallic Chem.*, 46 (1972) p. 21.
306. F. R. Jensen and J. A. Landgrebe, *J. Am. chem. Soc.*, 82 (1960) p. 1004.
307. V. I. Buzulukov, V. P. Maslennikov and Yu. A. Aleksandrov, *Zhur. Obshch. Khim.*, 42 (1972) p. 2583.
308. M. Dadić and D. Grdenić, *Croat. chem. acta*, 32 (1960) p. 39.
309. J. L. Wardell, R. D. Taylor and T. J. Lillic, *J. organometallic Chem.*, 33 (1971) p. 25.
310. R. M. G. Roberts, *J. organometallic Chem.*, 47 (1973) p. 359.
311. T. N. Mitchell, *J. organometallic Chem.*, 71 (1974) p. 27.
312. A. J. Bloodworth, *J. organometallic Chem.*, 23 (1970) p. 27.
313. D. Grdenić and F. Zado, *J. chem. Soc.* (1962) p. 521.
314. J. Lorberth and F. Weller, *J. organometallic Chem.*, 32 (1971) p. 145.
315. D. Breitinger and N. Quy Dao, *J. organometallic Chem.*, 15 (1968) p. P21.
316. R. J. Kline, L. F. Sytsma and D. R. Shackle, *Inorg. nucl. Chem. Lett.*, 7 (1971) p. 331.

317. A. S. Peregudov, D. N. Kravtsov and L. A. Fedorov, *J. organometallic Chem.*, 71 (1974) p. 347.
318. D. N. Krabtsov and A. N. Nesmeyanov, *Izv. Akad. Nauk SSSR, Ser. Khim.* (1967) p. 1487.
319. L. F. Sytsma and R. J. Kline, *J. organometallic Chem.*, 54 (1973) p. 15.
320. D. N. Kravtsov and A. N. Nesmeyanov, *Izv. Akad. Nauk SSSR Ser. Khim.* (1967) p. 1747.
321. S. F. Zhil'tsov and V. I. Shcherbakov, *Izv. Akad. Nauk SSSR Ser. Khim.* (1973) p. 474.
322. A. J. Bloodworth and M. E. Loveitt, Unpublished work; '*N*-phenylmercurio-pyrrole is a pale yellow solid m.p. 158°'.
323. W. Thiel, F. Weller, J. Lorberth and K. Dehnicke, *Z. anorg. allg. Chem.*, 381 (1971) p. 57.
324. J. Lorberth, *J. organometallic Chem.*, 71 (1974) p. 159.
325. J. Lorberth and G. Lange, *J. organometallic Chem.*, 54 (1973) p. 165.
326. I. A. Koten and R. Adams, *J. Am. chem. Soc.*, 46 (1924) p. 2764.
327. V. S. Petrosyan and O. A. Reutov, *J. organometallic Chem.*, 76 (1974) p. 123.
328. F. A. L. Anet, J. Krane, W. Kitching, D. Doddrell and D. Praeger, *Tetrahedron Letters* (1974) p. 3255.
329. W. Adcock, B. D. Gupta, W. Kitching, D. Doddrell and M. Geckle, *J. Am. chem. Soc.*, 96 (1974) p. 7360.
330. W. Kitching, D. Praeger, D. Doddrell, F. A. L. Anet and J. Krane, *Tetrahedron Letters* (1975) p. 759.
331. K. Kashiwabara, S. Konaka, T. Iijima and M. Kimura, *Bull. Soc. Chem. Japan*, 46 (1973) p. 407.
332. D. Grdenić, *Quart. Rev.*, 19 (1965) p. 303.
333. A. G. Robiette (1973), *Molecular Structure by Diffraction Methods*, vol. 1, ed. G. A. Sim and L. E. Sutton (Chem. Soc., London, 1973) p. 160.
334. L. V. Vilkov, M. G. Anashkin and G. I. Mamaeva, *Zhur. strukt. Khim.*, 9 (1968) p. 372.
335. C. Walls, D. G. Lister and J. Sheridan, *J. chem. Soc. Faraday II*, 71 (1975) p. 1091.
336. L. V. Vilkov and M. G. Anashkin, *Zhur. Strukt. Khim.*, 9 (1968) p. 690.
337. I. A. Ronova, O. Yu. Okhlobystin, Yu. T. Struchkov and A. K. Prokof'ev, *Zhur. Strukt. Khim.*, 13 (1972) p. 195.
338. L. G. Kuz'mina, N. G. Boki, Yu. T. Struchkov, D. N. Kravtsov and E. M. Rokhlina, *Zhur. strukt. Khim.*, 15 (1974) p. 491.
339. B. Ziolkovska, R. M. Myasnikova and A. I. Kitaigorodskii, *Zhur. strukt. Khim.*, 5 (1964) p. 737.
340. M. Matthew and N. R. Kunchur, *Canad. J. Chem.*, 48 (1970) p. 429.
341. N. R. Kunchur and M. Matthew, *Chem. Commun.* (1966) p. 71.
342. D. Grdenić, *Chem. Ber.*, 92 (1959) p. 231.
343. K. W. Küpper and H. J. Lindner, *Z. anorg. allg. Chem.*, 359 (1968) p. 41.
344. P. Luger and G. Ruben, *Acta Cryst. B*, 27 (1971) p. 2276.
345. D. Grdenić, *Acta Cryst.*, 5 (1952) p. 367.
346. D. Grdenić and A. I. Kitaigorodskii, *Zhur. fiz. Khim.*, 23 (1949) p. 1161.
347. J. C. Mills, H. S. Preston and C. H. L. Kennard, *J. organometallic Chem.*, 14 (1968) p. 33.
348. U. Müller, *Z. Naturforsch. B.*, 28 (1973) p. 426.
349. G. Dittmar and E. Hellner, *Angew. Chem.*, 81 (1969) p. 707.
350. Y. S. Wong, P. C. Chieh and A. J. Carty, *Chem. Commun.* (1973) p. 741.
351. Y. S. Wong, P. C. Chieh and A. J. Carty, *Canad. J. Chem.*, 51 (1973) p. 2597.

352. Y. S. Wong, N. J. Taylor, P. C. Chieh and A. J. Carty, *Chem. Commun.* (1974) p. 625.
353. T. A. Babushkina, E. V. Bryukhova, F. K. Velichko, V. I. Pakhomov, and G. K. Semin, *Zhur. strukt. Khim.*, 9 (1968) p. 207.
354. V. I. Pakhomov and A. I. Kitaigorodski, *Zhur. strukt. Khim.*, 7 (1966) p. 860.
355. R. D. Bach, A. T. Weibel, W. Schmonsees and M. D. Glick, *Chem. Commun.* (1974) p. 961.
356. L. G. Kuz'mina, N. G. Bokii, M. I. Rybinskaya, Yu. T. Struchkov and T. V. Popova, *Zhur. strukt. Khim.*, 12 (1971) p. 1026.
357. V. I. Pakhomov, *Kristallografiya*, 7 (1962) p. 456.
358. P. W. Jennings, S. K. Reeder, J. C. Hurley, C. N. Caughlan and G. D. Smith, *J. org. Chem.*, 39 (1974) p. 3392.
359. V. I. Pakhomov, A. V. Medvedev, V. I. Bregadze and O. Yu. Okhlobystin, *J. organometallic Chem.*, 29 (1971) p. 15.
360. D. Grdenić, B. Kamenar, B. Korpar-Colig, M. Sikirica and G. Jovanovski, *Chem. Commun.* (1974) p. 646.
361. V. I. Pakhomov, *Zhur. strukt. Khim.*, 4 (1963) p. 594.
362. B. Kamenar and M. Penavić, *Inorg. Chim. Acta*, 6 (1972) p. 191.
363. L. G. Kuz'mina, N. G. Bokii, Yu. T. Struchkov, D. N. Kravtsov and L. S. Golovchenko, *Zhur. strukt. Khim.*, 14 (1973) p. 508.
364. L. G. Kuz'mina, N. G. Bokii, Yu. T. Struchkov, V. I. Minkin, L. P. Olekhnovich and I. E. Mikhailov, *Zhur. strukt. Khim.*, 15 (1974) p. 659.
365. V. I. Pakhomov, *Zhur. strukt. Khim.*, 5 (1964) p. 873.
366. Y. Kobayashi, Y. Iitaka and Y. Kido, *Bull. chem. Soc. Japan*, 43 (1970) p. 3070.
367. V. I. Pakhomov, *Kristallografiya*, 8 (1963) p. 789.
368. A. J. Canty and B. M. Gatehouse, *Acta Cryst. B*, 28 (1972) p. 1872.
369. A. J. Canty and B. M. Gatehouse, *J. chem. Soc. Dalton* (1972) p. 511.
370. A. D. Redhouse, *Chem. Commun.* (1972) p. 1119.
371. E. D. Hughes, C. K. Ingold, F. G. Thorpe and H. C. Volger, *J. chem. Soc.* (1961) p. 1133.
372. M. H. Abraham, D. Dodd, M. D. Johnson, E. S. Lewis and R. A. More O'Ferrall, *J. chem. Soc. B* (1971) p. 762.
373. M. H. Abraham, P. L. Grellier and M. J. Hogarth, *J. chem. Soc. Perkin II* (1974) p. 1613.
374. E. D. Hughes, C. K. Ingold and R. M. G. Roberts, *J. chem. Soc.* (1964) p. 3900.
375. I. P. Beletskaya, K. P. Butin and O. A. Reutov, *Organometallic Chem. Rev. A*, 7 (1971) p. 51.
376. B. Praisnor, I. P. Beletskaya, V. I. Sokolov and O. A. Reutov, *Izv. Akad. Nauk SSSR Ser. Khim.* (1963) p. 970.
377. V. A. Kalyavin, T. A. Smolina and O. A. Reutov, *Doklady Akad. Nauk SSSR*, 156 (1964) p. 95.
378. I. P. Beletskaya, I. I. Zakharycheva and O. A. Reutov, *Doklady Akad. Nauk SSSR*, 195 (1970) p. 837.
379. H. Arzoumanian and J. Metzger, *Synthesis* (1971) p. 527.
380. L. G. Makarova, *Organometallic Reactions*, ed. E. I. Becker and M. Tsutsui (Wiley-Interscience, New York, 1971), vol. 1, p. 119 and vol. 2, p. 335.
381. J. L. Atwood, B. L. Bailey, B. L. Kindberg and W. J. Cook, *Austral. J. Chem.*, 26 (1973) p. 2297.
382. G. E. Coates and M. Tranach, *J. chem. Soc. A* (1967) p. 615.

383. S. W. Breuer, F. G. Thorpe and J. C. Podestá, *Tetrahedron Letters* (1974) p. 3719.
384. A. Storr and V. G. Wiebe, *Canad. J. Chem.*, 47 (1969) p. 673.
385. A. N. Nesmeyanov, A. E. Borisov and A. N. Abramova, *Izv. Akad. Nauk SSSR Otdel. Khim. Nauk* (1949) p. 570.
386. D. May, J. P. Oliver and M. T. Emerson, *J. Am. chem. Soc.*, 86 (1964) p. 371.
387. H. E. Zieger and J. D. Roberts, *J. org. Chem.*, 34 (1969) p. 2826.
388. V. I. Sokolov, V. V. Bashilov, L. M. Anishchenko and O. A. Reutov, *J. organometallic Chem.*, 71 (1974) p. C41.
389. J. Yamamoto and C. A. Wilke, *Inorg. Chem.*, 10 (1971) p. 1129.
390. A. J. Hart, D. H. O'Brien and C. R. Russell, *J. organometallic Chem.*, 72 (1974) p. C19.
391. I. E. Paleeva, N. I. Sheverdina, M. A. Zemlyanichenko and K. A. Kocheshkov, *Doklady Akad. Nauk SSSR*, 210 (1973) p. 1134.
392. J. J. Eisch and J.-M. Biedermann, *J. organometallic Chem.*, 30 (1971) p. 167.
393. K. Margiolis and K. Dehnicke, *J. organometallic Chem.*, 33 (1971) p. 147.
394. G. B. Deacon and J. C. Parrot, *Austral. J. Chem.*, 24 (1971) p. 1771.
395. A. Finch, P. J. Gardner, N. Hill and K. S. Hussian, *J. chem. Soc. Dalton* (1973) p. 2543.
396. A. N. Nesmeyanov, T. P. Tolstaya and V. V. Korol'kov, *Doklady Akad. Nauk SSSR*, 209 (1973) p. 1113.
397. V. A. Chauzov, V. M. Vodolazskaya, N. S. Kitaeva and Yu. I. Baukov, *Zhur. obshch. Khim.*, 43 (1973) p. 597.
398. C. Santini-Scampucci and J. G. Riess, *J. chem. Soc. Dalton* (1973) p. 2436.
399. A. Z. Rubezhov, A. S. Ivanov and S. P. Gubin, *Izv. Akad. Nauk SSSR Ser. Khim.* (1973) p. 951.
400. R. J. Cross and R. Wardle, *J. organometallic Chem.*, 23 (1970) p. C4.
401. R. J. Cross and R. Wardle, *J. chem. Soc. A* (1970) p. 841.
402. R. D. Brown, A. S. Buchanan and A. A. Humffray, *Austral. J. Chem.*, 18 (1965) pp. 1507, 1513.
403. S. Winstein and T. G. Traylor, *J. Am. chem. Soc.*, 77 (1955) p. 3747; 78 (1956) p. 2597.
404. M. M. Kreevoy and R. L. Hansen, *J. Am. chem. Soc.*, 83 (1961) p. 626.
405. J. R. Coad and M. D. Johnson, *J. chem. Soc. B* (1967) p. 633.
406. D. Dodd and M. D. Johnson, *J. chem. Soc. B* (1969) p. 1071.
407. I. P. Beletskaya, L. V. Savinykh and O. A. Reutov, *Doklady Akad. Nauk SSSR*, 197 (1971) p. 1325.
408. K. C. Bass, *Organometallic Chem. Rev.*, 1 (1966) p. 391.
409. T. Sakikibara and Y. Odaira, *J. org. Chem.*, 36 (1971) p. 3644.
410. W. R. R. Park and G. F. Wright, *Canad. J. Chem.*, 35 (1957) p. 1088.
411. J. J. Périé and A. Lattes, *Bull. Soc. Chim. France* (1971) p. 1378.
412. W. L. Weinberg, *Tetrahedron Letters* (1970) p. 4835.
413. J. H. Robson and G. F. Wright, *Canad. J. Chem.*, 38 (1960) p. 21.
414. F. G. Bordwell and M. L. Douglass, *J. Am. chem. Soc.*, 88 (1966) p. 993.
415. G. M. Whitesides and J. San Filippo, *J. Am. chem. Soc.*, 92 (1970) p. 6611.
416. W. R. Jackson, G. A. Gray and V. M. A. Chambers, *J. chem. Soc. C* (1971) p. 200.
417. V. M. A. Chambers, W. R. Jackson and G. W. Young, *J. chem. Soc. C* (1971) p. 2075.
418. C. L. Hill and G. M. Whitesides, *J. Am. chem. Soc.*, 96 (1974) p. 870.
419. R. P. Quirk, *J. org. Chem.*, 37 (1972) p. 3554.

420. R. P. Quirk and R. E. Lea, *Tetrahedron Letters* (1974) p. 1925.
421. B. Giese, S. Gantert, and A. Schulz, *Tetrahedron Letters* (1974) p. 3583.
422. D. Seyferth and R. J. Spohn, *Trans. New York Acad. Sci. Ser. II*, 33 (1971) p. 625.
423. G. T. Rodeheaver and D. F. Hunt, *Chem. Commun.* (1971) p. 818.
424. D. E. Bergbreiter and G. M. Whitesides, *J. Am. chem. Soc.*, 96 (1974) p. 4937.
425. (a) I. P. Beletskaya, O. A. Maksimenko, V. B. Vol'eva and O. A. Reutov, *Zhur. Org. Khim.*, 2 (1966) p. 1132. (b) I. P. Beletskaya, O. A. Maksimenko and O. A. Reutov, *Izv. Akad. Nauk SSSR Ser. Khim.* (1966) p. 662; *Doklady Akad. Nauk SSSR*, 168 (1966) p. 333.
426. A. L. Fridman, T. N. Ivshina, N. Ya. Volkova, V. P. Ivshin, V. A. Tartakovskii and S. S. Novikov, *Izv. Akad. Nauk SSSR Ser. Khim.* (1970) p. 1894.
427. M. A. Kazankova, L. I. Petrovskaya and I. F. Lutsenko, *Zhur. org. Khim.*, 7 (1971) p. 58.
428. G. A. Artamkina, I. P. Beletskaya and O. A. Reutov, *J. organometallic Chem.*, 42 (1972) p. C17.
429. G. Ahlgren, B. Akermark and M. Nilsson, *J. organometallic Chem.*, 30 (1971) p. 303.
430. E. Vedejs and P. D. Weeks, *Tetrahedron Letters* (1974) p. 3207.
431. M. O. Unger and R. A. Fouty, *J. org. Chem.*, 34 (1969) p. 18.
432. R. F. Heck, *J. Am. chem. Soc.*, 90 (1968) pp. 5518, 5526, 5531, 5535, 5538, 5542; 91 (1969) p. 6707; 93 (1971) p. 6896; 94 (1972) p. 2712; *J. organometallic Chem.*, 33 (1971) p. 399.
433. G. A. Shvekhgeimer, N. I. Sobtsova and A. Baranski, *Rocz. Chem.*, 47 (1973) p. 1243.
434. U. Blaukat and W. P. Neumann, *J. organometallic Chem.*, 49 (1973) p. 323.
435. T. N. Mitchell, *J. organometallic Chem.*, 71 (1974) p. 39.
436. G. A. Razuvaev, S. F. Zhil'tsov, Yu. A. Alexandrov and O. N. Drushkov, *Zhur. obshch. Khim.*, 35 (1965) pp. 1152, 1440 and earlier papers.
437. E. V. Uglova, V. D. Makhaev and O. A. Reutov, *Zhur. org. Khim.*, 9 (1973) p. 1304.
438. C. C. Lee, A. J. Cessna, E. C. F. Ko and S. Vassie, *J. Am. chem. Soc.*, 95 (1973) p. 5688.
439. W. Kitching, T. Sakakiyama, Z. Rappoport, P. D. Sleezer, S. Winstein and W. G. Young, *J. Am. chem. Soc.*, 94 (1972) p. 2329.
440. M. Matsuo and Y. Saito, *J. org. Chem.*, 37 (1972) p. 3350.
441. G. T. Rodeheaver and D. F. Hunt, *Chem. Commun.* (1971) p. 818.
442. W. L. Waters, P. E. Pike and J. G. Rivera, *A.C.S. Advances in Chemistry*, 112 (1972) p. 78.
443. R. C. Larock, *J. org. Chem.*, 39 (1974) p. 834.
444. W. Reid and W. Merkel, *Angew. Chem.*, 81 (1969) p. 400.
445. G. Marr and T. M. White, *J. chem. Soc. C*, (1970) p. 1789.
446. I. P. Beletskaya, O. A. Reutov and T. P. Gur'yanova, *Izv. Akad. Nauk SSSR Otdel. Khim. Nauk* (1961) pp. 1589, 1997.
447. O. A. Reutov, I. P. Beletskaya and T. P. Fetisova, *Doklady Akad. Nauk SSSR*, 166 (1966) p. 861.
448. O. A. Reutov, I. P. Beletskaya and T. A. Azizyan, *Izv. Akad. Nauk SSSR Otdel. Khim. Nauk* (1962) p. 424.
449. A. Lord and H. O. Pritchard, *J. phys. Chem.*, 70 (1966) p. 1689.
450. S. Winstein and T. G. Traylor, *J. Am. chem. Soc.*, 78 (1956) p. 2596.
451. I. P. Beletskaya, O. A. Reutov and T. A. Azizyan, *Izv. Akad. Nauk SSSR Otdel. Khim. Nauk* (1962) p. 223.

452. I. P. Beletskaya, O. A. Reutov and T. P. Gur'yanova, *Izv. Akad. Nauk SSSR Otdel. Khim. Nauk* (1961) p. 2178.
453. O. A. Reutov, E. V. Uglova, I. P. Beletskaya and T. B. Svetlanova, *Izv. Akad. Nauk SSSR Ser. Khim.* (1964) p. 1383.
454. F. R. Jensen and L. H. Gale, *J. Am. chem. Soc.*, 81 (1959) p. 1261; 82 (1960) p. 148.
455. F. R. Jensen, L. D. Whipple, D. K. Wedegaertner and J. A. Landgrebe, *J. Am. chem. Soc.*, 81 (1959) p. 1262; 82 (1960) p. 2466.
456. C. P. Casey, G. M. Whitesides and J. Kurth, *J. org. Chem.*, 38 (1973) p. 3406.
457. I. P. Beletskaya, L. V. Savinykh and O. A. Reutov, *J. organometallic Chem.*, 26 (1971) p. 13.
458. I. P. Beletskaya, L. V. Savinykh and O. A. Reutov, *Izv. Akad. Nauk SSSR Ser. Khim.* (1971) p. 1585.
459. I. P. Beletskaya, L. V. Savinykh, V. N. Gulyachkina and O. A. Reutov, *J. organometallic Chem.*, 26 (1971) p. 23.
460. E. Vedejes and M. F. Saloman, *Chem. Commun.* (1971) p. 1582.
461. A. J. Bloodworth and I. M. Griffin, *J. chem. Soc. Perkin I* (1975) p. 695.
462. H. C. Brown and P. Geoghegan, *J. Am. chem. Soc.*, 89 (1967) p. 1522; *J. org. Chem.*, 35 (1970) p. 1844.
463. H. C. Brown and W. J. Hammar, *J. Am. chem. Soc.*, 89 (1967) p. 1524.
464. H. C. Brown, J. H. Kawakami and S. Ikegami, *J. Am. chem Soc.*, 89 (1967) p. 1525.
465. D. H. Ballard, A. J. Bloodworth and R. J. Bunce, *Chem. Commun.* (1969) p. 815; see references 225 and 9.
466. H. C. Brown and J. T. Kurek, *J. Am. chem. Soc.*, 91 (1969) p. 5647.
467. C. H. Heathcock, *Angew. Chem.*, 81 (1969) p. 148.
468. S. C. Misra and G. Chandra, *Indian J. Chem.*, 11 (1973) p. 613.
469. J. W. Clark-Lewis and E. J. McGarry, *Austral. J. Chem*, 26 (1973) p. 2447.
470. P. Wilder, A. R. Partis, G. W. Wright and J. M. Sheppard, *J. org. Chem.*, 39 (1974) p. 1636.
471. A. J. Bloodworth and R. J. Bunce, *J. chem. Soc. Perkin I* (1972) p. 2787.
472. R. Gelin, S. Gelin and M. Albrand, *Bull. Soc. Chim. France* (1972) p. 1946.
473. L. A. Paquette and G. L. Thompson, *J. Am. chem. Soc.*, 94 (1972) p. 7118.
474. H. Hodjat, A. Lattes, J. P. Laval, J. Moulines and J. J. Périé, *J. heterocyclic Chem.*, 9 (1972) p. 1081.
475. V. Gómez Aranda, J. Barluenga, G. Arsensio and M. Yus, *Tetrahedron Letters* (1972) p. 3621.
476. A. J. Bloodworth and G. S. Bylina, *J. chem. Soc. Perkin I* (1972) p. 2433.
477. F. R. Jensen, J. J. Miller, S. J. Cristol and R. S. Beckley, *J. org. Chem.*, 37 (1972) p. 4341.
478. D. Seyferth, J. M. Burlitch and J. K. Heeren, *J. org. Chem.*, 27 (1962) p. 1491.
479. D. Seyferth, *Acc. Chem. Res.*, 5 (1972) p. 65.
480. D. Seyferth, J. Y.-P. Mui and J. M. Burlitch, *J. Am. chem. Soc.*, 89 (1967) p. 4953.
481. D. Seyferth, J. Y.-P. Mui and R. Damrauer, *J. Am. chem. Soc.*, 90 (1968) p. 6182.
482. D. Seyferth and J. M. Burlitch, *J. Am. chem. Soc.*, 86 (1964) p. 2730.
483. D. Seyferth and C. K. Haas, *J. organometallic Chem.*, 39 (1972) p. C41.
484. D. Seyferth, H. D. Simmons and G. Singh, *J. organometallic Chem.*, 3 (1965) p. 337.
485. D. Seyferth and S. P. Hopper, *J. org. Chem.*, 37 (1972) p. 4070.

486. D. Seyferth and G. J. Murphy, *J. organometallic Chem.*, 49 (1973) p. 117.
487. D. Seyferth and G. J. Murphy, *J. organometallic Chem.*, 52 (1973) p. C1.
488. D. Seyferth and D. C. Mueller, *J. organometallic Chem.*, 25 (1970) p. 293.
489. D. Seyferth and E. M. Hanson, *J. organometallic Chem.*, 27 (1971) p. 19.
490. D. Seyferth and R. A. Woodruff, *J. org. Chem.*, 38 (1973) p. 4031.
491. D. Seyferth, W. Tronich and H.-M. Shih, *J. org. Chem.*, 39 (1974) p. 158.
492. D. Seyferth, W. Tronich, W. E. Smith and S. P. Hopper, *J. organometallic Chem.*, 67 (1974) p. 341.
493. D. Seyferth, W. Tronich, R. S. Marmor and W. E. Smith, *J. org. Chem.*, 37 (1972) p. 1537.
494. D. Seyferth and H. Shih, *J. Am. chem. Soc.*, 95 (1973), 8464; *J. org. Chem.*, 39 (1974) pp. 2329, 2336.
495. D. Seyferth and Y. M. Cheng, *J. Am. chem. Soc.*, 95 (1973) p. 6763.
496. D. Seyferth, Y. M. Cheng and D. D. Traficante, *J. organometallic Chem.*, 46 (1972) p. 9.
497. L. H. Sommer, L. A. Ullard and G. A. Parker, *J. Am. chem. Soc.*, 94 (1972) p. 3469.
498. D. Seyferth, H. Shih, J. Dubac, P. Mazerolles and B. Serres, *J. organometallic Chem.*, 50 (1973) p. 39.
499. D. Seyferth, F. M. Armbrecht and B. Schneider, *J. Am. chem. Soc.*, 91 (1969) p. 1954.
500. D. Seyferth, J. K. Heeren, G. Singh, S. O. Grim and W. B. Hughes, *J. organometallic Chem.*, 5 (1966) p. 267.

PART 4

The Biochemistry and Toxicology of Mercury[†]

Kenneth H. Falchuk[‡], Leonard J. Goldwater and Bert L. Vallee

Biophysics Research Laboratory of the Department of Biological Chemistry, the Department of Medicine, Harvard Medical School;

Peter Bent Brigham Hospital, Boston, Massachusetts;

Department of Community Health Science, Duke University Medical Center, Durham, North Carolina

† This work was supported by Grant-in-Aid GM-15003 from the National Institute of Health, of the Department of Health, Education and Welfare.

‡ Investigator of the Howard Hughes Medical Institute.

24 Introduction

Mercury is distributed widely in the earth's crust, in sea, ground and rain water. Importantly, all phyla and species naturally contain traces, present either as inorganic or organometallic compounds, or both.[1,2] A biological role for the element is thus far undefined; however, it is present in rat liver chromatin[3] and methyl mercury induces hepatic protein synthesis in this species.[4] It has been known that mercury enters into biological life cycles; however, the awareness that inorganic mercury can be converted into organometallic compounds by bacteria and higher organisms has recently stimulated further interest both in the chemistry and toxicological potential of this element.[5-9] This biological conversion of inorganic mercury into organic mercury is particularly significant since extensive industrial and agricultural usage of mercurials affects and increases its distribution in specific regions. Moreover, the burning of fossil fuels generates environmental mercury in amounts comparable to those from industrial processes.[10]

The biochemical basis for the toxicological effects of mercury, both in its inorganic and organic forms, as well as a role in normal metabolism, if any, are probably dependent not only on dose but also on its interaction, *inter alia*, with thiol, selenide, phosphate, amino and carboxyl groups of such cellular components as amino acids, proteins, enzymes, nucleic acids and lipids. Detailed studies of the toxicology of mercury and the biochemistry of its interaction with cell constituents have recently been facilitated by the capability to quantitate its content in biological samples. This section will deal with the chemistry of this element, its analysis in biological samples, and a possible biochemical basis for its effect on proteins, nucleic acids, cells, tissues, and whole organisms.

Pertinent properties of this element are shown in table 24.1. Mercury, as the other group II B elements, is regarded as a non-transition element, being more electropositive, softer, having lower melting point and hence greater volatility. It is limited to oxidation states no higher than II due to the fact that its third ionisation potential is so high that solvation or lattice formation do not suffice to make oxidation state III stable. Mercury, as other metals, forms strong covalent bonds with atoms that donate electron pairs, thus, it readily complexes

Table 24.1

Boiling point ($^{\circ}$C)	357
Melting point ($^{\circ}$C)	−38.4
Density (g/ml)	13.6
Outer electron configuration	$(5d)^{10}(6s)^2$
Ionisation potentials (eV)	
1st	10.41
2nd	18.74
3rd	34.3
Ionic radii (nm)	
Pauling	0.110
Goldschmidt	0.093
Favoured geometry (high spin)	linear > tetrahedral > others

with ammonia, amines, halide and hydroxyl ions, and cyanide; and it reacts
with sulphur and other non-metals such as phosphorus and selenium. The mercury
ion has a strong tendency to complex formation, particularly two-coordinate
linear and four-coordinate tetrahedral complexes. Consequently, mercury would
be expected to bind strongly to biological substances containing free ammonia,
phosphates, sulphydryl and carboxyl groups, such as amino acids and proteins.
This has been well established and the stability constants for various ligands are
shown in table 24.2.

The analysis of inorganic mercury has been difficult, owing particularly to its
volatility, which may result in loss during separation of the metal from its matrix.
Moreover, until recently, methylmercury was both unsuspected and undetected
in biological materials owing to lack of awareness of its significance and the time

Table 24.2. Stability Constants of 1:1 Mercuric Complexes

Ligand	Log k_1
Cl^-	6.74[11]
Br^-	9.05[11]
I^-	12.87[11]
OH^-	10.3[12]
NH_3	8.8[13]
Imidazole	3.57[13]
Ethylenediamine (N–N)	14.3[13]
Cysteine (N–S)	45.4[13]
Glycine (N–O)	10.3[13]
Histidine	7.9[13]
Tris(2-amino-2-hydroxymethyl-1,3 propanediol)	8.3[14]

lag required for the development of adequate analytical methods. A variety of methods are now available. These include ashing and digestion of biological samples, followed by extraction into solvents, such as dithizone, for subsequent analysis by atomic absorption spectroscopy, colorimetric reactions—including measurement with —SH groups—gas chromatography, or mass spectroscopy.[15–17] Each method differs in sensitivity, in the capability to differentiate between inorganic and organic mercury, and in preparative losses such as those associated with ashing. Gas chromatography has proved suitable as a quantitative procedure for identifying specifically methyl-, ethyl-, methoxyethyl-, methylthiol- and phenyl-mercury, obviating to some extent many previous analytical problems. Dimethylmercury must be converted to methylmercury halides to be identified by the electron-capture detector, but it can be determined either directly by thermal combustion, or by flameless atomic-absorption measurement of the resultant vapour.

Recently, microwave induced emission spectrometry has been utilised as well to measure mercury in both inorganic and organic forms.[18,19] Mercury shows a strong affinity for ligands such as phosphates, cysteinyl and histidyl side chains of proteins, purines, pteridines and porphyrins (table 24.2). Thus, complexes of great stability are formed between mercury and —SH groups of amino acids, proteins, nucleotides, and nucleic acids. This remarkable affinity for such complexes has served analytical biochemistry since the complexes can be quantitatively followed by spectrophotometric methods.

25 Interaction of Mercury with Amino Acids, Proteins and Enzymes

The affinity of mercury for –SH groups is greater than for any other single ligand (table 24.2). Reactions with these groups can result in different complexes depending on whether the metal is present as Hg^{2+} or $R–Hg^{+}$, whether the ligand is a mono- or di-thiol and dependent on the relative concentrations of mercury and –SH groups. Importantly, complexes with $R–Hg^{+}$ bind thiols stoichiometrically.[20,21] These reactions can be followed polarographically and have been used for quantitative analysis of thiols. One such method for thiol determination utilises spectrophotometric titration with pCMB (parachloro-mercury benzoate). Though this approach is used for both amino acids and proteins, it principally identifies cysteine residues.[22] Reaction of this monofunctional mercurial with –SH groups results in an increase in the absorbance at 250 nm at pH 7. The change in absorbance is proportional to the number of –SH groups reacting with the mercurial. While hydrogen-ion concentration can affect formation of Hg–S complexes, the resulting mercaptides are usually stable over the entire pH range. The presence of other ligands competing for mercury also affect the binding of mercury with thiols. The existence of the S–Hg–S bond in proteins can, therefore, be demonstrated by the simultaneous elimination of mercury by EDTA and carboxymethylation of the –SH groups thus released.[23]

Mercury also interacts with tryptophan and characteristically alters its absorption spectrum.[24] These altered absorption spectra are broader, less structured and red shifted as compared with controls. Moreover, on complexing with tryptophan, mercury completely quenches fluorescence and phosphorescence. This loss of phosphorescence of proteins has been attributed to energy transfer. Analogous fluorescence quenching of phenylalanine and tyrosine has also been reported.[25] Mercury also interacts with phosphoryl groups of cell membranes[26] and with amino and carboxyl groups of enzymes.[27–29]

Most proteins contain various ligands for mercury, including sulphydryl groups.

The interaction with albumin is a model for studies of mercury binding to proteins in general, including enzymes. Fresh human albumin contains from 0.65 to 0.70 sulphydryls per mole and that of bovine serum albumin from 0.50 to 0.75 —SH groups per mole, as determined by reaction with organic mercurials, Hg^{2+}, and other —SH-group reagents.[30,31]

Mercaptalbumin contains one mercury ion and two albumin molecules.[32] As a monomer it differs from albumin only in its sulphydryl content. It has been studied in great detail in terms of its interaction with mercurials and illustrates that mercury binding to protein —SH groups is more complex than to similar groups in amino acids such as, for example, cysteine. Thus in albumin, as in other proteins, the reactivity of —SH groups varies, some groups binding readily while others react only when the protein is denatured. On addition of the mono-functional organic mercurials such as, for example, pMB, to mercaptalbumin a 1 : 1 complex forms. Further increases in mercury concentration result in the formation of mercaptalbumin dimers. The amount of dimerisation increases to a molar ratio of mercury to protein of 0.5, as expected from a protein with a single sulphydryl group. Dimerisation, in turn, is reversed on increasing the mercurial concentration further.[33] Halides also dissociate the dimer in the order of $I^- > Br^- > Cl^-$. The interaction of mercury with a second mercaptalbumin monomer is considerably weaker than that expected for a mercuric mercaptide, suggesting that reactivity may be influenced by vicinal protein structure.

Such studies with albumin or mercaptalbumin have delineated the effect of pH, temperature, other ligands, etc. on mercurial interaction with proteins. Furthermore, the presence of 4–5 grams per cent of albumin in mammals, including man, capable of binding ingested mercury suggests that this protein can affect the distribution of this metal in organs and, as a consequence, its toxic effects.[34] In this regard, other sulphur-containing proteins have also been studied in terms of mercury binding. Thus, metallothionein is found in human kidney,[35,36] rabbit liver and kidney.[37] It contains exceptionally high amounts of sulphur[38] and variable amounts of metal, for example 6 per cent cadmium and 0.2 per cent zinc and small amounts of copper and iron; depending on the source.[36,39,40] Humans treated with mercurial diuretics contain mercury bound more firmly than either zinc or cadmium to kidney metallothionein[35,40] suggesting that this protein may have a role in the metabolism of metals, including mercury, by the kidney. Data on binding of inorganic and organic Hg^{2+} are consistent with these studies.[41,42]

While —SH ligands in proteins have the greatest affinity for mercurials, proteins and enzymes contain other ligands that complex mercury. Thus, pCMB (parachloromercury benzoate) binds more rapidly to a number of sites other than to —SH groups in hemerythrin.[43] For example, mercurials bind and inactivate apocarbonicanhydrase, an enzyme that does not contain an —SH group at the active site. Moreover, mercurials also bind to this enzyme at the single —SH group located 1.5 nm from the zinc binding site.[44-47] Similarly, pMB binds to the active site of apocarboxypeptidase though there is no —SH group.[29,48] As a consequence, the kinetics of the reaction of mercury binding to proteins and enzymes, in terms of specificity for Hg—S reaction, must be confirmed by alternative methods before conclusions can be drawn regarding the role of —SH groups in the function of particular proteins.

There are numerous enzymes that react with organic and inorganic mercurials with resultant changes either in spectral characteristics or in enzyme activity. As examples, *E. coli* DNA-dependent RNA polymerase is inactivated by *p*MB. The inhibition results from an inability to bind to the DNA template which retains the capability of binding substrate.[49] In addition, most bacterial- and plant proteases have —SH groups essential for function and most are inactivated by *p*MB.[50] Other examples of the effects of mercury on enzyme activities are shown in table 25.1. These data illustrate that not only inhibition but also activation of certain enzymes follows administration of mercurials to animals.

The reactivity of different —SH groups in particular enzymes varies considerably, probably due to their spatial relationship to other groups in the three-dimensional enzyme structure, the presence of metals, pH, etc. Thus, *p*MB inactivates and disaggregates glutamine synthetase if the reaction is carried out in the absence of divalent ions. In the presence of Mn^{2+} such treatment does not affect enzyme activity.[59] Native beef-heart supernatant malic dehydrogenase contains six sulphydryl groups, three of which react with *p*MB, inactivating the enzyme. In 2.6 M urea, the enzyme remains active and can be inactivated by *p*MB, but under these conditions all six —SH groups react. This loss of activity, however, is not due to actual blocks of —SH groups but to a change in protein conformation.[60] In addition, interaction of —SH-containing enzymes with mercurials varies

Para mercury benzoate

Table 25.1. Alteration of Enzymatic Activities Following Administration of Mercury to Animals

Increased activities

alkaline phosphatase	liver, lung (rat)[51]
β-glucuronidase	kidney lysosomes (mice)[52]
N-acetylglucosaminidase	kidney lysosomes (mice)[52]
acid phosphatase	liver (rat)[51]
lactic dehydrogenase	cerebellum (rat)[53]
glutamic oxaloacetic transaminase	urine (rat)[54]
aspartate aminotransferase	urine (rat)[55]

Decreased activities

alkaline phosphatase	leukocytes (rat)[56]
acid phosphohydrolase	kidney lysosomes (mice)[52]
glucose 6-phosphatase	kidney lysosomes (mice)[52]
cytochrome *c* oxidase	kidney mitochondria (mice)[52]
glutaminase	brain (guinea pig)[57]
lactic dehydrogenase	kidney (rat)[58]

greatly, not only as a function of reaction conditions, but also of their sources. Thus, phosphoglyceromutase from adult human and chicken skeletal muscle is more sensitive to Hg^{2+} than the enzyme obtained from fetal muscle.[61] Mercury reversably transforms the specificity of human, rabbit and chicken muscle phosphoglyceromutase into that of a 2,3 diphosphoglycerate, but not that of an enzyme purified from swine kidney, beef liver and brain, and human red cells.[62]

26 Interaction with Purines, Pyrimidines and Nucleic Acids

Mercury can derange numerous cellular functions at all levels of nucleic-acid metabolism, including transcription and translation and it can cause chromosomal aberrations such as C-mitosis, polyploidy and aneuploidy.[63-65] Mercury binds to single-stranded tobacco mosaic virus RNA in a combining ratio of 1 : 1.[66] Methylmercury induces hepatic protein synthesis in rats, accompanied by increases in ribosomes, ribosomal sub-units and polyribosomes.[4] Binding of mercury to the nuclear components, including chromatin and nucleic acids of viruses, cells and tissues is the likely basis of such effects. The recent observation that mercury binds to the chromatin of normal rat liver has suggested that the element may have a role in normal nucleic-acid metabolism.[3] Thus, while reactions of mercury with purine and pyrimidine bases, nucleosides, nucleotides and nucleic acids are clearly of toxicological importance, they may be of biological significance, but have been studied less intensively than reactions with proteins.

Mercury can bind to nucleic acids through the hydroxyl groups of the sugar moieties, the phosphate, and oxygen or nitrogen donor groups of the bases. However, binding probably occurs principally through complexes with the bases.[67] Thus, Hg^{2+} is known to form 1 : 1 stable complexes with adenine, guanine, and cytidine. Hg^{2+} has a particularly high affinity for the adenine base, the stability constant of the 1 : 1 Hg–adenosine complex is 4.3.[68] Importantly, on forming a mercury complex the absorption spectrum of nucleoside bases is shifted drastically up to 11 nm from that characteristic of the 260 nm absorption band.[69] The shift is reversed by the addition of mercury complexing agents. Complex formation with adenosine is sensitive to pH and occurs at an amino group of the base through proton displacement, as evidenced by release of H^+ proportional to Hg^{2+} binding. Similarly, interaction with guanosine is pH dependent. Thus, at high pH Hg^{2+} binds to guanosine at the N1 group and to the NH_2 group at low pH.[69-74] Significantly, on complexation of Hg^{2+} with the 5′GMP, a novel absorption maximum appears.[75] As with mercury binding to nucleosides, a shift in the absorption spectrum of the various nucleosides and nucleic acids also follows Hg^{2+} binding. In fact, mercury complexes affect optical rotatory dispersion of nucleotides characteristically and differentially.[76] Based on these observations, for example, shift in the typical absorption peak, stoichiometry of complexation, and release of H^+, it has been concluded that mercury interacts with nucleic acids through the bases, not just the phosphate groups. Following such metal complexation, the intrinsic viscosity, spectra, thermal melting point and optical rotatory dispersion curves of the nucleic acids are all affected.[69,70,77-79] On binding to DNA, mercury interpolates in stoichiometric quantities between each DNA strand. This results in a rigid Hg^{2+}-DNA complex, which does not cause the double strands to unwind. Mercury binding to DNA appears to occur through linkage of the metal to every other base of the molecule. Despite marked effects on the spectra and other physical properties described above, removal of the metal by complexing agents regenerates native DNA.[77,80] This reversible binding, and the resulting effects on specific rotation, are more pronounced for A–T- than for G–C-rich DNA. These differences have been utilised to separate DNA molecules of varying base composition.[81-83]

27 Interaction with Cells and Cellular Components

Extra- and intracellular proteins, nucleic acids, membranes, mitochondria, mitotic apparatus, etc. contain numerous ligands for mercury, including –SH groups. As a consequence, addition of mercury to these systems has resulted in derangements of their function. The multiplicity of variables involved in the *in vitro* interaction of mercury and its derivatives with ligands of amino acids, proteins, nucleic acids, and phospholipids also pertain to interactions of mercurials with sub-cellular and cellular components. However, the biological effects on cells and their components are more complex since additional variables such as, for example, the capability of the mercurial to cross the cell membranes, distribute in cellular compartments, to bind to ligands whose availability may be altered by cellular metabolism, etc. must also be considered. Hence, in sea-urchin eggs undergoing cytokinesis there are cyclical increases in free –SH groups.[84] An effect of mercury on this process would depend on whether mercury can enter the fertilised egg and bind to the free –SH groups available during a particular phase of the cell cycle.

Such considerations have direct bearing on the different effects of inorganic and organic mercury on the same biological system. Thus, while all mercury compounds are cytotoxic to cells in culture, the organic mercurials are more effective than inorganic ones.[85] Thus, any biological effects of mercurials must be interpreted in terms of the chemical state of the mercury and its distribution, as well as the metabolism and composition of the cells involved.

Mercurials have significant effects on over-all cellular processes. Thus the alkyl mercury compounds decrease glucose transport across membranes,[86] decrease phosphate transport in the myocardial membrane,[87] Na–K-ATPase[88] and water permeability of red-cell membranes.[89] In addition, mercurials increase the lag in ATP-driven reduction of NAD by succinate in phosphorylating sub-mitochondrial particles from beef heart, while dithiothreitol eliminates the lag.[90] In addition, accumulation of inorganic mercury in liver-cell lysosomes causes release of hydrolytic enzymes, a response that probably accounts for the resultant cellular toxicity.[91] Mercury compounds produce chromosomal abnormalities and induce genetic and teratogenic effects.[92] Both phenyl- and methylmercury derivatives can inhibit spindle formation during mitosis.[63,64,93,94] Moreover, and consistent with the high sulphydryl content of these structures,[95] on exposure to mercurials, the microtubules of isolated mitotic apparati from sea-urchin eggs selectively disappear, as monitored morphologically. Specific proteins, closely similar to those found in sea-urchin sperm tails, can be isolated subsequently.[96] The offspring of Drosophila fed 0.25 ppm methylmercury contain extra chromosomes.[63,64,97]

In addition, organic mercurials, for example, salyrgan, can alter chromosomal structure and chromosome agglutination, inhibit root growth and induce both weak C-mitotic and radiomimetic effects in bulbs of *Allium sepa* and *Vicia fava* and in larvae of *Drosophila melanogaster*.[63-65,97] Methyl- and phenylmercury compounds at a concentration of 2.5×10^{-7} M (0.05 ppm) cause C-mitosis in *A. sepa* while concentrations of methyoxyethylmercury about ten times higher are required for the same effects; these are reversed by BAL in both instances.[63,64]

Significantly, chromosomal aberrations have not been reported in humans consuming a regular diet of mercury-laden fish[98] nor were they found in victims of Minamata disease in Japan.[99] However, 10 ppm of mercury (but not copper, silver, lead, zinc, nickel or chromium) induces human lymphocyte transformation and mitosis *in vitro*.[100-102] Mercurials also bind to viruses with functional consequence to their infectivity. Thus, Hg^{2+} reacts both with the brome grass mosaic virus itself and its RNA, resulting in decreased infectivity without apparent degradation of the virus. EDTA, mercaptoethanol, CN^- and other complexing agents reverse this effect.[103] Hg^{2+} also interacts with turnip yellow mosaic virus, in this case dissociating virions into RNA-free protein and nucleoprotein.[104]

28 Toxicological Aspects of Mercury Metabolism

As is true of all metals, the toxic effects of mercury depend, *inter alia*, on the chemical form, the size of the dose, and the route of entry into the body. Its relatively high volatility at ordinary temperatures, a unique feature of mercury, results in potential health hazards from inhalation of its vapours, a threat not presented to the same degree by other metals.

28.1 Metallic Mercury

In its liquid form, metallic mercury can enter the body by ingestion, through the skin, and by injection. If swallowed it passes through the gastrointestinal tract without being absorbed and hence causes little or no adverse effect. Passage of mercury through the unbroken skin was the basis for the inunction treatment for syphilis, which was in vogue from the late fifteenth to the early twentieth century. It is still significant as a route of absorption among industrial workers who handle the metal. Parenteral administration of metallic mercury is not common but has occurred accidentally and as a means of attempted suicide. The use of a mercury seal in drawing blood for gas analysis, as, for example, during cardiac catheterisation, has resulted in accidental entry of the metal into the blood stream[105] with occasional fatal outcome from embolisation.

28.2 Mercury Vapour

Inhalation of mercury vapour can cause either acute or chronic effects. If metallic mercury is heated it can liberate vapours of sufficient concentration to produce a severe or even fatal pneumonitis.[106,107] Pulmonary absorption of mercury vapours at concentrations that occur at ordinary temperatures, and of durations measured in months or years, can produce toxic effects. The classical manifestations in such instances are a fine intention tremor, gingivitis and erethism. A variety of subjective complaints such as headache, insomnia, bad dreams,

metallic taste in the mouth and impaired memory have been associated with chronic exposure to mercury vapours. A few cases of nephrotic syndrome have been reported;[108,109] milder degrees of albuminuria are more common.[111]
As a general rule all toxic manifestations are accompanied by elevations in the mercury levels in blood and urine above the 'normal' concentrations of 5.0 μg/100 ml and 25 μg/l respectively.[112,113]

28.3 Inorganic Salts

Best known for its toxic effects is mercuric chloride (corrosive sublimate). Before barbiturates became readily available, bichloride of mercury was a popular suicidal agent. A gram or two taken by mouth will cause corrosion of the oesophagus and stomach, resulting in vomiting and diarrhoea as well as bleeding from the intestinal tract. On reaching the kidneys, the absorbed compound produces severe tubular injury ultimately causing total renal failure and death from uraemia. Although relatively common in the past, such cases are now extremely rare.

Mercurous chloride (calomel), being much less soluble than the sublimate, is consequently less drastic in its action if swallowed. In fact, it has been widely used as a therapeutic agent, as a cathartic, anti-syphilitic, and diuretic. Administered orally to young children as a teething powder, calomel has been responsible for numerous cases of poisoning and has been implicated as a cause of acrodynia or 'pink disease'.[114] Other than the chlorides, inorganic salts of mercury are of little toxicological importance.

When absorbed into the body, inorganic forms of mercury tend to be deposited in the kidneys in higher concentrations than in any other organ and tissue.[110,115] Ordinarily these deposits have little or no pathological effect. They may result in the appearance of elevated mercury levels in the urine for a number of years, even when the exposure to the metal has been terminated.[116]

In the mild and moderate cases of chronic poisoning due to inorganic mercurials, such as are seen as a result of occupational exposures, the toxic manifestations ordinarily clear up without any specific treatment. In severe cases it may be desirable to administer 2,3-dimercapto-1-propanol (British AntiLewisite, BAL) intramuscularly. This drug frequently causes moderate or severe local and systemic side reactions and so should be used with reserve and caution. Another thiol-containing compound, N-acetyl-d,1-penicillamine, has been found useful in enhancing the excretion of inorganic mercury.

28.4 Organic Mercurials

From a toxicological point of view it is necessary to distinguish between three classes of organomercurials: aryl, alkyl, and alkoxyalkyl. Among them, these three groups span the entire toxicological spectrum from life-saving medicines to highly lethal poisons. Obviously it is improper to look upon all organic compounds as a single class.[117,118] Of the three, the alkyls are by far the most toxic.

28.4.1 Arylmercurials

This class embraces all mercury derivatives of aromatic hydrocarbons including such groups as benzyls, tolyls, napthyls and many others, as well as their halogenated and other forms. Little toxicological information is available except for the phenylmercurials, especially phenylmercuric acetate (PMA). There is some evidence, however, that other arylmercurials such as the benzoate, lactate, oleate, and propionate resemble PMA in their toxicological effects[119]

Extensive studies of human exposure to phenylmercurials in an industrial setting have shown that these compounds are of relatively low toxicity. They can be absorbed through the skin as well as through the lungs and gastrointerestinal tract. Strong solutions in the per cent range will regularly cause second-degree chemical burns of the skin with blister formation, while weaker solutions are only slightly irritating and rarely cause skin sensitisation. Chronic poisoning from phenylmercurials is practically unknown but if it were to occur the manifestations would be expected to resemble those of mercury vapour as described above.[119]

28.4.2 Alkylmercurials

Massive outbreaks of human poisoning due to methylmercury compounds in Japan,[120] Iraq[121] and elsewhere have focused world-wide attention on the toxic effects of these compounds. There has been ample documentation of the major differences between the alkyl compounds on the one hand, and the inorganic and aryl compounds on the other.[122]

Alkylmercury compounds can be absorbed through the skin, by inhalation, and via the gut. They can also pass through the placenta and cause damage to the fetus.[99] As a result of their affinity for lipid tissues, the alkylmercurials select the central nervous system as the principal site of injury. Permanent brain damage is frequently found in cases of poisoning and common manifestations are disturbances in vision, hearing, sensation, muscle function, and mentality. Congenital poisoning produces a clinical picture very similar to that of cerebral palsy. Absorbed alkylmercurials are liberated slowly, the 'half-time' for humans being about seventy days for the whole body. The average retention time is longer for the brain than for other tissues. There is no known effective treatment for alkylmercury poisoning, but its effects may be mitigated by selenium.[123]

28.4.3 Alkoxyalkylmercurials

Compared with the aryl- and alkylcompounds of mercury, relatively little is known about the toxicity of the alkoxyalkyls. Those that are related chemically to the mercurial diuretics are, of course, practically non-toxic.[124] Limited animal experimental work has shown that methoxyethylmercuric chloride resembles PMA in its behaviour, and that it is distinctly less toxic than its alkyl equivalent.[125]

28.5 Mercury–Selenium Interaction

It has become clear recently that the distribution and toxic manifestation of both inorganic and organic mercurials in mammals, including man, is affected by, and can be obviated by, the presence of ligands for mercury. Thus, selenium or selenium compounds protect animals from lethal doses of mercuric compounds.[123,126-131] The mechanism by which selenium exerts its protective effect is unknown; however, it increases the mercury content of blood, causes retention of mercury in animals and in humans, and accumulation in specific organs such as the thyroid, pituitary, brain, and kidneys.[132,133] The presence of selenium and its interaction with mercury thus constitutes a natural mechanism for the detoxification of compounds such as the mercurials, cadmium, and others.[132,134] In support of this thesis,[133] a 1:1 ratio of selenium to mercury has been reported in marine mammals,[135] in tuna,[123] as well as in humans, while the mercury content of various marine fish has been found associated with a forty-fold greater selenium content.[136,137]

Thus, while methylation of mercury by biological organisms, particularly fresh-water fish, represents a pathway of potential toxicological significance, variables such as the role of selenium must be considered in toxicological studies. Several useful reviews on the general aspects of the toxicology of mercury have been written.[122,138,139]

28.6 Methylation of Mercury by Biological Systems

There has been considerable controversy over the role and behaviour of methyl-mercury in aquatic ecosystems and particularly in food chains. The metal may be present not only as a result of man-made pollution but also from natural biological processes. While it is clear that industrial discharge can account for the high mercury content of exposed fresh-water fish and those in contaminated estuaries, mercury in ocean water may well derive nearly entirely from natural deposits and it seems possible that the mercury content of tuna and swordfish are no higher than they were many decades ago.[140] The question of whether or not salt-water fish themselves can convert Hg^{2+} to methylmercury by an enzymatic or organic mechanism has not been answered. It is known that marine algae can accumulate mercury up to 100-fold the amount present in sea water, providing a ready entrance to the fish food chain.[141] Gas-chromatographic analysis carried out on many forms of fish indicate that the bulk of the metal, 85–90 per cent, is in the form of methylmercury whether due to natural or man-made exposure.[142,143] Once accumulated in fish, methylmercury appears to be retained efficiently and is excreted only slowly, with a half-time of several years.[144] Such findings indicate that in those animals the presence of mercury *per se* does not result in toxic manifestations.

Methylation of mercury can occur by several processes. Methylpentacyano-cobaltate, a model compound for vitamin B_{12}, can react with Hg^{2+} yielding methylmercury.[145] The mechanism of methylmercury synthesis by this model compound has focused on pathways in living organisms that are possibly involved in similar reactions. Thus, S-adenosylmethionine, N^5-methyltetrahydrofolate derivatives, and methyl corrinoids are involved in methyl-group transfer of biological systems.[6] The first two transfer the methyl group as a carbonium ion (CH_3^+) and hence are incapable of methyl transfer to the metal of Hg^{2+} salts. However, the methyl corrinoids could generate methyl groups as carbanions (CH_3^-), carbonium ions (CH_3^+), or radicals $(\cdot CH_3)$ and, therefore, are potentially capable of methylation of inorganic mercury salts (with CH_3^-) or of mercury metal (with $\cdot CH_3$).

Microorganisms and mammals contain methylcobalamin[146] and synthesis of both monomethylmercury and dimethylmercury by enzymatic or non-enzymatic pathways has been shown.[5] The relative proportions of the mono- or dimethyl forms are determined by the Hg^{2+} concentration.[147]

Nuclear magnetic resonance has facilitated studies of the occurrence and mechanisms of non-enzymatic synthesis by methylcobalamin. The methyl group is transferred as CH_3^- since Co^{3+} does not change valence.[148] The reaction proceeds by electrophilic attack of Hg^{2+} on methylcobalamin. The pH-dependent methyl transfer requires a $2:1$ ratio of Hg^{2+} to methylcorrinoid. Two components are evident in this reaction. A rapid component involves reaction of the first Hg^{2+} ion with the nitrogen of 5,6-dimethylbenzimidazole; followed by a second, slower step involving electrophilic attack by a second Hg^{2+} ion on the CH_3^-. In aerobic organisms in which methyl corrinoids are involved in intermediary metabolism, this reaction may generate methylmercury since the non-enzymatic methyl transfer does not occur in the presence of Hg^+ or Hg^0, the forms of mercury that predominate under anaerobic conditions. Non-enzymatic methyl transfer from methylcobalamin to mercury compounds, for example, $HgCl_2$, also occurs.[149] The reaction with organic mercury compounds in aqueous solution was found to be slower than with $HgCl_2$, in the order chloride > hydroxide > dicyandiamide. Tris and phosphate buffers as well as thiols are inhibitory.

Mechanisms for the enzymatic synthesis of methylmercury compounds have been suggested for three systems in which methylcorrinoids are implicated in methyl transfer reactions.[6] A cobalamin-dependent N^5 methyltetrahydrofolate-homocysteine transmethylase (methionine synthetase),[150,151] possibly capable of methylmercury synthesis, has been found in a number of aerobic microorganisms, facultative and strict anaerobes, as well as mammalian liver. Methylcobalamin serves as a coenzyme, methyl transfer to homocysteine occurs in the presence of catalytic amounts of S-adenosylmethionine, and the enzyme-coenzyme complex is re-methylated by N^5-methyltetrahydrofolate. In the presence of Hg^{2+}, electrophilic attack on CH_3^- of the enzyme–conezyme complex should yield methylmercury and the aquocobalamin–enzyme complex. The complex would be restored by reduction of Co^{3+} to Co^+ with $FADH_2$, followed by remethylation from N^5-methyltetrahydrofolate. Methylmercury interacts with the substance, homocysteine, giving methylmercurihomocysteine, and additional methyl transfer may also produce some methylmercurythiomethyl.

The reaction may occur slowly both in mammalian liver[152] and in micro-organisms.[153]

Certain species of *Clostridia*, as well as other anaerobes, utilise methyl cobalamin in acetate synthesis from carbon dioxides.[154] Methyl cobalamin is the substrate for the C_2 carbon of acetic acid and CO_2 the source of carbon C_1. The mechanism of action of the methylcorrinoids in the acetate synthetase system is not yet understood. Methylmercury is generated from Hg^{2+} and methylcobalamin is again regenerated by N^5-methyltetrahydrofolate.[6]

The mechanism of methylation of mercury in methane synthesis is thought to involve a methyl radical ($\cdot CH_3$).[6] Methanogenic bacteria containing large amounts of methylcobalamins,[155] are very numerous in the anaerobic sediments of rivers and canals and in sludge of sewage beds, and are also present in the rumen of herbivores. They synthesise methane from CO_2 and H_2, the latter being evolved by other organisms dwelling in a symbiotic relationship with those containing the synthetase.[156,157]

The low redox potential at which anaerobic organisms thrive reduces inorganic mercury to Hg^0 and dimethylmercury is readily formed by methyl radical addition. Thus, Hg^{2+} may be the form of mercury that enters the bacterial cell.[5] Once dimethylmercury is formed it may diffuse from the cell, and at alkaline pH would be released as such into the water or, at acid pH, possibly converted to monomethylmercury and methane. This may be a detoxification reaction since bacteria may thus free themselves of mercury. In this regard, mercury-tolerant strains of *N. crassa* have a higher methylating capacity than do normal strains.[153]

Dimethylmercury may be the primary product of biological methylation of mercury, monomethylmercury being its decomposition product,[5,8,9] though the biological significance of the enzymatic reactions has been questioned.[149]

The identity of the substrate, which is generally provided either by industrial waste or mercury derivatives from natural sources, either favours or hinders the synthesis of methylmercury by microorganisms. Mercury appears in industrial discharges in five principal forms: (i) inorganic divalent mercury, Hg^{2+}; (ii) metallic mercury, Hg^0; (iii) phenyl mercury, $C_6H_5Hg^+$; (iv) alkoxyalkyl mercury, $CH_3O-CH_2-CH_2-Hg^+$; (v) methylmercury CH_3Hg^+. In nature the first four can be converted to methylmercury (see below). (HgS) is converted to methylmercury in the presence of oxygen when it is oxidised to sulphate (SO_4^{2-}), permitting methylation of mercury.[143]

In addition to synthesising organic mercury compounds, bacteria can also decompose them. *Pseudomonas* K-52 and a variety of other bacteria decompose various types of organic mercurials, yielding metallic mercury.[158−160,163] Cell-free extracts of a mercury-resistant *Pseudomonas* also catalyse the decomposition of phenylmercuric acetate into benzene and volatilise mercury.[162,163] Such extracts also catalyse the decomposition of ethylmercuric phosphate to ethane and methylmercuric chloride to methane, albeit more slowly. More recently it has been shown that demethylating species of bacteria are prevalent in the environment and that they are highly effective in preventing progressive build-up of methylmercury in waters.[164]

Conclusion

The biological effects of mercury, in both its organic and inorganic forms, depend on its interaction with amino acids, proteins, enzymes, nucleic acids, as well as cellular components such as membranes, mitochondria, etc. The capability to measure mercury in biological materials and the appreciation that living organisms can methylate mercury has led to an increasing awareness of its presence and distribution in living organisms as well as its toxicological importance in terms of human disease. While increasing quantities of mercury are both accumulating in the environment and entering the food chain of a variety of animals, the significance of this accumulation to the normal function of these animals must be assessed not only in terms of the absolute quantities of metal involved, but the functional consequences of its presence in their tissues and organs. In this regard the high mercury content in apparently normal animals suggests that biochemical processes involved in detoxification, including interaction with proteins such as albumin or perhaps metallothionin, and, as recently described, with selenium, may be critical to an over-all understanding of the biology and toxicology of this element.

References

1. R. A. Wallace, W. Fulkerson, W. D. Shults and W. S. Lyons, *Mercury in the Environment–The Human Element. Oak Ridge Tenn: Nat. Lab. NSF-EP-1.* (1971).
2. H. V. Weiss, M. Koide and E. D. Goldberg, *Science,* 174 (1971) p. 692.
3. S. E. Bryan, A. L. Guy and K. J. Hardy, *Biochem.,* 13 (1974) p. 313.
4. P. E. Brubaker, G. W. Lucier and R. Klein, *Biochem. Biophys. Res. Commun.,* 44 (1971) p. 1552.
5. J. M. Wood, F. S. Kennedy and C. G. Rosen, *Nature,* 220 (1968) p. 173.
6. J. M. Wood, *Advan. Sci. Technol II* eds R. Metcalf and J. N. Pitts, (Wiley-Interscience, New, York, 1971) pp. 39–55.
7. S. Jensen and A. Jernelöv, *Nordforsk. Biocid. Inform.,* 10 (1967) p. 4.
8. S. Jensen and A. Jernelöv, *Nordforsk. Biocid. Inform.,* 14 (1968) p. 3.
9. S. Jensen and A. Jernelöv, *Nature,* 223 (1969) p. 753.
10. O. I. Joensuu, *Science,* 172 (1971) p. 1027.
11. K. G. Sillén, *Acta Chem. Scand.,* 3 (1949) p. 539.
12. S. Hietanen and K. G. Sillén, *Acta Chem. Scand.,* 6 (1952) p. 747.
13. K. G. Sillen and A. E. Martell, *Stability Constants of Metal-Ion Complexes,* (Chemical Society, London, 1964).
14. D. P. Hanlon, D. S. Watt and E. W. Westhead, *Anal. Biochem.,* 16 (1966) p. 225.
15. T. W. Clarkson and M. R. Greenwood, *Anal. Biochem.,* 37 (1970). p. 236.
16. R. Von Burg, F. Farris and J. C. Smith, *J. Chromatography,* 97 (1974) p. 65.
17. T. Giovanoli-Jakubczak, M. R. Greenwood, J. Crispin Smith and T. W. Clarkson, *Clin. Chem.,* 20 (1974) p. 222.
18. H. Kawaguchi and B. L. Vallee, *Anal. Chem.* 47 (1975) p. 1029.
19. I. Atsuya, H. Kawaguchi and B. L. Vallee, *Anal. Biochem.* (1976) in press.
20. R. Benesch and R. E. Benesch, *Arch. Biochem. Biophys.,* 38 (1952) p. 425.
21. F. L. Hoch and B. L. Vallee, *Arch. Biochem. Biophys.,* 91 (1960) p. 1.
22. P. D. Boyer, *J. Am. chem. Soc.,* 76 (1954) p. 4331.
23. Y. Burnstein and R. Sperling, *Biochem. Biophys. Acta,* 221 (1970) p. 410.
24. L. K. Ramachandran and B. Witkop, *Biochemistry,* 3 (1964) p. 1603.
25. R. F. Chen, *Arch. Biochem. Biophys.,* 142 (1971) p. 552.
26. H. Bassow, A. Rothstein and T. W. Clarkson, *Pharmacol. Rev.,* 13 (1961) p. 185.
27. M. C. Battigelli, *J. occup. Med.,* 2 (1960) p. 337.
28. E. S. Cook and G. Parisutti, *J. biol. Chem.,* 167 (1947) p. 827.

29. W. N. Lipscomb, J. A. Hartsuck, G. N. Reeke, Jr, F. A. Quiocho, P. H. Bethge, M. L. Ludwig, T. A. Steitz, H. Muirhead and J. C. Coppola, *Brookhaven Symp. Biol.*, 21 (1968) p. 24.
30. W. L. Hughes, Jr., *J. Am. chem. Soc.*, 69 (1947) p. 1836.
31. I. M. Kolthoff, A. Anastasi, W. Stricks, B. H. Tan and G. S. Deshmukh, *J. Am. chem. Soc.*, 79 (1957) p. 5102.
32. W. L. Hughes, Jr. and H. M. Dintzis, *J. biol. Chem.*, 239 (1964) p. 845.
33. H. Edelhoch, E. Kathalski, R. H. Maybury, W. J. Hughes, Jr. and J. T. Edsall, *J. Am. chem. Soc.*, 75 (1953) p. 5058.
34. J. L. Webb, *Enzyme and Metabolic Inhibition* vol. 2 (Academic Press, New York and London, 1969).
35. P. Pulido, J. H. R. Kägi and B. L. Vallee, *Biochemistry*, 5 (1966) p. 1768.
36. J. H. R. Kägi, S. R. Himmelhoch, P. Whanger and J. L. Bethune, *J. biol. Chem.*, 249 (1974) p. 3537.
37. M. Piscator, *Nord. Hyg. Tidskr.*, 44 (1963) p. 17.
38. M. Margoshes and B. L. Vallee, *J. Am. chem. Soc.*, 79 (1957) p. 4813.
39. J. H. R. Kägi and B. L. Vallee, *J. biol. Chem.*, 235 (1960) p. 3188.
40. J. H. R. Kägi and B. L. Vallee, *J. biol. Chem.*, 236 (1961) p. 2435.
41. M. Jakubowski, J. Piotrowski and B. Trojanowska, *Toxicol. Appl. Pharmacol.*, 16 (1970) p. 743.
42. J. Piotrowski and W. Bolanowska, *Med. Pracy.*, 21 (1970) p. 338.
43. M. H. Klapper, *Biochem. Biophys. Res. Commun.*, 38 (1970) p. 172.
44. E. E. Rickli and J. T. Edsall, *J. biol. Chem.*, 237 (1962) p. PC258.
45. B. G. Malmstrom, P. O. Nyman, B. Strandberg and B. Tilander, *Structure and Activity of Enzymes*, eds T. W. Gordon, J. L. Harris, B. S. Hartley, (Academic Press, London; 1964) p. 121.
46. B. Tilander, B. Strandberg and K. Fridborg, *J. mol. Biol.*, 12 (1965) p. 740.
47. K. Fridborg, K. K. Kannan, A. Liljas, G. Lundin, B. Strandberg, R. Strandberg, B. Tilander and G. Wirén, *J. mol. Biol.*, 25 (1967) p. 505.
48. B. L. Vallee, T. L. Coombs and F. L. Hoch, *J. biol. Chem.*, 235 (1960) p. PC45.
49. A. Ishihama and J. Hurwitz, *J. biol. Chem.*, 244 (1969) p. 6680.
50. B. L. Vallee and J. Riordan, *Ann. Rev. Biochem.*, 38 (1969) p. 733.
51. B. Nowak, *Med. Pracy.*, 20 (1969) p. 333.
52. M. A. Verity and W. J. Brown, *Am. J. Pathol.*, 61 (1970) p. 57.
53. A. Teraoka, T. Inoue, N. Hotta, N. Udo and K. Miyakawa, *Seibutsu Butsuri Kagaku*, 14 (1970) p. 291.
54. L. F. Prescott and S. Ansari, *Toxicol. Appl. Pharmacol.*, 14 (1969) p. 97.
55. M. Sternberg, I. Szlamka, M. Moisy, P. Rebeyrotte and G. Lagrue, *C. R. Acad. Sci. D*, 271 (1970) p. 1134.
56. K. Lutomska, J. Pawlak and E. Witek, *Czas. Stomatol.*, 23 (1970) p. 917.
57. K. Takahama, *Kumamoto med. J.*, 23 (1970) p. 84.
58. S. Ringoir, *Nephron.*, 7 (1970) p. 538.
59. B. M. Shapiro and E. R. Stadtman, *J. biol. Chem.*, 242 (1967) p. 5069.
60. A. Guha, S. Englard and I. Listowsky, *J. biol. Chem.*, 243 (1968) p. 609.
61. J. Grisolia Ripoll, D. A. Diederich and S. Grisolia, *Biochem. Biophys. Res. Commun.*, 41 (1970) p. 1238.
62. D. Diederich, A. Khan, I. Santos and S. Grisolia, *Biochem. Biophys. Acta*, 212 (1970) p. 441.
63. C. Ramel, *Nord. Hyg. Tidskr.*, 50 (1969) p. 135.
64. C. Ramel, *Hereditas*, 61 (1969) p. 208.
65. R. Lorente Albinana, *Rev. Biol. Lisbon.*, 6 (1968) p. 313.
66. S. Katz and V. Santilli, *Fed. Proc.*, 21 (1962) p. 376.

67. L. S. Kan and N. C. Li, *J. Am. chem. Soc.*, 92 (1970) p. 4823.
68. H. T. Writh and N. Davidson, *J. Am. chem. Soc.*, 86 (1964) p. 4325.
69. C. A. Thomas, Jr., *J. Am. chem. Soc.*, 76 (1954) p. 6032.
70. T. Yamane and N. Davidson, *J. Am. chem. Soc.*, 83 (1961) p. 2599.
71. T. Yamane and N. Davidson, *Biochem. Biophys. Acta*, 55 (1962) p. 609.
72. G. L. Eichhorn and P. Clark, *J. Am. chem. Soc.*, 85 (1963) p. 4020.
73. G. L. Eichhorn, J. J. Butzow, P. Clark and E. Tarren, *Biopolymers*, 5 (1967) p. 283.
74. R. B. Simpson, *J. Am. chem. Soc.*, 86 (1964) p. 2059.
75. P. Y. Cheng, D. S. Honbo and J. Rozsnyai, *J. Biochem.*, 8 (1969) p. 4470.
76. P. Y. Cheng, *Biochem. Biophys. Res. Commun.*, 33 (1968) p. 746.
77. S. Katz, *J. Am. chem. Soc.*, 74 (1952) p. 2238.
78. D. W. Gruenwedel and N. Davidson, *J. mol. Biol.*, 21 (1966) p. 129.
79. R. N. Izatt, J. J. Christensen and J. H. Rytting, *Chem. Revs.*, 71 (1971) p. 439.
80. G. L. Eichhorn, *Inorganic Biochemistry*, vol. 2., ed. G. L. Eichhorn, (Elsevier Scientific Publishing Co., Amsterdam, London, New York, 1973) p. 1226.
81. N. Davidson, J. Widholm, U. S. Nandi, R. Jensen, B. M. Olwera and J. C. Wang, *Proc. Natn. Acad. Sci. US.*, 53 (1965) p. 111.
82. U. S. Nandi, J. C. Wang and N. Davidson, *Biochemistry*, 4 (1965) p. 1687. Niigata Report, (Tokyo, Ministry of Health and Welfare, 1967).
83. G. Luck and G. Zimmer, *Eur. J. Biochem.*, 18 (1971) p. 140.
84. H. Sakai, *Biophys. Biochem. Cytol.*, 8 (1960 (1)) p. 609.
85. M. Umeda, K. Saito, K. Hirose and M. Saito, *Jap. J. exp. Med.*, 39 (1969) p. 47.
86. W. D. Stein, *The movement of molecules across cell membranes*, (Academic Press, New York, 1967) pp. 289–95.
87. D. D. Tyler, *Biochem. J.*, 111 (1969) p. 65.
88. J. C. Skou, *Progress Biophys. Biophys. Chem.*, 14 (1964) p. 131.
89. R. I. Macay and R. E. L. Farmer, *Biochem. Biophys. Acta*, 211 (1970) p. 104.
90. S. F. Schuurmans, B. P. Sani and D. R. Sanadi, *Biochem. Biophys. Res. Commun.*, 39 (1970) p. 1026.
91. T. Norseth, *Biochem. Pharmacol.*, 17 (1968) p. 581.
92. H. V. Malling, J. S. Wasson and S. S. Epstein, *Newslett. Environ. Mutagen Soc.*, 3 (1970) p. 7.
93. G. Fiskesjo, *Hereditas*, 62 (1969) p. 314.
94. G. Fiskesjo, *Hereditas*, 64 (1970) p. 142.
95. D. Mazia, The role of thiol groups in the structure and function of the mitotic apparatus 367 Sulfur in Proteins, ed. R. Benesch (Academic Press, 1959).
96. T. Bibring and J. Baxandall, *J. cell Biol.*, 48 (1971) p. 324.
97. C. Ramel, *Hereditas*, 57 (1967) p. 445.
98. S. Skerfving, K. Hansson and J. Lindsten, *Archs envir. Hlth*, 21 (1970) p. 133.
99. Kumamoto University, *Pathological, Clinical and Epidemiological Research About Minamata Disease: 10 Years Later (Japanese)* (1973).
100. E. Schoepf, *Naturwissenshaften*, 56 (1969) p. 464.
101. E. Schoepf, K. H. Schulz and I. Isensee, *Arch. klin. Exp. Dermatol.*, 234 (1969) p. 420.
102. E. Schoepf and G. Nagy, *Acta Haematol.*, 43 (1970) p. 73.
103. P. Pfeiffer and B. Dorne, *Biochem. Biophys. Acta*, 228 (1971) p. 456.
104. B. Dorne, G. Jonard, J. Witz and L. Hirth, *Virology*, 43 (1971) p. 279.

105. J. T. Buxton, J. C. Hewitt, R. H. Gadsden and G. B. Bradham, *J. Am. med. Ass.*, 193 (1965) p. 573.
106. D. T. Teng and J. C. Brennan, *Radiology*, 73 (1959) p. 354.
107. J. T. Hallee, *Am. Rev. Respirat. Diseases*, 99 (1969) p. 430.
108. L. Friberg, S. Hammarstrom and A. Hystrom, *AMA Arch. Ind. Hyg.*, 8 (1953) p. 149.
109. L. J. Goldwater, *AMA Arch. Ind. Hyg.*, 8 (1953) p. 558.
110. M. M. Joselow, L. J. Goldwater and S. B. Weinberg, *Arch. environ. Hlth*, 15 (1967) p. 64.
111. M. M. Joselow and L. J. Goldwater, *Arch. environ. Hlth*, 15 (1967) p. 155.
112. L. J. Goldwater, A. C. Ladd and M. B. Jacobs, *Arch. environ. Hlth*, 9 (1964) p. 735.
113. M. B. Jacobs, A. C. Ladd and L. J. Goldwater, *Arch. environ. Hlth*, 9 (1964) p. 454.
114. J. Warkany and D. M. Hubbard, *Lancet*, 254 (1948) 829.
115. K. Liebscher and H. Smith, *Arch. environ. Hlth*, 17 (1968) p. 881.
116. L. J. Goldwater and A. Nicolau, *Arch. environ. Hlth*, 12 (1966) p. 196.
117. L. J. Goldwater, *Mercury: A History of Quicksilver* (Baltimore, York Press, 1972).
118. W. L. Hughes, Jr., *Ann. N.Y. Acad. Sci.*, 65 (1957) p. 454.
119. A. C. Ladd, L. J. Goldwater and M. B. Jacobs, *Arch. environ. Hlth*, 9 (1964) p. 43.
120. M. Kutsuna, ed., *Minamata Disease, Kumamoto, Japan*, [Kumamoto University, (English), 1968].
121. F. Bakir, S. F. Damluji, L. Amin-Zaki, M. Murtadha, A. Khalidi, N. Y. Al-Rawi, S. Tiktriti, H. I. Dhahir, T. W. Clarkson, J. C. Smith and R. A. Doherty, *Science*, 181 (1973) p. 230.
122. L. Friberg and J. Vostal, (eds), *Mercury in the Environment: A Toxicological and Epidemiological Appraisal*, (Stockholm: Karolinska Institute, 1971).
123. H. E. Ganther, C. Goudie, M. L. Sunde, M. J. Kopecky, P. Wagner, O. Sange Hwan and W. G. Hoekstra, *Science*, 175 (1972) p. 1122.
124. J. E. Baer and K. H. Beyer, *Ann. Rev. Pharmacol.*, 6 (1966) p. 261.
125. J. W. Daniel and J. C. Gage, *Biochem. J.*, 111 (1969) p. 20P.
126. J. Părízek and I. Oštádalová, *Experientia*, 23 (1967) p. 142.
127. J. Părízek, I. Beneš, I. Oštádalová, A. Babický, J. Beneš and J. Lener, *Physiologia bohemoslov*, 18 (1969) p. 95.
128. O. A. Levander and L. C. Argrett, *Toxic. Appl. Pharmac.*, 14 (1969) p. 308.
129. M. M. El-Begearmi, C. Goudie, H. E. Ganther and M. L. Sunde, *Fed. Proc.*, 32 (1973) p. 886.
130. B. Stillings, H. Lagally, J. Soares and D. Miller, *Abstracts of Short Communications, 9th International Congress of Nutrition, Mexico, D.F.*, (1972) p. 206.
131. S. D. Potter and G. Matrone, *Fed. Proc.*, 32 (1973) p. 929.
132. J. Părízek, I. Oštádalová, J. Kalousková, A. Babický and J. Beneš, *Newer Trace Elements in Nutrition*, ed. W. Mertz and W. E. Cornatzer (Marcel Dekker, New York, 1971) p. 85.
133. L. Kosta, A. R. Byrne and V. Zelenko, *Nature*, 254 (1975) p. 238.
134. J. Părízek, I. Beneš, I. Ostádalová, A. Babický, J. Beneš and J. Pither, *Mineral Metabolism in Paediatrics*, eds D. Bartrop and W. L. Burland, (Blackwell, Edinburgh, 1969).

135. J. H. Koeman, W. H. M. Peeters and C. H. M. Koudstaal-Hol, *Nature*, 245 (1973) p. 386.
136. D. E. Robertson, L. A. Rancitelli, J. C. Langford and R. W. Perkins, *Workshop Sponsored by the National Science Foundation's Office for the International Decade for Oceanographic Exploration* (Brookhaven National Laboratory, Upton, New York, May 24–26, 1973).
137. G. Lunde, *J. Sci. Fd. Agric.*, 19 (1968) p. 432.
138. T. W. Clarkson, 'Recent Advances in the Toxicology of Mercury with Emphasis on the Alkylmercurials'. Cleveland: *CRC Reviews in Toxicology*, March 1972, p. 205.
139. T. W. Clarkson, 'The Pharmacology of Mercury Compounds', *Ann. Rev. Pharmacol.*, 12 (1972) p. 375.
140. R. Kishore and V. Guinn, *Chem. Eng. News*, 14 (Aug. 30, 1971).
141. H. T. Shacklette, *U.S. Geol. Surv. Prof. Pap.*, 713 (1970) p. 35.
142. K. Noren and G. Westöö, *Var. Foeda.*, 19 (1967) p. 13.
143. A. Jernelöv, *Chemical Fallout*, eds M. W. Miller and G. G. Berg, Springfield, Ill: Thomas, 1969, pp. 71, 93, 240.
144. T. Jarvenpää, M. Tillander and J. K. Miettinen, *Suom. Kemistilehti B*, 43 (1970) p. 439.
145. J. A. Halperin and J. P. Maher, *J. Am. chem. Soc.*, 86 (1964) p. 2311.
146. K. Lindstrand, *Nature*, 204 (1964) p. 188.
147. N. Imura, E. Sukagawa, S.-K. Pan, K. Nagao, G.-Y. Kim, T. Kivan and T. Ukita, *Science*, 172 (1971) p. 1248.
148. H. A. O. Hill, J. M. Pratt, S. Risdale, F. R. Williams and R. J. P. Williams, *Chem. Commun.*, 124 (1970) p. 341.
149. L. Bertilsson and H. Y. Neujahr, *Biochemistry*, 10 (1970) p. 2805.
150. J. R. Guest, C. W. Helleiner, M. J. Cross and D. D. Woods, *Biochem. J.*, 76 (1960) p. 396.
151. G. T. Burke, J. H. Mangum and J. D. Brodie, *Biochem.*, 9 (1970) p. 4297.
152. G. Westöö, *Var Foeda*, 19 (1967) p. 121.
153. L. Landner, *Nature*, 230 (1971) p. 452.
154. L. Ljungdahl, E. Irion and H. G. Wood, *Biochemistry*, 4 (1965) p. 2771.
155. B. A. Blaylock and T. C. Stadtman, *Arch. Biochem. Biophys.*, 116 (1966) p. 138.
156. H. A. Barker and A. van Leenwenhoch, *Microbiol. Serol.*, 6 (1940) p. 20.
157. M. P. Bryant, E. A. Wolin, M. J. Wolin and R. S. Wolfe *Arch. Mikrobiol.*, 59 (1967) p. 20.
158. T. Suzuki, K. Furukawa and K. Tonomura, *J. Ferment. Technol.*, 46 (1968) p. 1048.
159. T. Suzuki and K. Tonomura, *Kogyo Gijutsuin Biseibutsu Kogyo Gijutsu Kenkyusho Kenkyu Hokoku* 36 (1969) p. 11.
160. K. Furukawa, T. Suzuki and K. Tonomura, *Agr. biol. Chem.*, 33 (1969) p. 128.
161. K. Tonomura, *Kagaku, Kyoto*, 24 (1969) p. 508.
162. K. Tonomura and F. Kanzaki, *Hakko Kogaku Zasshi*, 47 (1969) p. 430.
163. K. Tonomura and F. Kanzaki, *Biochim. Biophys. Acta*, 184 (1969) p. 227.
164. W. J. Spangler, G. L. Spigarelli, G. M. Rose and H. M. Miller, *Science*, 180 (1973) p. 192.

Index